METALLOPROTEINS
Theory, Calculations, and Experiments

METALLOPROTEINS
Theory, Calculations, and Experiments

Edited by
Art E. Cho
Korea University, Korea

William A. Goddard III
California Institute of Technology, Pasadena, USA

CRC Press
Taylor & Francis Group
Boca Raton London New York

CRC Press is an imprint of the
Taylor & Francis Group, an **informa** business

CRC Press
Taylor & Francis Group
6000 Broken Sound Parkway NW, Suite 300
Boca Raton, FL 33487-2742

First issued in paperback 2017

© 2015 by Taylor & Francis Group, LLC
CRC Press is an imprint of Taylor & Francis Group, an Informa business

No claim to original U.S. Government works

ISBN-13: 978-1-4398-1318-8 (hbk)
ISBN-13: 978-1-138-89438-9 (pbk)

Visit the Taylor & Francis Web site at
http://www.taylorandfrancis.com

and the CRC Press Web site at
http://www.crcpress.com

We dedicate this volume to the authors who took time from their very busy fundamental research activities to provide an up-to-date summary of the recent advances in both the experimental and theoretical understanding of the fundamental processes in Metalloproteins. We hope that this will help engage new generations in the ongoing adventure of explaining and harnessing the powerful life processes mediated by these systems.

Contents

Preface

Metalloproteins play essential critical functional roles in enzymatically catalyzing reactions difficult to achieve without metals, in signal transduction, and in storage and transport of proteins constituting over 1/3 of the proteome of living organisms. Recent breakthroughs in crystallography have provided insights into understanding their mechanisms, but challenges remain in identifying the particularly redox states of the metals and in the chemical mechanisms for the complex reactions they catalyze. With these advances in structural information, theory and computation are starting to play a more significant role, particularly in identifying the reaction mechanism. This volume summarizes some of the recent progresses in both experiment and theory/computation showing the synergy that is now developing.

Two papers review the progress in understanding the Oxygen Evolution Reaction (OER) accomplished by Photosystem II (PSII), where four thrusts have converged to provide something close to an atomistically based mechanism:

- X-ray structures of some (but unfortunately not all) of the intermediate states in the catalytic reactions and characterization of the active sites through techniques such as EXAFS and XANES
- Model systems synthesized to reproduce the active states to enable clear-cut experimental studies of the details of the structures and redox states of the metals
- EPR and NMR studies of many of the active states to clarify the steps
- Quantum mechanical calculations of the active states to determine the structures and reaction barriers

This remarkable progress is summarized in the papers from the Britt and Agapie groups.

A driving force in trying to understand the mechanisms of the metalloenzymes is to use them as a model for developing biomimics that catalyze such important reactions as CH_4 activation and the OER. The paper by Chan and coworkers describes the Odyssey from characterizing the complex structures and chemistry in methane monooxygenase (pMMO) to determining its unique structure involving three distinct Cu sites and then to developing a functioning biomimic catalyst that is very effective for selective oxidation of CH_4 and alkanes.

A particularly important class of metalloenzymes is the radical S-adenosyl-l-methionine (SAM) superfamily in which 4Fe–4S complexes carry out important anaerobic transformations as explained in the paper by Betz, Shepard, and Broderick.

With the development of accurate structural models, it has become possible to use first principle quantum mechanics (DFT) to develop an understanding of the electron processes of the catalytic and electron transfer reactions involved as summarized by the Klein group. Cho then shows how to model the binding of ligands to metalloproteins by including the role of the more remote parts of the protein on these electronic processes through coupling of QM with a force field description (QM/MM).

The advances in our understanding of the electrochemical processes essential to metalloproteins through protein film electrolysis (PFE) are summarized by Armstrong, showing how this provides detailed mechanisms underlying how these proteins play essential roles in life.

Ferritins play an essential role in storing iron in bioavailable assembly, providing an example of the interplay between metal transport and storage of well-characterized structural detail as explained in the paper by Kim and Kim showing the unique characteristics that are important in the functioning of *Helicobacter pylori*.

At the heart of essentially all metalloenzymes is the role of electron transfer in controlling the rates of their reactions. The elucidation of how these electron transfer processes are mediated by the structures of the metalloenzymes involves a combination of theory and experiment as explained by the Gray group.

We hope that this volume will ignite further collaborations between experiment and theory in elucidating and exploiting the metalloproteins. We envision that the volume will be valuable to researchers in the general field of bioinorganic chemistry and also as a sourcebook for courses at the biology/chemistry interface.

We hope that this comprehensive collection of the recent research by top researchers will stimulate accelerated developments in combining experimental and theoretical studies of metalloproteins.

William A. Goddard III
Materials and Process Simulation Center,
California Institute of Technology

Art E. Cho
Department of Bioinformatics, Korea University

Editors

William A. Goddard III is the Charles and Mary Ferkel Professor of chemistry, materials science, and applied physics, and the director of the Materials and Process Simulation Center at the California Institute of Technology. He earned his BS degree in engineering from the University of California, Los Angeles, in 1960 and his PhD degree in engineering science with a minor in physics from the California Institute of Technology in 1965. After passing his PhD exam in October 1964, he joined the Chemistry department at Caltech in November 1964, where he continues his activities in research and teaching.

Dr. Goddard has pioneering methods for quantum mechanics (generalized valence bond theory, first principle pseudo-potentials) and reactive fields molecular dynamics (ReaxFF and eFF), and complete sampling for protein structure predicting and docking (GEnSeMBLE and DarwinDock), which he has applied to many areas of chemical reaction theory, catalysis, materials science, and selective ligand design. He has been a member of the US National Academy of Sciences since 1984 and is a fellow (IAQMS, APS, AAAS, Am Acad, Arts Science) or a member (ACS, MRS, Protein Society) of many other organizations. He was a cofounder of Materials Simulation Inc. (now Accelrys) and Schrodinger Inc., and continues with recent startups in electron etching of semiconductors (Systine) and design of therapeutics (GIRx).

Art E. Cho is a professor in the Department of Bioinformatics, Korea University. He graduated from the University of California at Berkeley in 1988, with a double major in physics and mathematics. After completing the master's program in mathematics at the University of Chicago, he pursued a PhD degree in physics at Brown University. Before finishing his PhD degree, he took time off and returned to his home country of Korea to fulfill mandatory military duty. As a substitution for service in the army, he worked as a senior research scientist at the supercomputing center of Samsung Advanced Institute of Technology affiliated with Samsung Electronics Co. After earning a PhD degree in physics in 2001, he worked as a postdoctoral scholar at Caltech and then as a research scientist at Columbia University. He also worked as an applications scientist at Schrodinger Inc. Since 2007, he has been with Korea University. In 2010, while taking up a departmental duty as the chair, he founded a startup company, named Quantum Bio Solutions, which specializes in computational drug design.

Contributors

Theodor Agapie
Department of Chemistry
California Institute of Technology
Pasadena, California

Mercedes Alfonso-Prieto
Institute for Computational Molecular
    Science
Temple University
Philadelphia, Pennsylvania

Fraser A. Armstrong
Department of Chemistry
University of Oxford
Oxford, United Kingdom

Jeremiah N. Betz
Department of Chemistry and
    Biochemistry
Montana State University
Bozeman, Montana

R. David Britt
Department of Chemistry
University of California
Davis, California

Joan B. Broderick
Department of Chemistry and
    Biochemistry
Montana State University
Bozeman, Montana

Sunney I. Chan
Institute of Chemistry
Academia Sinica
Taipei, Taiwan

Maraia E. Ener
Beckman Institute
California Institute of Technology
Pasadena, California

Harry B. Gray
Beckman Institute
California Institute of Technology
Pasadena, California

Jacob S. Kanady
Department of Chemistry
California Institute of Technology
Pasadena, California

Kyung Hyun Kim
Department of Bioinformatics
Korea University
Sejong, Korea

Sella Kim
Department of Bioinformatics
Korea University
Sejong, Korea

Michael L. Klein
Institute for Computational Molecular
    Science
Temple University
Philadelphia, Pennsylvania

Suman Maji
School of Physical Sciences
Lovely Professional University
Punjab, India

Jose Mendoza-Cortes
Department of Chemical and
    Biomedical Engineering
Florida State University
Tallahassee, Florida

Eric M. Shepard
Department of Chemistry and
    Biochemistry
Montana State University
Bozeman, Montana

Troy A. Stich
Department of Chemistry
University of California
Davis, California

Jeffrey J. Warren
Beckman Institute
California Institute of Technology
Pasadena, California

Jay R. Winkler
Beckman Institute
California Institute of Technology
Pasadena, California

Steve S.-F. Yu
Institute of Chemistry
Academia Sinica
Taipei, Taiwan

1 Advanced Electron Paramagnetic Resonance Studies of the Oxygen-Evolving Complex

Troy A. Stich and R. David Britt

CONTENTS

CONSPECTUS

Electronic structure models of metalloenzyme active sites offer invaluable insight into determining atom-level details of the corresponding enzyme mechanism. Results from density functional theory (DFT) and other levels of computational chemistry can afford such models. However, these computational results must first be spectroscopically validated—meaning they must be shown to accurately predict electronic structure properties. In the case of the tetranuclear manganese oxygen-evolving complex (OEC), which is the active site for water oxidation in photosystem II (PSII), DFT predictions have afforded several geometric models that were consistent with various x-ray diffraction and scattering results, although evaluation of the corresponding electronic structures would be more discriminating. To this end, many groups have used results from electron paramagnetic resonance (EPR) spectroscopy—in particular, advanced pulse EPR techniques—to probe the distribution of the unpaired electron spin about the OEC and the strength of the hyperfine interaction between magnetic nuclei and this electron spin. This work has yielded a sea of magnetic parameters that collectively serve as a perspicacious judge of the computer-generated structural models of the OEC. In this chapter, we summarize these EPR-derived magnetic parameters for various states and inhibited forms of the OEC. These findings have ruled out several proposed structures for the OEC and, with them, certain reaction pathways. Recently, however, one model (informed by the high-resolution crystal structure coordinates) has emerged that performs remarkably well, especially in predicting the magnetic parameters of the constituent ^{55}Mn ions. Further refinement of this model requires continued spectroscopic characterization of the OEC, and EPR spectroscopy will be a key tool to do so.

KEYWORDS

photosystem II; oxygen-evolving complex; electron paramagnetic resonance; electron spin echo envelope modulation; electron–nuclear double resonance

1.1   INTRODUCTION

Oxygenic photosynthesis is the process by which green plants, algae, and cyanobacteria use sunlight to power the production of cell carbon from carbon dioxide via the Calvin cycle. These actions work in concert with the four-electron oxidation of two waters to yield one molecule of dioxygen and four protons catalyzed by the multicomponent system of proteins termed photosystem II (PSII). Two pseudo-C_2 symmetry-related polypeptide chains, D1 and D2, span thylakoid membranes [1,2]. The D1 peptide provides a majority of the ligands to the tetranuclear manganese core of the oxygen-evolving complex (OEC; Figure 1.1) [3–6]. Surrounding this catalytic center are several proteins that include cytochrome b_{559}, whose function may be that of a photoprotectant [7]; a set of small extrinsic proteins involved in stabilizing

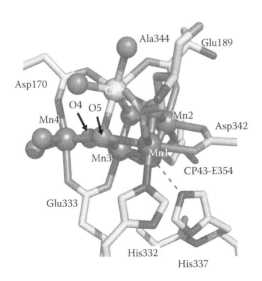

FIGURE 1.1 Structure of the PSII active site highlighting possible protein-derived ligands to the Mn cluster. (Adapted from crystal structure reported by Y. Umena et al., *Nature*, 473, 55, 2011.)

the OEC [8]*; and CP43 and CP47, a pair in a series of light-harvesting complexes (LHCs). CP43 and CP47 shuttle photonic energy to P_{680}, a pair of chlorophyll *a* molecules found at the interface of the D1 and D2 polypeptides, to generate the charge-separated state designated as $P_{680}^{+\bullet}Q_A^{-\bullet}$ [9].

This last process begins as visible-wavelength (400–680 nm) photons and are absorbed by an array of pigment molecules (e.g., carotenoids and chlorophylls) found in peripheral antenna proteins, the LHCs, and throughout the D1 and D2 polypeptides [10]. Through a series of very fast ($t_{1/2} < 100$ ps) energy transfer steps, this exciton ultimately leads to oxidation of P_{680} to $P_{680}^{+\bullet}$ [11,12]. It is noted, however, that multiple pathways exist to effect P_{680} oxidation involving pheophytin and/or chlorophyll bound to the D1 protein [13]. A plastoquinone molecule (Q_A) bound tightly to the D2 protein accepts the liberated electron forming $Q_A^{-\bullet} \cdot Q_A^{-\bullet}$ then transfers an electron, this time across a nonheme iron/bicarbonate mediator to another plastoquinone, Q_B ($t_{1/2} \approx 200$ μs), to yield $Q_B^{-\bullet}$ [14,15]. A second reduction of Q_A along with proton transfer to the Q_B-binding site ultimately leads to a quinol form of Q_B that diffuses into a pool of exchangeable plastoquinones within the thylakoid membrane [16,17]. This plastoquinol is oxidized back to quinone upon docking with the cytochrome b_6f complex, transferring electrons further along the chloroplast electron transport chain [18]. The $P_{680}^{+\bullet}$ cation that results from the initial electron transfer is reduced by a nearby tyrosine residue, D1-Tyr161 (Y_Z) [19,20]. It is postulated that oxidation of Y_Z is coincident with the transfer of the phenolic proton to the imidazole

---

* These extrinsic proteins include PsbO (the 33 kDa or manganese-stabilizing protein), PsbP (23 or 24 kDa protein), PsbQ (16–18 kDa protein), PsbR, PsbU, PsbV (cytochrome c_{550}, not found in plants), and Psb27.

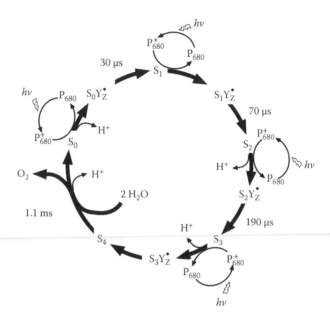

FIGURE 1.2 Modified Kok cycle, showing the light-induced oxidation of chlorophyll P_{680} that is quickly reduced by Y_Z to generate $Y_Z^\bullet$. This organic radical oxidizes the OEC by one electron, thus storing oxidizing equivalents for the water-splitting reaction. (From B. Kok et al., *Photochem. Photobiol.*, 11, 457, 1970.)

nitrogen of D1-His190 [21–26]. This neutral $Y_Z^\bullet$ radical, in turn, abstracts an electron from the nearby OEC ($r_{Y_Z\text{-OEC}} \approx 5$–11 Å)* [27–29].

This cycle repeats until a total of four oxidizing equivalents are stored on the OEC, which then dumps these electron holes into two water molecules, converting them to dioxygen and protons. The nomenclature (S_0–S_4) for these five photooxidized intermediates or S-states of the OEC was first introduced by Bessel Kok and is presented in Figure 1.2 [30,31]. In the absence of all light, PSII possesses natural mechanisms to return OECs that are at any of the various stages of the Kok cycle back to the S_1 state, also called the dark-adapted state. The oxidation state assignment for the four Mn centers in S_1 is believed to be $Mn(III)_2Mn(IV)_2$, a finding based on results from x-ray absorption spectroscopic (XAS) studies as well as the lack of a perpendicular-mode (⊥-mode) electron paramagnetic resonance (EPR) signal [32–34]. Resonances are found in the parallel-mode (∥-mode) EPR spectrum (see Section 1.9). This behavior generally implies that the system has an integer value for the electron spin quantum number S. A single, saturating flash of light ($t_{flash} < 1$ μs) upon dark-adapted PSII particles at ambient temperatures, or continuous white-light illumination at 195–200 K, advances the OEC to the S_2 state. The positive Mn edge shift (1.5 eV) observed in the XAS spectrum of PSII in the S_2 state compared with that in the S_1 state indicates Mn-centered oxidation [35]. The S_2 state is thus formulated as being $Mn(III)Mn(IV)_3$, i.e., with an odd number of electrons and,

---

* The broad range of reported distances derives from the manner in which this metric is calculated: from the phenolic oxygen to the nearest Mn ion or based on the interspin distance.

accordingly, "EPR active." One more light flash brings the OEC to the S_3 state; however, whether a similar oxidation state change occurs during the $S_2 \rightarrow S_3$ transition remains a hotly debated subject [33,36–38] (see Section 1.7 for more discussion). PSII particles poised in the S_3 state and flashed once more rapidly decay ($t_{1/2} = 1.1$ ms) to S_0, the most reduced state in the Kok cycle, coincident with release of molecular oxygen. On the basis of Joliot's observation of an oscillating pattern for flash-driven O_2 evolution [39], Kok originally suggested that an S-state, known as S_4, could transiently exist between S_3 and S_0 [30]. Attempts to determine the nature of this intermediate have been largely unsuccessful; although, results from time-resolved (10-µs resolution) room-temperature x-ray fluorescence studies have suggested that no Mn-centered oxidation occurs following administration of the third flash (starting from S_1) [40]. Instead, Dau and coworkers observed a 200-µs "lag" following the formation of S_3-$Y_Z^{\bullet}$, during which they postulate that a proton is released from a substrate water, a protonated µ-oxido bridge in the OEC, or from surrounding proteinaceous ligands [41,42]. After this lag period, the XAS edge position shifts to lower energy ($t_{1/2} \approx 1$ ms), signaling that the Mn-centered reduction and formation of the S_0 state have occurred along with release of a second proton [40,43–45].

EPR spectroscopy has long been an essential tool in the characterization of the numerous redox-active cofactors bound within the PSII supercomplex [46,47]. Advanced EPR studies, in particular, have been invaluable in understanding geometric and electronic structures of carotenoid pigment [48,49], P_{680} [50,51], chlorophyll [51,52], pheophytin [53,54], cytochrome b_{559} [55], quinone [56–60], the nonheme Fe site [61–64], and the stable radical $Y_D^{\bullet}$, a redox-active tyrosine residue located in the D2 peptide [65–68]. In native preparations of the enzyme, $Y_Z^{\bullet}$ is too quickly reduced by the Mn cluster to trap for study with advanced EPR methods. Notably, however, rapid formation and decay of an EPR signal (historically known as signal IIvf) attributed to $Y_Z^{\bullet}$ was monitored using time-resolved continuous-wave (CW)-EPR techniques [69,70]. Chemical inhibition of Mn to $Y_Z^{\bullet}$ electron transfer (Section 1.5) as well as very low-temperature illumination methods (Section 1.6) have allowed for some study of $Y_Z^{\bullet}$. Owing to the limitations in studying $Y_Z^{\bullet}$, many researchers have looked to characterize $Y_D^{\bullet}$ that resides at a site that is related to $Y_Z^{\bullet}$ by a C_2 rotation [27,67,71–73]. The precise role of Y_D is unclear. However, it has been implicated as being operative in the assembly of the OEC, OEC protection via oxidation of S_0 to the more stable S_1 state, and directing the primary electron transfer events such that they proceed along the D1 rather than the D2 branch of PSII [74,75].

In general, insight provided by EPR spectroscopic results is a necessary complement to even the highest-resolution crystal structure data, as redox heterogeneity induced by the x-ray beam tends to plague the interpretation of the resulting structure [76]. More important, EPR spectroscopy is able to characterize both electronically excited and otherwise short-lived states that crystallographic techniques rarely observe. Finally, EPR results afford great insights into the ground-state electronic structure.

Results from biochemical, kinetic, and spectroscopic assays have established that the inorganic cofactors that compose the OEC include the aforementioned four Mn ions as well as one Ca^{2+} and one Cl^- ion [77,78]. However, despite considerable research efforts, the precise geometric arrangement of the atoms found within the OEC and the evolution of this structure during the events of the Kok cycle have remained elusive. Of particular note is the vital role played by extended x-ray absorption fine structure (EXAFS) spectroscopy in determining the Mn⋯Mn and Mn⋯Ca internuclear distances for PSII trapped in all S-states but S_4. Recent reviews of these studies can be found in Refs. [33,79–81]. The many x-ray structures culminating with he 1.9 Å resolution structure presented in 2011 [82] show that the four manganese ions are distributed in a 3 + 1 arrangement wherein a trimer of oxido-bridged Mn centers are linked to the fourth, "dangling" Mn ion. Notably, this architecture was first posited on the basis of results from ^{55}Mn electron spin-echo-detected electron nuclear double-resonance spectroscopy (pulse ENDOR), a pulse EPR technique [83]. This is one example, among many, that illustrates how results from EPR spectroscopic studies have contributed to our understanding of the OEC structure and mechanism. A summary of these and many other EPR spectral studies that attempt to piece together the nature of the OEC and its protein environment is the subject of this review.

1.2   EPR CHARACTERIZATION OF THE S-STATES

Since it was first discovered that green laser light flashes or continuous white-light illumination of chloroplast membranes give rise to a CW-EPR signal that suggested the presence of redox-active Mn ions in PSII [84–86], EPR spectroscopy has been vigorously employed to study the OEC at all stages of the Kok cycle [38,46,87–92]. Unfortunately, interpretation of this flood of EPR spectral data is nontrivial owing to the presence of multiple unpaired electrons distributed among four exchange-coupled ^{55}Mn ions, each having a nuclear spin of $I = 5/2$ (^{55}Mn is 100% naturally abundant). The interaction of unpaired electrons with an ^{55}Mn nucleus leads to a 6-fold ($2I + 1$) increase in the number of EPR-allowed transitions. The number of hyperfine transitions increases further, in multiplicative fashion, upon the inclusion of additional magnetic nuclei. Therefore, for a single unpaired electron interacting with the four Mn ions of the OEC, one would expect an imbroglio of 1296 EPR lines. However, many of these resonances overlap and/or are obscured in the inhomogenously broadened EPR lineshapes, as exemplified by the mere 18–20 lines visible in the spectrum of PSII particles in the S_2 state (Figure 1.3a, Section 1.3). A final complication is encountered when one considers the potential for any unpaired electrons to couple to other nearby magnetic nuclei (e.g., the ^{14}N nuclei in the imidazole ring of histidine [93]). Very rarely are these ligand hyperfine couplings (HFCs) strong enough to be apparent in typical CW-EPR spectra as they also tend to be concealed within the inhomogeneously broadened EPR lineshape. Yet, advanced EPR techniques such as ENDOR and electron-spin-echo envelope modulation (ESEEM) spectroscopies provide access to these unresolved features. Results from these experiments can not only discern whether a certain ligand is coordinating to the OEC—a subject of enduring debate despite the ever-increasing quality of x-ray diffraction data—but can also report directly on the strength of this metal–ligand bond (i.e., covalency).

The interpretation of EPR spectra of the multiple paramagnetic species present in the vicinity of the OEC requires use of the following *uncoupled* spin Hamiltonian.

$$\hat{H} = \sum_{i}^{n}\sum_{j}^{m}\left[\mu_B \vec{B}_0 g_i \hat{S}_i + \hat{S}_i A_{ij} \hat{I}_j + \hat{S}_i D_i \hat{S}_i + \hat{I}_j P_j \hat{I}_j + \mu_N g_{N,j} \vec{B}_0 \hat{I}_j\right] - \sum_{i>k}^{n} \hat{S}_i J_{ik} \hat{S}_k \quad (1.1)$$

In this most-generalized effective spin Hamiltonian for n electron and m nuclear magnetic dipoles, the matrices that couple the angular momenta on a single site (e.g., g_i)

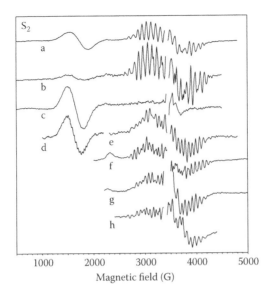

FIGURE 1.3 Low-temperature (4–10 K) X-band CW-EPR spectra of various PSII preparations poised in the S_2 state by continuous white-light illumination. In all cases, the spectrum of the nonilluminated or dark-adapted species was subtracted. (a) Untreated; (b) 5% (v/v) methanol added; (c) 100 mM NaF added; (d) 100 mM NH_4Cl added with PSII-containing particles oriented on Mylar sheets; (e) 100 mM NH_4Cl treatment followed by 273 K annealing; (f) Ca^{2+} depleted (NaCl wash); (g) Ca^{2+} depleted followed by reconstitution with Sr^{2+}; (h) D1-H332E mutant (*Synechocystis* sp. 6803).

do not correspond to more directly observable quantities such as the effective molecular g_{eff} matrix presented in Equation 1.8. Rather, they describe the electronic structure properties of an imaginary isolated paramagnet within the spin-coupled system. The first term in Equation 1.1 describes the electron Zeeman interaction with the applied magnetic field B_0. In the uncoupled formalism, a separate g-tensor (g_i) exists for each moiety that possesses unpaired electrons. The electron–nuclear hyperfine interaction is given by the second term and can be further decomposed into an isotropic contribution (a_{iso}) governed by unpaired electron density localized at the nucleus (i.e., unpaired electron density in s-orbitals; Fermi contact)* and a dipolar coupling tensor (T) reflecting the through-space interaction of the electron and nuclear spins [94,95].

$$A = a_{iso}(1) + T \tag{1.2}$$

where

$$T_{ij} = \frac{\mu_0}{4\pi\hbar}\mu_B g_N \mu_N \left\langle \psi \left| \frac{3g_{ij}r_ir_j - \delta_{ij}g_{ij}r^2}{r^5} \right| \psi \right\rangle \tag{1.3}$$

---

* The mechanisms of spin polarization—unpaired electron density in p- or d-orbitals mixed into s-orbitals by configuration interaction—and spin-orbit coupling are especially important contributions to a_{iso} in the case of transition metal complexes.

The electron and nuclear g-factors are given by g and g_N, respectively. μ_B and μ_N are the Bohr and nuclear magnetons, respectively. The term in bracket notation represents the integral of the $1/r^3$ operator (expanded to include one- and two-center terms) over the ground-state wave function ψ.

If anisotropy in the g-matrix can be ignored and the center containing the unpaired electron spin and magnetic nucleus are rather distant from one another ($r > 2.5$ Å) and orbital contributions to the HFC are small, Equation 1.3 can be simplified to the oft-used spin-only point-dipole approximation:

$$T = \frac{\mu_0}{4\pi\hbar} g_e\mu_B g_N\mu_N\rho\left(\frac{3\cos^2\theta-1}{r^3}\right)$$

(1.4)

The spatial separation between the electron and nuclear spin is represented by r, and the angle of this vector with respect to $\vec{B}_0$ is defined as θ. The unpaired spin population on the central ion is given by ρ.

When multiple unpaired electrons are present—as is the case with all high-spin Mn ions (e.g., the oxidation states of Mn relevant to the OEC are as follows: Mn(II) with $S = 5/2$, Mn(III) with $S = 2$, and Mn(IV) with $S = 3/2$)—their interactions with one another induces further fine structure in the EPR spectra as the degeneracy of spin multiplets at zero applied magnetic field is broken.* The third term in Equation 1.1 represents this zero-field splitting (ZFS) interaction. Such magnetic anisotropy is often understood in terms of an axial D and rhombic E distortions away from cubic symmetry. D and E are defined such that their ratio E/D is always between 0 (i.e., at the axial limit) and 1/3 (i.e., at the rhombic limit).

$$\hat{S}D\hat{S} = D\left(\hat{S}_z^2 - \frac{S(S+1)}{3}\right) + E(\hat{S}_x^2 - \hat{S}_y^2)$$

(1.5)

On the basis of EPR spectroscopic and magnetization studies of mononuclear Mn-containing compounds, the dependence of the magnitude of the ZFS interaction on the Mn ion oxidation state trends as $|D_{Mn(II)}| < |D_{Mn(IV)}| \ll |D_{Mn(III)}|$, with Mn(III)-containing complexes exhibiting ZFS constants almost an order of magnitude greater ($|D| = 1$–4 cm^{-1}) than either Mn(IV) or Mn(II) [96]. Therefore, the presence of Mn(III) ions in the exchange-coupled OEC can have significant influence on the observed spectroscopic properties, while ZFS contributions from Mn(IV) and Mn(II) can often be ignored.

The final two terms within the double sum in Equation 1.1 represent the nuclear quadrupole (for nuclei with $I > 1/2$) and nuclear Zeeman interactions, respectively. While these contributions to the spin Hamiltonian are not usually manifested when interpreting field-swept EPR spectra, they are essential to our understanding of

---

* Of course, ZFS effects cannot overcome Kramers degeneracy.

ESEEM and ENDOR results. Much like ZFS, the quadrupolar term in the effective spin Hamiltonian can be rewritten as

$$\hat{I}P\hat{I} = P\left(3\hat{I}_z^2 - I(I+1)\right) + \eta\left(\hat{I}_x^2 - \hat{I}_y^2\right) \tag{1.6}$$

where $P = \dfrac{e^2 q_{zz} Q}{4I(2I-1)\hbar}$, with Q being the nuclear quadrupole coupling constant, and q_{zz} is defined as the largest magnitude of the electric field gradient (EFG) centered at the nucleus. Identification of this principal quadrupole axis is of particular importance as its orientation, relative to the molecular frame, usually coincides with the direction of the dominant bonding interaction. The asymmetry parameter $\eta = \left|\dfrac{q_{xx} - q_{yy}}{q_{zz}}\right|$ gives the deviation of the EFG from axial symmetry (η can range from 0 to 1 describing an axial and rhombic EFG, respectively). These nuclear quadrupole interaction (NQI) parameters have been shown to be predictive of the coordination environment for hyperfine-coupled ^{14}N nuclei. Specifically, knowledge of the NQI can allow one to distinguish between imidazole, amino, and amido nitrogens [97,98].

The final term in Equation 1.1 couples the electrons formally localized on one metal center to those on adjacent metal centers. While the J matrix has isotropic, symmetric, and antisymmetric contributions [99], EPR data of the OEC have only been interpreted using the isotropic exchange-coupling parameter J_{ik}. In this approximation, the Heisenberg–Dirac–van Vleck Hamiltonian is used.

$$H_{\mathrm{HDvV}} = -2\sum_{i<k} J_{ik}\hat{S}_i\hat{S}_k \tag{1.7}$$

Six such pairwise electronic exchange-coupling interactions are possible in the tetranuclear OEC, and their magnitudes serve as a measure of the degree of electron sharing between every set of two Mn ions. This interaction tends to be strongest* through the covalent bonds provided by single atom bridges such as the μ-oxido atoms that are prevalent in the OEC [100–103]. The strength of the exchange-coupling interaction is also affected by the number of bonds in the exchange pathway and the angle formed by the two metal ions and the bridging atom(s) [104].

In the presence of exchange interactions, all electron spin momenta couple to yield a new ladder of possible spin states S_T (an example for an Mn(III)Mn(IV) dimer is presented in Figure 1.4), where $S_T = \sum_i |S_i|, \left(\sum_i |S_i|\right) - 1, \ldots \geq 0$. Temperature- and microwave power-dependent saturation CW-EPR studies can provide an estimate of the energy gap between the spin-coupled ground state and the first "rung" of this spin ladder (as can magnetic susceptibility and variable-temperature variable-field

---

* "Strong" meaning large, negative J_{ik}, and thus antiferromagnetic using the convention described by the effective spin Hamiltonian equation (Equation 1.7).

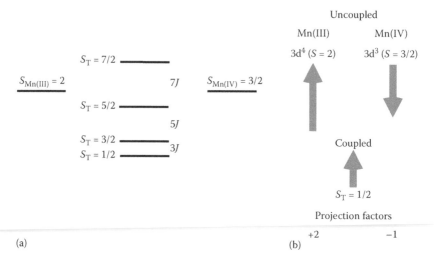

FIGURE 1.4 Spin ladder (a) illustrating the new system of spin manifolds that results from coupled two-spin angular momenta. In this example, $S = 3/2$ from Mn(IV) and $S = 2$ from Mn(III) couple to give rise to a net $S_T = 1/2$ spin ground state (b).

magnetic circular dichroism) [105]. For pulse EPR applications, the inversion recovery pulse sequence [106] can be used to distinguish between Raman [107] and Orbach [108] relaxation kinetics. If the latter mechanism is operative, then the temperature dependence of the spin-lattice relaxation rate (T_1) can be used to determine the energy separation between the ground and first excited states. This parameter is an essential constraint when searching for possible values for the six isotropic exchange-coupling constants in the OEC (Section 1.3.1).

In the coupled or molecular spin-state representation, Equation 1.1 can be rewritten as

$$\hat{H} = \mu_B \vec{B}_0 \boldsymbol{g}_{eff} \hat{S}_T + \hat{S}_T \boldsymbol{D}_{eff} \hat{S}_T + \sum_j^m \left[\hat{S}_T \boldsymbol{A}_{eff,j} \hat{I}_j + \hat{I}_j \boldsymbol{P}_j \hat{I}_j + \mu_N g_{N,j} \vec{B}_0 \hat{I}_j \right] \quad (1.8)$$

where $\boldsymbol{g}_{eff}$ is the effective g-matrix for the total coupled molecular spin S_T, and, if $S_T > 1/2$, $\boldsymbol{D}_{eff}$ is the corresponding molecular ZFS tensor. The interaction of this net spin S_T with nearby magnetic nuclei j is governed by new effective hyperfine matrices, $\boldsymbol{A}_{eff,j}$. To interpret these measured hyperfine parameters, site-specific values must be calculated through use of the projection factors c_i.

$$A_{eff,j} = \sum_i c_i a_{ij} \quad (1.9)$$

The values c_i are a function of the oxidation state of site i, the total spin of the system (S_T), the individual site ZFS tensors, and the magnitude of the exchange-coupling parameter (J_{ik}) to the other centers of unpaired spin. Alternatively, c_i can be considered as the ratio between real or intrinsic (site-specific) and effective unpaired spin vectors. a_{ij} can be thought of as the intrinsic HFC of the magnetic nucleus to site i, and it is only from a_{ij} that the degree of covalency and internuclear distance can be calculated.

More commonly, EPR spectroscopy is performed in "perpendicular mode" ($\perp$) wherein the oscillating magnetic field vector of the incident microwave radiation ($\vec{B}_1$) is aligned perpendicularly to the direction of the static magnetic field ($\vec{B}_0$). The selection rules for this technique allow for only the $\Delta m_S \pm 1$ transitions to be observed in the corresponding spectra. Such experiments are employed to study paramagnetic systems with odd numbers of unpaired electrons ($S = 1/2$, $3/2$, etc.), as there will always be sets of levels whose Kramers degeneracy can only be lifted by the applied magnetic field. Therefore, one will inevitably find a magnitude of $\vec{B}_0$ that induces a splitting in the spin levels that matches the energy of the microwave photon (i.e., creating a resonance condition). As is illustrated in the following, $\perp$-mode EPR experiments are central to investigations of PSII poised in the S_0, S_2, and $S_2Y_Z^{\bullet}$ states.

The sequential photooxidation of the OEC is also expected to lead to non-Kramers-type systems for S_1 and S_3 states. This assumes that Mn-centered oxidation is coincident with each S-state advancement, a point that is not universally accepted. In these cases, there is an even number of unpaired electrons and the ground spin state depends on the nature of the exchange-coupling interactions between metal centers. Owing to significant ZFSs (D and E from Equation 1.5), the electron spin energy levels for states of integer-spin species whose m_S differ by ± 1 are often too distant from one another for a transition to be induced by the commonly available microwave frequencies. Occasionally, such resonances can be observed, although most often by using higher-frequency microwaves. Some signals are found in the X-band (≈ 9 GHz) $\perp$-mode CW-EPR spectra of PSII in the S_3 state, suggesting that the corresponding ZFS energy gaps between levels is on the order of 0.3 cm^{-1} [38,109]. Possible origins of these features are discussed in Section 1.7.

Alternatively, the electronic structure of integer-spin species can be probed using parallel-polarization ($\parallel$-mode) EPR experiments. With the orientations of the $\vec{B}_0$ and $\vec{B}_1$ fields parallel to each other, new selection rules take effect and only $\Delta m_S = 0$ transitions can be observed. In cases where the energy splitting between the spin multiplets at zero field is less than the energy of the microwave photon, a resonance condition can be attained in a $\parallel$-mode experiment. At low values of $\vec{B}_0$, m_S is not a good quantum number and the pure Zeeman spin-state wave functions (high-field limit) mix. The rhombic ZFS term E can further mix the spin levels, helping satisfy the $\Delta m_S = 0$ selection rule (see Equation 1.5). More advanced EPR techniques such as ESEEM and ENDOR have been applied to a select few non-Kramers-type spin systems [110–114]; however, no such studies of the OEC have been undertaken.

A recent review summarizes the appearance and properties of CW-EPR signals that have been observed for PSII prepared in S_0, S_1, S_2, and S_3 states, as well as those for a variety of split-signals S_n–$Y_Z^{\bullet}$ [92]. Therefore, here we will focus on the discussion of ESEEM and pulse ENDOR results, and will address the recent use of $\parallel$-mode and multifrequency EPR methods applied to the study of the OEC. CW-EPR spectra will only be discussed when they aid in the interpretation of the pulse EPR data.

A key tool needed in the interpretation of the measured magnetic parameters is the density functional theory (DFT) calculations. Until recently, most computational studies of PSII attempted to determine the structure of the OEC during each stage of the water oxidation mechanism [115–119]. These calculations have been able to produce several models of the OEC that have structural and thermodynamic properties consistent with XAS and kinetic results, respectively. More discerning, however, is computation of the EPR parameters of the OEC, which, based on the spectroscopic evidence discussed herein, are keenly sensitive to its redox and chemical environment. Progress on this front has been realized very recently with the development of a convincing computational OEC model [120] later informed by the high-resolution crystal structure [82]. Using broken-symmetry DFT [121–124], molecular g-values and magnetic nuclei HFCs can be routinely computed for exchange-coupled spin systems [125], and successfully so for models of the OEC [126–134].

As it is the most-often studied, results from spectroscopic and computational studies of the S_2 state of PSII will be presented first.

1.3 S_2-MULTILINE SIGNAL

The S_2 state can be prepared simply by several minutes of low-temperature (195 K) continuous white-light illumination without risk of further advancement through the Kok cycle. Oxidation of the OEC to the S_3 state during continuous illumination below 210 K is blocked as a proton translocation event on the acceptor side is necessary to prevent charge recombination [135,136]. This attribute and the fact that it is relatively long lived even at room temperature ($t_{1/2} \approx 200$ s) [137] and gives rise to an intense EPR spectrum have allowed S_2 to become the best spectroscopically characterized S-state. As discussed previously, a number of other species present in PSII give rise to EPR signals; however, these contributions can be removed by subtracting the spectrum of dark-adapted PSII particles from the signal of illuminated particles. Features in such "light-minus-dark" spectra are interpreted as arising exclusively from the photooxidized OEC. (One should note that there can also be a light-induced signal that arises from reduction at the Q_A-site, $Fe^{2+}Q_A^-$ [63,138] as well as additional intensity from cytochrome b_{559} [139,140], $Y_D^\bullet$, and other pigment-derived radicals with narrow EPR signals centered at $g \approx 2$.) The X-band $\perp$-mode CW-EPR spectrum of the S_2 state of the OEC exhibits two distinct features that have been attributed to the Mn cluster: an 18–20 line multiline signal (MLS) centered at $g = 1.98$, and a broad derivative-shaped feature at $g = 4.1$ (see Figure 1.3a). The origin of each of these features is discussed in the following.

The MLS is spread across ≈ 1800 G and exhibits line spacings ranging from 85 to 90 G, which arise from ^{55}Mn hyperfine interactions. The large spectral breadth associated with the MLS is a product of numerous factors, including anisotropy of g-values (while not noticeable at X-band excitation frequencies, the effect of the rather small g-anisotropy is noticeable using higher-field EPR spectroscopy; see Table 1.1); the number of spin-coupled metal centers (four in this case); their individual oxidation states (mononuclear Mn(II) sites tend to have larger hyperfine constants than analogous Mn(III) or Mn(IV) sites); and, most important, the values of the site-specific

TABLE 1.1
EPR Parameters and ^{55}Mn HFC Constants (in MHz) for OEC Poised in Different S-States

PSII Preparation					Method (Freq.)	References
S_2 (spinach; 3% ROH)	$S = \frac{1}{2}$					
	$g = [1.97, 1.97, 1.99]$					[83]
	$g_x = 1.988, g_y = 1.981, g_z = 1.965$				CW-EPR (W)	[141]
	$g = [1.997, 1.970, 1.965]$				ESE-EPR (W)	[142]
	Nucleus	a_{iso}	T	$P_\parallel$		
	^{55}Mn$_A$	−245	−13	−3	Pulse ENDOR (X)	[89]
	^{55}Mn$_B$	+217	+17	−3		
	^{55}Mn$_C$	−297	+14	+8		
	^{55}Mn$_D$	+200	+20	+1		
	^{55}Mn$_A$	245	−10		Pulse ENDOR (Q)	[143]
	^{55}Mn$_B$	205	−20			
	^{55}Mn$_C$	295	15			
	^{55}Mn$_D$	195	−20			
	^{55}Mn$_A$	248	−13		Pulse ENDOR (Q)	[144]
	^{55}Mn$_B$	205	−20			
	^{55}Mn$_C$	298	12			
	^{55}Mn$_D$	193	−23			
S_2 (*T. elongatus*; 3% ROH)	$S = \frac{1}{2}$					
	$g = [1.971, 1.948, 1.985]$					[130]
	Nucleus	a_{iso}	T	$P_\parallel$		
	^{55}Mn$_A$	251	−13		Pulse ENDOR (Q)	[130]
	^{55}Mn$_B$	208	−17			
	^{55}Mn$_C$	312	18			
	^{55}Mn$_D$	191	−36			
S_2 (spinach; 100 mM NH$_4$Cl)	$S = \frac{1}{2}$					
	$g = [1.99, 1.99, 1.96]$					[83]
	Nucleus	a_{iso}	T	$P_\parallel$		
	^{55}Mn$_A$	+191	−17	−3	Pulse ENDOR (X)	[83]
	^{55}Mn$_B$	−137	+13	−3		
	^{55}Mn$_C$	+205	−17	1		
	^{55}Mn$_D$	−333	−28	8		
S_2 (*T. elongatus*; 2 mM NH$_3$)	$S = \frac{1}{2}$					
	$g = [1.993, 1.974, 1.964]$					
	Nucleus	a_{iso}	T	$P_\parallel$		
	^{55}Mn$_A$	321	+98		Pulse ENDOR (Q)	[126]
	^{55}Mn$_B$	246	−31			
	^{55}Mn$_C$	209	−44			
	^{55}Mn$_D$	170	−58			
	$S = 5/2$					
	$g_{eff} = 4.1$					

(Continued)

TABLE 1.1 (CONTINUED)

EPR Parameters and ^{55}Mn HFC Constants (in MHz) for OEC Poised in Different S-States

PSII Preparation	Nucleus	a_{iso}	T	$P_{\parallel}$	Method (Freq.)	References
	$^{55}Mn_A$	132			CW-EPR (X)	[145]
	$^{55}Mn_B$	104				
	$^{55}Mn_C$	95				
	$^{55}Mn_D$	45				
S_0 (spinach; 3% ROH)	$S = \frac{1}{2}$					
	$g = [1.99, 1.99, 1.89]$					[143]
	$g = [2.009, 1.855, 1.974]$					[144]
	Nucleus	a_{iso}	T	$P_{\parallel}$		
	$^{55}Mn_A$	245	25		Pulse ENDOR (Q)	[143,144]
	$^{55}Mn_B$	220	−30			
	$^{55}Mn_C$	345	−35			
	$^{55}Mn_D$	195	−25			
S_1 (−17 and −23 kDa)	$S = 2$					
	$g = [2.0\ 2.0\ 1.942]$		$D = 0.805\ cm^{-1},$ $E = 0.26\ cm^{-1}$			[146]
	Nucleus	a_{iso}	T	$P_{\parallel}$		
	$^{55}Mn_A$	186	10		CW-EPR (X)	
	$^{55}Mn_B$	186	10			
	$^{55}Mn_C$	186	10			
	$^{55}Mn_D$	186	10			

projection factors. For example, the projection factors for an Mn(II,III) dimer ($c_{Mn(II)}$ = +7/3 and $c_{Mn(III)}$ = −4/3) are greater than those for an Mn(III,IV) dimer ($c_{Mn(III)}$ = +2 and $c_{Mn(IV)}$ = −1); thus, one would expect the EPR spectrum of the former to be broader than that of the latter, an inveterate fact based on experimental EPR data of such complexes [147].

1.3.1 Electronic Structure of the Tetranuclear Mn Cluster

Attempts to simulate the MLS using Equation 1.8 resulted in a number of satisfactory fits of the CW-EPR spectrum, although with disparate spin Hamiltonian parameters [29,83,96,148–151]. These simulations differed most prominently in the values for the ^{55}Mn HFC and the number of coupled Mn nuclei. Confirming earlier predictions of the g-values for the S_2 state [83], recent high-frequency (W-band, 94 GHz) CW [141] and pulse EPR [142] studies of PSII single crystals have shown that the anisotropy in g_{eff} is rather small ($\Delta g_{eff} < 0.04$; see Table 1.1), precluding this as an origin of the observed large spectral width. Furthermore, on the basis of the lack of a prominent edge feature in XAS spectra, which is characteristic of manganous ions, it seems unlikely that Mn(II) is present in the S_2 form of the OEC [152].

One of the more intriguing aspects of the MLS is its apparent equilibrium with the signal at $g = 4.1$ (for more information on the origin of the $g = 4.1$ signal, see Section 1.4). This equilibrium can be manipulated to favor the MLS by the addition of 3%–5% (v/v) methanol (or any of a number of other small terminal alcohols; Figure 1.3b), 50% glycerol, or 30% polyethylene glycol to the buffering solution. Alternatively, additions of sucrose [153,154], certain amines [155], F^- (Figure 1.3c) [156], or other inhibitors of Cl^- binding lead to formation of the $g = 4.1$ signal in greater yield [77,157,158]. A few site-directed mutants have been shown to enhance production of the $g = 4.1$ signal, including mutation of the D1 C-terminus, D1-Ala344, to glycine [159]. Illumination conditions also seem to play a large role. The $g = 4.1$ signal is generated upon white-light illumination of dark-adapted PSII particles or near-infrared (NIR) illumination of PSII in the S_2 state at 130–140 K. Yet, as previously stated, the MLS is most easily achieved by white-light illumination at temperatures around 195 K.

Methanol treatment provides an attractive method for generating the MLS almost exclusively while not inhibiting oxygen-evolving activity [160]. Thus, most of the pulse EPR studies cited throughout Section 1.3.1 used methanol (or ethanol)-treated preparations of PSII. Notably, however, results from ESEEM investigations indicate that methanol likely coordinates directly to the Mn cluster and perhaps in a substrate water-binding site [160,161]. Furthermore, methanol-treated PSII from spinach (but not from cyanobacteria [162]) exhibits no sensitivity to NIR light, while 40% of centers in untreated PSII convert to the $g = 4.1$ signal upon NIR illumination [163–165]. These findings indicate that methanol may also perturb the exchange interactions between some of the Mn centers (and by extension, the effective ^{55}Mn HFCs; cf. Equation 1.9). Indeed, comparisons of pulse ENDOR spectra of methanol-treated and untreated PSII isolated from spinach show some small (<5%) differences in measured ^{55}Mn frequencies (cf. Figure 1.5a,b) [28,83,130]. Interestingly, the difference between methanol-treated and untreated PSII from the thermophilic cyanobacterium *Thermosynechococcus elongatus* are even smaller [130]. Through simulation of just the CW-EPR data, Charlot et al. derived two distinct sets of ^{55}Mn HFC values for methanol-treated and untreated PSII trapped in the S_2 state [166]. Their results indicate that the isotropic contribution to the hyperfine matrix for each ^{55}Mn nucleus changes by <2% upon addition of methanol, whereas the anisotropic components change by much more (cf. Table 1.1).

Through application of saturation and inversion recovery methods [106], the MLS of alcohol-treated PSII has been shown rigorously to arise from an antiferromagnetically coupled $S_T = 1/2$ ground state [130,167], and, in agreement with earlier CW-EPR studies [168,169], the energy of an excited spin-state manifold ($S_T = 3/2$) was found to be $\approx$30–40 cm^{-1} (Δ) above the ground state [170–172].* This excited state energy is a crucial parameter in the evaluation of possible exchange-coupling schemes in models of the OEC as Δ is a function of the values for J_{ik}. Belinskii

---

* Su et al. observed that methanol-treated PSII from both *T. elongatus* and spinach exhibit very similar values for Δ (22.4 and 24.7 cm^{-1}, respectively) [130]. For untreated spinach PSII, however, they found $\Delta = 2.7$ cm^{-1}. This small energy gap between the $S_T = 1/2$ and $S_T = 3/2$ states led to significant state mixing and rationalized the dramatic changes in the ^{55}Mn ENDOR spectrum (Figure 1.5) observed upon treating spinach PSII with methanol (see Table 1.1).

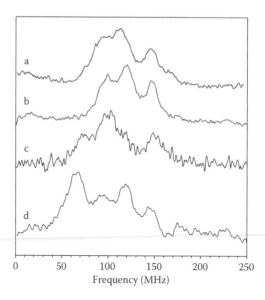

FIGURE 1.5 Light-minus-dark [55]Mn ESE-Davies ENDOR spectra of various preparations of spinach PSII: (a) untreated; (b) 5% (v/v) methanol added; (c) 100 mM NH_4Cl treatment followed by 273 K annealing; (d) 400 mM acetate treatment followed by 5 s illumination at 273 K. (From J.M. Peloquin et al., *J. Am. Chem. Soc.*, 120, 6840, 1998; J.M. Peloquin et al., *J. Am. Chem. Soc.*, 122, 10926, 2000; J.M. Peloquin, R.D. Britt, *Biochim. Biophys. Acta*, 1503, 96, 2001.)

has derived all matrix elements for the general Heisenberg Hamiltonian for an arbitrary $Mn(IV)_3Mn(III)$ cluster with an $S = 1/2$ ground state [173]. These calculations indicate that an accurate treatment of the exchange-coupling problem in the Mn tetramer will ultimately require the inclusion of excited spin states as these are expected to admix into the ground state (as in the case described in footnote on page 15). However, given the currently available data, it seems that our understanding of exchange coupling within the Mn cluster will have to endure certain approximations.

As described in Section 1.1, only when [55]Mn X-band pulse ENDOR data (Figure 1.5) became available [83,174] was a unifying model proposed that could be used to simulate both CW-EPR and ENDOR data for the S_2 state while remaining consistent with structural results from EXAFS spectroscopy. These early results have since been confirmed by analogous studies at Q-band [143]. The measured effective [55]Mn HFC constants are presented in Table 1.1 and were used to constrain the calculation of the *relative* magnitudes of the six possible intermanganese exchange-coupling parameters (see Equation 1.7). The energy of the first excited spin manifold (Δ)—30–40 cm^{-1} in this case—is used to calculate the *absolute* magnitude of each J-coupling parameter. From this initial [55]Mn ENDOR data, the best fit was achieved when, of these six couplings, three were set to zero; two couplings (J_{AB} and J_{BC}) were strongly antiferromagnetic ($J > -100$ cm^{-1}); and the link to the fourth Mn was much weaker, ranging from $J_{CD} = -20$ cm^{-1} to $+20$ cm^{-1}. This so-called 3-J

coupling scheme is consistent with a monomer–trimer type of arrangement for the Mn in the OEC and became known as the "dangler" model (see Mn4 in Figure 1.1). Once in hand, this set of J-values allows for the determination of the site-specific projection values using Equation 1.9 and the numerical recipes given in Bencini and Gatteschi [99]. In following this procedure, Peloquin et al. computed intrinsic ^{55}Mn HFC values that were consistent with a single Mn(III) ion ($d = -1.25$ to -2.25 cm^{-1}) being present in the S$_2$ form of the OEC [83,89]. Importantly, the simulation strongly suggests that the Mn(III) ion is located within the trimer. The value of the axial ZFS parameter implies that the Mn(III) site exists in either a five-coordinate square pyramidal or tetragonally elongated octahedral geometry.

Alternatively, the analysis can proceed by first assuming certain intrinsic ^{55}Mn HFC constants for Mn(III) and Mn(IV) [175–177].* Using these site-specific HFC values, the set of c_i needed to calculate the effective HFC parameters determined by either CW-EPR or, more preferably, ^{55}Mn ENDOR experiments can be derived, and from these, the J-couplings can be extracted (Figure 1.6). This second approach was employed in the J-space and projection factor calculations conducted by Charlot et al. [166], in which it is argued that four quite strong J-couplings (4-J scheme; -130 to -290 cm^{-1}) are required to simulate the MLS and to compute the energy of an excited spin state just 30 cm^{-1} above the ground state. However, the influence on effective HFC parameters due to ZFS of the Mn(III) ion(s) is not taken into account, nor were the simulations constrained by experimentally determined effective ^{55}Mn HFC constants from the variety of ENDOR studies discussed previously. These omissions give rise to purely isotropic (and, thus, perhaps unrealistic) projection factors (see Equation 1.9) and too large effective ^{55}Mn HFC, respectively.

Importantly, by only modifying the magnitudes of the exchange-coupling parameters and the locality of the Mn(III) within the trimer, an $S_T = 5/2$ ground state can be achieved that models the form of S$_2$ responsible for the $g = 4.1$ signal (see Section 1.4) [158,178]. Further manipulation of the J couplings can give rise to an $S_T = 7/2$ ground state. Such a formulation for the S$_2$ state is relevant as it is believed to be generated by the NIR illumination-induced decay of PSII particles poised in the S$_3$ state [179,180].

1.3.2 LIGANDS TO THE OEC

While the combination of results from EXAFS, x-ray crystallographic, and ^{55}Mn ENDOR spectroscopic studies have provided a framework for the generation of models of the OEC architecture, the precise nature of the ligand set that binds the Mn cluster to the protein matrix remains undetermined. EXAFS spectroscopy cannot distinguish between O and N coordinating atoms, and it is now generally accepted that the several reported x-ray structures suffer from significant photodamage leading to Mn reduction and Mn-ligand bond breakage. However, the x-ray diffraction data can be relied on to provide candidates for amino acid-derived ligands. These

---

* In most cases, intrinsic ^{55}Mn HFC values have been derived on the basis of results from EPR spectroscopic studies of monomeric Mn(III) [175] and Mn(IV) [176] systems as well as Mn(III,IV) dimers (see Supporting Information of Ref. [177]).

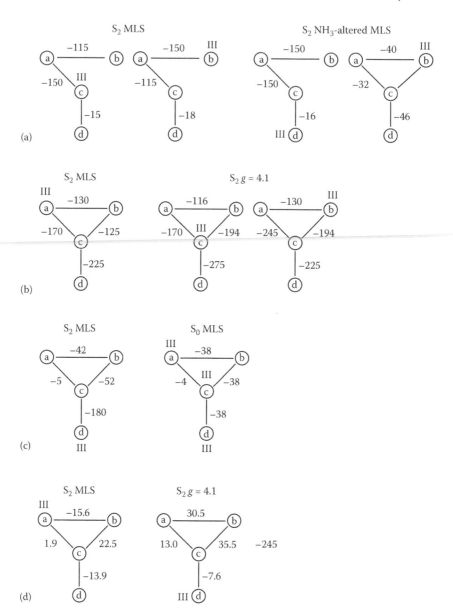

FIGURE 1.6 Proposed exchange-coupling schemes for the tetranuclear Mn cluster from (a) J.M. Peloquin et al., *J. Am. Chem. Soc.*, 122, 10926, 2000; J.M. Peloquin, R.D. Britt, *Biochim. Biophys. Acta*, 1503, 96, 2001; (b) M.-F. Charlot et al., *Biochim. Biophys. Acta Bioenerget.*, 1708, 120, 2005; (c) L.V. Kulik et al., *J. Am. Chem. Soc.*, 129, 13421, 2007; and (d) D.A. Pantazis et al., *Angew. Chem. Int. Ed.*, 51, 9935, 2012. The Mn ion labeled D is considered to be the dangling Mn ion. Computed isotropic exchange-coupling parameters (J_{ij}) are given between the two corresponding spin centers in units of cm^{-1}. "III" labels the Mn(III) ion; all other ions are in the +4 oxidation state.

include D1-Asp170, D1-Glu189, D1-His332, D1-Glu333, D1-Asp342, CP43-Glu354, and the carboxy terminus of the D1 polypeptide D1-Ala344 (Figure 1.1). Furthermore, mutation of these residues (we refer the interested reader to the following examples: D1-Asp170 [181], D1-Glu189 [182], D1-His332 [183], D1-Ala344 [184–187], and CP43-Glu354 [188]) can retard or even halt oxygen-evolving activity [189,190].

As a definitive determination of which amino acids are ligated to specific Mn nuclei is unavailable from crystallographic studies, our group and others have looked to ESEEM and ENDOR spectroscopies to probe the nature of the OEC ligand set [191,192]. X-band (9 GHz) ESEEM data collected in resonance with the MLS of S_2 show one broad peak centered at ≈5 MHz (see Figure 1.7B.a) that disappears upon global ^{15}N-labeling of histidine [193,194]. In assigning this hyperfine-coupled N nucleus to a specific nearby nitrogen-containing amino acid (i.e., His, Arg, Gln, etc.), we note that only mutations of CP43-Arg357 [195], D1-His332 [196], and D1-His190 [182] deleteriously effect O_2-evolution activity. D1-His190 is postulated to be a hydrogen bond partner with Y_Z and thus is too far away to coordinate directly to Mn, and CP43-Arg357 is positioned in the second coordination sphere. Therefore, the peak observed in the X-band ESEEM spectrum is attributed to D1-His332. Indeed, later results from analogous ESEEM studies showed that the peak at ≈5 MHz disappears as the D1-His332 is mutated to a glutamate (see also Section 1.3.4.4) [197].

Unfortunately, this single peak in the ESEEM spectrum is insufficient to determine the hyperfine and nuclear quadrupole coupling terms of the corresponding ^{14}N atom (or atoms). Therefore, analogous studies (Figure 1.7B.b through B.e) were performed at the higher excitation frequencies of P (14 GHz) up to Ka (30.5 GHz) band. These results indicate that at ≈30 GHz, the system is very near the so-called exact-cancellation regime wherein the hyperfine-induced shift of the nuclear energy levels in one electron spin manifold (by notational convenience, we assume this manifold to be the α-electron spin manifold) is nearly exactly offset by the field-dependent nuclear Zeeman effect (i.e., $A_{iso} \approx \nu_N/2$; cf. Figure 1.7A). This results in an ESEEM spectrum where up to three sharp peaks that correspond to pure quadrupolar (NQI) transitions in the α manifold can be observed at low frequencies and one broader peak at higher frequency corresponds to the double quantum (dq) transition in the β manifold (see Figure 1.7) [198–200]. In the limit where the anisotropic HFC (T) is small relative to A_{iso}, the quadrupolar coupling parameters described in Equation 1.6 can be calculated for an $I = 1$ nucleus directly from the frequency of these NQI transitions as follows:

$$\nu_{\alpha,0} = \frac{1}{2}\eta e^2 q_{zz}Q, \quad \nu_{\alpha,-} = \frac{1}{4}(3-\eta)e^2 q_{zz}Q, \quad \nu_{\alpha,+} = \frac{1}{4}(3+\eta)e^2 q_{zz}Q \quad (1.10)$$

and an upper bound of the frequency of the dq transition is given by

$$\nu_{\beta,dq} = 2\left[\left(\nu_I + \left|\frac{A_{iso}}{2}\right|\right)^2 + \frac{(e^2 q_{zz}Q)}{16}(3+\eta^2)\right]^{\frac{1}{2}} \quad (1.11)$$

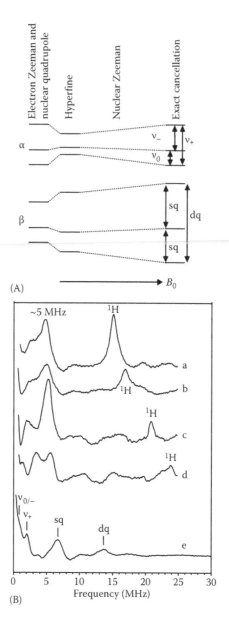

(A)

(B)

FIGURE 1.7 (A) Qualitative nuclear spin energy-level diagram describing the effect of "hyperfine cancellation" wherein the field-dependent nuclear Zeeman contribution (third column) in one electron spin manifold (the top or α manifold in this case) cancels out the shift in energy due to HFC (second column). (B) Multifrequency cosine Fourier-transformed frequency-domain two-pulse ESEEM spectra of PSII trapped in the S_2 state (light minus dark). All data were acquired in resonance with $g = 1.98$ using the following field positions and excitation frequencies, respectively: (a) 3525 G, 9.749 GHz; (b) 3955 G, 10.995 GHz; (c) 4875 G, 13.530 GHz; (d) 5603 G, 15.517 GHz; (e) 11,079 G, 30.757 GHz. (From G.J. Yeagle et al., *Inorg. Chem.*, 47, 1803, 2008; G.J. Yeagle et al., *Philos. Trans. R. Soc. B*, 363, 1157, 2008.)

On the basis of simulations of these multifrequency ESEEM data sets (and confirmed by separate ENDOR studies [201]), Stich and Yeagle and coworkers have determined the following HFC and NQI parameters for the coupled ^{14}N nucleus: $A_{iso} = 6.95$ MHz, $A_{aniso} = [0.2, 1.3, -1.5]$ MHz, $e^2Qq_{zz}/h = 1.98$ MHz, and $\eta = 0.82$ [202–204]. Importantly, the major features in ESEEM data sets acquired at all excitation frequencies are satisfactorily simulated using only these HFC parameters, meaning only one N nucleus is strongly magnetically coupled to the Mn cluster in the S_2 state. Furthermore, the rather large value of A_{iso} derived from these studies would seem to rule out the coupling of D1-His332 to the Mn cluster via a hydrogen bond network of any length. Rather, Stich et al. propose that the τ-N of the imidazole ring is directly coordinated to Mn. Our analysis of the HFC and what we believe the projection factors are for the S_2 state point to this coordinating Mn ion to be the lone Mn(III) center. On the basis of the multitude of x-ray crystal structures available, which consistently show D1-His332 bound to Mn1 (see Figure 1.1), we assign this Mn as the manganic ion at low temperature. Results from broken-symmetry DFT studies essentially agree with this oxidation state assignment [126,134], although the A_{iso} for ^{14}N tends to be underestimated.

As an interesting side note, the D1-H332Q and D1-H332S mutant is modestly active, exhibiting 10%–15% of wild-type activity for PSII from *Synechocystis* [196]. This may suggest that whatever the role of the D1-His332 nitrogen—be it Mn ligand or participant in an essential hydrogen-bonding network—the amido nitrogen from glutamine or the serine alcohol can substitute, albeit poorly. However, we are not aware of any EPR spectroscopic studies that have attempted to measure possible magnetic interactions between the Mn cluster and the amido group of glutamine-332. Notably, these same mutations of PSII from the thermophile *T. elongatus* maintain 80% of wild-type oxygen-evolution activity and do not affect the X-band ESEEM spectrum [205]. In fact, several phototrophically grown *T. elongatus* mutants of PSII exhibit properties that are very similar to wild-type enzyme, whereas those of analogous heterotrophically grown *Synechocystis* mutants are dramatically altered (cf. results in Ref. [206] to those from Refs. [11,207,208]). Two other classes of weakly coupled nitrogen-containing moieties have been proposed based on X-band hyperfine sublevel correlation (HYSCORE) spectra of PSII poised in the S_2 state [209]. HYSCORE spectra are populated by correlation peaks or ridges that cross at the frequencies of nuclear transitions of the hyperfine-coupled nucleus in opposite electron spin manifolds [210]. The reported isotropic hyperfine values for these nitrogens are very small ($A_{iso} < 0.3$ MHz), suggesting that they are not participating in a bond to Mn. Uncoordinated peptide nitrogens, however, could give rise to such spectroscopic features.

Of the many possible carboxylate ligands to the OEC that have been postulated, only one has been confirmed via advanced EPR studies: D1-Ala344. This amino acid is the carboxy terminus of the D1 polypeptide, and results from several studies indicated that D1-Ala344 plays an important role in maintaining the structure and reactivity of the OEC [185,211–214]. By ^{13}C-labeling the C-terminal α-COO$^-$ group as well as all alanine-derived peptide carbonyl groups in PSII from *Synechocystis* sp. PCC 6803, a single doublet was observed in the ENDOR spectrum, centered at the ^{13}C Larmor frequency (Figure 1.8) [215]. The corresponding magnitude of the

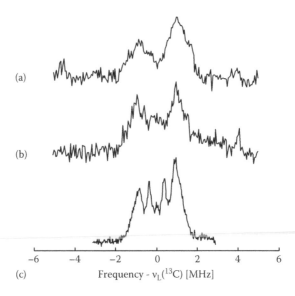

(a)

(b)

(c)   Frequency - $v_L(^{13}C)$ [MHz]

FIGURE 1.8 Q-band Mims ENDOR spectra (black) of (a) ^{13}C-Ac Mn(III,IV) TACN, (b) α-^{13}C Ala PSII, and (c) ^{13}C-PSII.

spin-projected a_{iso} was found to be at least 1.2 MHz, larger than that found for the carbon of a bridging acetate in a mixed valence dinuclear Mn(III,IV) model complex (cf. Figure 1.8a,b). These results were interpreted as indicating direct coordination of the D1-Ala344 carboxylate by at least one Mn of the OEC in the S_2 state. Similar studies of uniformly ^{13}C-labeled PSII indicate that several additional ^{13}C couplings exist, although these interactions have yet to be fully dissected and assigned to specific amino acids.

Thus far, no other proteinaceous ligand to the OEC has been identified by EPR studies. However, ‖-mode CW-EPR studies have indicated that an active-site aspartate residue D1-Asp170 in Mn-depleted PSII operates as a nucleation site for the photoassembly of the OEC [216]. During this process, an exogenous Mn(II) ion binds to D1-Asp170 and is then photooxidized to Mn(III) by the same light-harvesting machinery used for water oxidation [181,217]. The cycle presumably repeats as additional equivalents of Mn(II) as well as Ca^{2+} and Cl^- cofactors bind to yield the fully functional OEC. On the basis of the published crystal structure data, it seems likely that D1-Asp170 may be close enough to the "dangling" Mn ion ($r = 2.40$ Å) [218] to form a bond; however, Fourier transform infrared (FTIR) data suggest that if so, this Mn center does not undergo a change in oxidation state as a function of S-state [219,220]. Furthermore, multifrequency ESEEM spectra of the D1-D170H mutant show no new features that would indicate Mn coordination by the imidazole ring nitrogens of the mutated histidine residue [204,221]. There is also some evidence that D1-His332 and/or D1-His337 may participate in photoassembly by coordinating, with high-affinity, an Mn(II) ion [187,222].

Whatever the arrangement of amino acid-derived ligands about the OEC, it is clear these are insufficient to coordinatively saturate the Mn centers (Figure 1.1). As described previously, seven amino acids have been implicated as possible ligands to the OEC, yet the Ca^{2+} and the four Mn ions can be considered to have 31 possible coordination sites. Many (up to 13) of these vacancies are filled by the μ-oxido bridges between Mn centers; however, this still leaves several open sites. Some researchers have proposed that a bicarbonate molecule may ligate Mn, although it is believed that such coordination is perhaps more relevant to the process of OEC photoassembly than to water-splitting chemistry [223–231]. Also, the Cl^- ion that is known to be necessary for optimal oxygen evolution must bind somewhere. While both common isotopes of Cl are magnetic (^{35}Cl, 76% n.a., $I = 3/2$; ^{37}Cl, 24% n.a., $I = 3/2$), no ENDOR or ESEEM results have indicated any significant HFC of this nucleus to the Mn cluster in the S_2 state [232].* The lack of signal from a strongly coupled Cl^- ion could be due to a variety of reasons. The corresponding nuclear g-values for chlorine are small and, thus, any features due to a weakly coupled nucleus would appear in the congested low-frequency area of the nuclear frequency spectrum. The rather large quadrupole constant for Cl (the quadrupole moment for both isotopes is >4 times that of ^{14}N) would significantly broaden any peaks, perhaps beyond detection. Finally, the Cl^- ion could be too distant from the unpaired electron spin to afford significant HFC. This last point is made all the more credible by EXAFS studies on Br^--substituted PSII from spinach poised in the S_1 state [233]. These results indicate that a bromide ion, which can functionally replace chloride in the OEC, is at least 5 Å from manganese or calcium in the S_1 state. However, this behavior could be attributed to the fact that Cl^- is not required for the $S_0 \rightarrow S_1$ and $S_1 \rightarrow S_2$ transitions [78,234] and may instead occupy a prebinding site 5 Å away from the Mn cluster until it is needed for the later S-state transitions. Recent x-ray crystal structures confirm this distant binding of halide [82,235]; again, however, these structures represent an S-state that is likely reduced compared with S_0. Interestingly, Cl^- depletion by dialysis with chloride-free buffer abolishes the S_2 MLS, although the $g = 4.1$ signal remains [236,237]. This behavior suggests that the presence of Cl^- influences the S_2 electronic structure. Several other small anions (NO_3^-, NO_2^-, I^-) [237–240] are competitive with chloride and either retard or completely inhibit oxygen-evolving activity. No advanced EPR studies of these inhibited species have yet been reported.

Another inhibitor, azide, is also believed to bind in the chloride site [241]. ESEEM studies of Cl^--depleted PSII samples treated with terminally labeled N_3^- found that A_{iso} (^{15}N) = 0.7 MHz and suggest that the Cl^--binding site is in close proximity to the Mn cluster in the S_1 and S_2 states [242]. The disparity between the Br^- EXAFS results, which suggest a distant Cl^--binding site, and the N_3^- ESEEM results, which imply a more intimate interaction between the Cl^- ion and the Mn cluster, has not been resolved.

---

* It should be noted, however, that an ESEEM study of salt-washed (i.e., extrinsic proteins are removed) PSII from spinach that was reconstituted with either $CaCl_2$ or $CaBr_2$ may provide some evidence of HFC for each nucleus [232]. The difference (Cl^--substituted minus Br^--substituted) of the three-pulse ESEEM spectra indicated that the halide is very weakly coupled to the OEC, as evidenced by extremely weak positive and negative features near the Larmor frequency for Cl^- and Br^-, respectively. Boussac et al. noted that such signals could not be observed for similarly treated Cl^--depleted (via high pH treatment) or Ca^{2+}-depleted (via EGTA treatment) PSII particles [232].

More insight into the location of the chloride-binding site has been achieved via EPR studies of Cl^--depleted or acetate-treated PSII. As such treatments allow for the formation of the "split" CW-EPR signal attributed to the $S_2Y_Z^*$ state, these results are discussed in Section 1.5. Intriguingly, the binding of F^-, also a competitive inhibitor of Cl^-, to the OEC abolishes the MLS, leading to exclusive formation of a slightly narrower version of the $g = 4.1$ signal in the X-band CW-EPR spectrum (see Section 1.4). That some of these small anions both inhibit the water-splitting reaction and directly bind to the OEC in early S-states (supported by the aforementioned EPR results as well as FTIR studies) further emboldens claims that the role of the chloride ion is dynamic during the Kok cycle. At states S_0 through to S_2, it appears as if the chloride coordination site associated with the Mn cluster must be in rapid exchange (or at the very least, the presence of Cl^-, specifically, is innocuous). When this site is instead occupied by, for example, azide, S-state advancement is halted at S_2. At later S-states, the Cl^- ion is needed in much closer proximity to the OEC, perhaps to equalize built-up positive charge on the highly oxidized Mn centers [234]. Chloride has also been invoked as a participant in a proton relay mechanism, helping pump H^+ ions to the lumenal or donor side of the chloroplast membrane [243–245]. Chloride depletion or the binding of acetate to the chloride site would likely disrupt such a proton-pumping pathway. Analysis of x-ray structure data combined with results from DFT/quantum mechanics–molecular mechanics simulations indicated that the absence of Cl^- leads to formation of a salt bridge between D1-D61 (see also [246]) and D2-K317, and perturbs the waters bound to the dangling Mn and calcium [247]. Furthermore, the combined substitution of iodide for chloride and strontium for calcium (see also Section 1.3.4.3) dramatically slowed the rate of O_2 production by a factor of 40 [248]. These results also point to a through-space interaction between Cl^- and Ca^{2+} that is likely mediated by the intervening H-bonds and water molecules. Unfortunately, any of these hypotheses have yet to be proven by direct observation of the chloride ion at any true stage of the Kok cycle.

1.3.3 SUBSTRATE BINDING

As water is the substrate of the OEC, the most attractive candidate to fill all remaining open coordination sites on the Mn and Ca^{2+} ions in computational models of the OEC is water (or hydroxide). Yet, there are far too many possible combinations of the number of ligands and their respective protonation state of water molecules to discriminate solely on the basis of computed energy of a given structure. Magnetic parameters of these water ligands, on the other hand, provide a much more quantitative means to assess these computational models.

To this end, several groups turned to CW-ENDOR spectroscopy to determine the presence of exchangeable protons in the vicinity of the OEC in the S_2 state [249–251]. This was done by comparing the ENDOR spectra obtained using natural-abundance water to those of PSII membranes that had been exchanged in D_2O for as long as 24 h (example spectra are shown in Figure 1.9). The ^1H HFC values (A_{obs}) for resonances that either disappeared or significantly decreased in intensity upon D_2O incubation are summarized in Table 1.2. Protons that are hyperfine coupled to the Mn cluster were distinguished from those coupled to Y_D^* (whose EPR spectrum somewhat

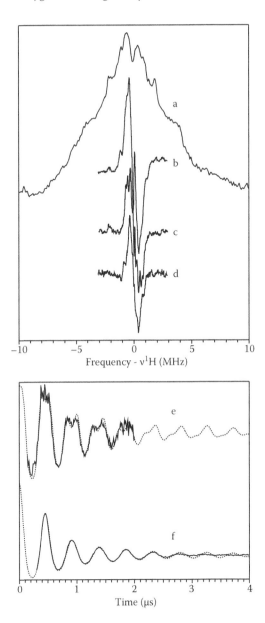

FIGURE 1.9 X-band (a) Davies and (b through d) CW-ENDOR spectra of PSII trapped in the S_2 state. (From C.P. Aznar, R.D. Britt, *Philos. Trans. R. Soc. Lond., Ser. B*, 357, 1359, 2002; H. Yamada et al., *Biochim. Biophys. Acta Bioenerget.*, 1767, 197, 2007.) The spectra are centered at the ^1H Larmor frequency (set to 0 MHz). The CW-ENDOR data were acquired following (b) 0 h, (c) 3 h, and (d) 24 h incubation in ^2H_2O-containing buffer. Ratioed $(^2H/^1H)$ (e) two-pulse and (f) three-pulse ESEEM time-domain spectra achieved following 30 min exchange in ^2H_2O-containing buffer. (From C.P. Aznar, R.D. Britt, *Philos. Trans. R. Soc. Lond., Ser. B*, 357, 1359, 2002.) Simulations of the time-domain data (dotted lines) were generated using the ^2H HFC parameters given in Table 1.2.

TABLE 1.2

HFC Parameters (in MHz) for Solvent-Exchangeable Protons (and Deuterons) and Oxygen Nuclei in the Vicinity of the OEC

PSII Preparation	Nucleus	$A_\perp$ (obs)	Method (Freq.)	References
S_2				
(Spinach: 50% ethylene glycol; pH 6.2)	^1H	4.016	CW-ENDOR (X)	[249]
	^1H	2.011		
(Spinach; 30% glycerol; 0.2 M sucrose; pH 6.8)	^1H	4.016	CW-ENDOR (X)	[249]
	^1H	2.412		
(Synechococcus; 50% glycerol; pH 6.5)	^1H	4.9	CW-ENDOR (X)	[191]
	^1H	2.4		
	^1H	1.0		
	^1H	0.53		
(Spinach: 0.4 M mannitol; pH 6.5)	^1H	4.0–4.2	CW-ENDOR (X)	[250]
	^1H	2.3		
	^1H	1.1		
	^1H	0.6		
(Spinach: 0.4 M sucrose; pH 6.5)	^1H	4.00	CW-ENDOR (X)	[251]
	^1H	2.19		
	^1H	1.38		
	^1H	1.10		
	^1H	0.69		
	^1H	0.43		

# of Equiv. Nuclei	Nucleus	A_{iso}	T	r_{eff} (Å)		
(Spinach: 5% ethanol; 0.4 M sucrose; pH 6.0)						
2	^1H (^2H)	2.93 (0.45)	4.17 (0.64)	2.67	Pulse ENDOR/ESEEM (X)[a]	[252,253]
2	^1H (^2H)	0.00	3.97 (0.61)	2.71		
2	^1H (^2H)	0.00	2.02 (0.31)	3.43		
1	^{17}O	1.5	1.00	2.2	ESEEM (X)	[253]

	n	Nucleus				Method	Ref
(Reported values are an average of results from studies of PSII from pea and spinach; see ref. for details)	1	^{17}O	0.0	0.54	2.7	ESEEM (X)[b]	[161]
	2	^{1}H (^{2}H)	2.08 (0.32)	6.31 (0.97)		ESEEM (X)	[254]
	2	^{1}H (^{2}H)	1.76 (0.27)	4.23 (0.65)		ESEEM (X)	[255]
	0–20	^{1}H (^{2}H)	0.39 (0.06)	2.86 (0.44)		EDNMR (W)	[128]
	8	^{1}H (^{2}H)	0.33 (0.05)	1.63 (0.25)		EDNMR (W)	[126]
NO-treated acetate inhibited		^{17}O	5.0				
	2	(^{2}H)	(0.80)	(0.62)			
	4	(^{2}H)	(0.00)	(0.45)			
(*T. elongatus*; pH 6.5)		^{17}O	9.7	2.2			
		^{17}O	4.5	0.6			
		^{17}O	1.4	0.6			
(*T. elongatus*; 2 mM NH_3, pH 7.6)		^{17}O	7.0	2.2			
		^{17}O	3.1	0.6			
$S_2 Y_Z$:							
(Acetate inhibited)	2	(^{2}H)	(0.00)	(0.57)		ESEEM (X)	[256]
(Acetate inhibited)	4	(^{2}H)	(0.00)	(0.42)		ESEEM (X)	[257]
(Ca^{2+} depleted)	2	(^{2}H)	(0.00)	(0.48)		ESEEM (X)	[258]
		Nucleus	$A_\perp$ **(obs)**				
S_0		^{1}H	4.01			CW-ENDOR (X)	[251]
		^{1}H	2.33				
		^{1}H	1.62				
		^{1}H	0.99				
		^{1}H	0.48				
		^{1}H	0.28				

(Continued)

TABLE 1.2 (CONTINUED)

HFC Parameters (in MHz) for Solvent-Exchangeable Protons (and Deuterons) and Oxygen Nuclei in the Vicinity of the OEC

PSII Preparation	# of Equiv. Nuclei	Nucleus	A_{iso}	T	r_{eff} (Å)	Method (Freq.)[b]	References
(Spinach FCCP treated; 1% methanol)	2	^{1}H(^{2}H)	2.60 (0.40)	5.53 (0.85)	2.43	ESEEM (X)[b]	[91]
	2	^{1}H(^{2}H)	(0.00)	4.56 (0.70)	2.59		
	2	^{1}H(^{2}H)	(0.00)	2.60 (0.40)	3.12		
	1	^{1}H(^{2}H)	1.95 (0.30)	4.88 (0.75)	2.53		
(Reported values are an average of	2	^{1}H(^{2}H)	1.76 (0.27)	5.86 (0.90)		ESEEM (X)[b]	[161]
results from studies of PSII from pea	2	^{1}H(^{2}H)	1.37 (0.21)	3.54 (0.58)			
and spinach; see ref. for details)	10–16	^{1}H(^{2}H)	0.33 (0.05)	2.73 (0.42)			
	8	^{1}H(^{2}H)	0.33 (0.05)	1.63 (0.25)			

[a] $A_\perp = A_{iso} + T$ for ^{1}H measured by pulse ENDOR spectroscopy at X-band and used to constrain the simulations of the corresponding D$_2$O-exchanged ESEEM data.

[b] Reported ^{1}H values were calculated from those measured for ^{2}H through scaling by the ratio of the nuclear g-factors (g(^{1}H)/g(^{2}H) = 6.51).

overlaps with the MLS) using the ENDOR-induced EPR technique. Resonances at 0.5–0.6, 1.0–1.1, 2.2–2.4, and 4.0–4.2 MHz seem to be apparent in each of the five CW-ENDOR studies listed in Table 1.2. Minor differences in the precise values of the ^{1}H ENDOR frequencies measured by different laboratories could be due to the different PSII preparatory methods employed (leftmost column, Table 1.2) or simply due to the manner in which interpeak distances were measured (e.g., peak maxima or inflection points). In the most recent of these studies, Mino and coworkers discovered a time dependence to the disappearance of ^{1}H peaks [251]. They observed that the resonance corresponding to $A_\perp = 2.19$ MHz went away after 3 h of incubation in D_2O, while the more strongly coupled proton ($A_\perp = 4.00$ MHz) exchanged with deuterium only after 24 h of incubation (cf. Figure 1.9b through d). Earlier results from time-resolved mass spectrometric studies indicated that two exchangeable water molecules exist in close proximity to the OEC in the S_2 state: one with an apparent exchange rate of 120 s^{-1} and the other with 2 s^{-1} [259–261]. It is clear that the proton that exchanges with bulk water after 24 h of incubation cannot correspond to either of the much faster exchanging protons observed in the mass spectrometry experiment. Therefore, Mino and coworkers assign the proton(s) responsible for the resonance at 4 MHz as belonging to an amino acid residue that is likely coordinated to the Mn cluster [251]. Alternatively, this slowly exchanging proton could be from a so-called structural, nonsubstrate water (hydroxide) ligand. The remaining low-frequency resonances whose intensities are sensitive to D_2O exchange are attributed to more distant water molecules not directly bound to the OEC.

Interpretation of results from CW-ENDOR studies present three important difficulties when trying to understand the nature of water substrate binding to the OEC. First, the intensity of a certain spectral peak is not exclusively a function of the number of equivalent protons contributing to that CW-ENDOR resonance, making "water counting" impossible. Second, the magnitude of splitting of the CW-ENDOR peaks centered approximately at the ^{1}H ($I = 1/2$) Larmor frequency (ν_I) is a function of both A_{iso} and T (the dipolar contribution to the HFC term)

$$\nu_{\alpha,\beta} = \sqrt{(\nu_I \pm A/2)^2 + (B/2)^2} \qquad (1.12)$$

where

$$A = A_{iso} + T(3\cos^2\theta - 1); \quad B = 3T\sin\theta\cos\theta \qquad (1.13)$$

Rudimentary analysis of the experimentally observed ENDOR frequency often neglects the B term and reduces Equation 1.12 to

$$\nu_{\alpha,\beta} = |\nu_I \pm A/2| \qquad (1.14)$$

In the axial approximation ($A_{xx} \approx A_{yy} \equiv A_\perp = A_{iso} + T$; $A_{zz} \equiv A_\| = A_{iso} - 2T$), it is often not possible to distinguish the resonances that correspond to $A_\|$ (i.e., $\theta = 0°$) from those for $A_\perp$ in the CW-ENDOR spectrum. Therefore, one has no way to solve for both A_{iso} and T for a specific hyperfine-coupled nucleus. Finally, CW-ENDOR is rather insensitive to strongly ($A_{obs} > 5$ MHz) coupled protons owing to the large spectral range over which the corresponding resonances are spread.

Fortunately, the Davies ENDOR pulse sequence can be used to selectively probe such strongly coupled nuclei [262]. Indeed, ^{1}H pulse ENDOR studies of PSII reveal some new features not previously observed in the aforementioned CW-ENDOR studies (cf. Table 1.2 and see Figure 1.9a) [252]. It is still difficult to unambiguously assign pulse ENDOR features to $A_\perp$ or $A_\parallel$. However, the modulation depth parameter $k = \left(\dfrac{\nu_I B}{\nu_\alpha \nu_\beta} \right)^2$ (shown here for an $I = 1/2$ nucleus) of the time-domain ESEEM spectrum is also a function of A_{iso} and T. Using this relation in conjunction with Equation 1.12 through 1.14 allows for the determination of a unique set of HFC parameters that can simultaneously simulate both ESEEM and ENDOR data.

Following just 30 min of incubation in D_2O, the three-pulse ESEEM spectrum of PSII in the S_2 state dramatically changes and new modulation patterns emerge (see Figure 1.9e and f). By taking the ratio of these data and the natural-abundance H_2O data, one can, in an ideal case, observe exclusively the modulation of the echo envelope due to the exchangeable deuterons within the vicinity of the unpaired electron.* The modulation depth parameter k for a weakly coupled deuteron is related to that for its proton isotopologue by a factor of about 8/3. NQI contributions for the $I = 1$ deuteron are generally small and thus ignored. Using the pulse ENDOR results to constrain the value of $A_\perp = A_{iso} + T$ and by extension B (see Equation 1.13), one can compute the modulation pattern for a single exchangeable deuteron. Integer multiples of these "unit" modulation patterns are multiplied together to match the observed modulation depth and provide an estimate for the number of equivalent deuterons.

This data analysis method was employed by both the Britt [91,252] and Pace [161, 165,263] groups to determine the number of exchangeable water molecules as well as the corresponding Mn–$^{1/2}$H internuclear distances within the vicinity of the OEC. The analysis presented in Aznar et al. found two classes of hydrogens that exchanged within 30 min,† each with a population of two (indicating that the H atom belongs

---

* For three-pulse ESEEM spectroscopy, the modulation observed in the time-domain spectrum is given by $V_{3p}(\tau,T) \propto \sum_\theta \left(\prod_q V_{3p}^\alpha(q,\theta) + \prod_q V_{3p}^\beta(q,\theta) \right)$. The term $V_{3p}^{\alpha(\beta)}(q,\theta)$ describes the contribution to the modulation by nucleus q with orientation θ. The modulations from all like-oriented nuclei contributing via the same coherence transfer pathway (α or β) are multiplied according to the product rule; then, the products from each coherence transfer pathway are added. Finally, this result is summed over all possible orientations to give the total observed echo envelope modulation. It is clear that this ratioing procedure is strictly valid only when the precessional frequencies of nucleus q in the α and β manifold (ν_α and ν_β) are identical and there is no orientation dependence of these frequencies (or the anisotropy of the g values allows for exclusive orientation selection). Yet, were these conditions held, there would be no ESEEM modulation at all (i.e., $k = 0$). In practice, the application of the ^{2}H/^{1}H ratioing procedure should be reserved for cases in which τ is chosen to suppress modulations due to protons and where the exchangeable protons are only weakly hyperfine coupled leading to even smaller (by a factor of 6.51 = $g_N(^1H)/g_N(^2H)$) deuteron HFC parameters.

† In these studies, S-state cycling was accomplished by 30 min of ambient light illumination to ensure incorporation of D_2O into the OEC active site. This was followed by 45 min of dark adaptation on ice before the samples were frozen in liquid N_2.

to a molecule of H_2O and not OH^-). These deuterons exhibited rather large dipolar HFC terms ($T = 4.17$ and 3.97 MHz; Table 1.2) indicating close proximity to the $S = 1/2$ species [252,253]. Intriguingly, only one class possessed significant isotropic hyperfine coupling ($A_{iso} = 2.93$ MHz), which suggests that the corresponding water molecule is covalently bound to a center with unpaired electron density. These $^1H/^2H$ ESEEM results coincide with those from the mass spectrometric studies in their suggestion that two water molecules are directly coordinated to the OEC: one water molecule bound to an Mn ion is expected to have significant unpaired spin density on the H atoms (and thus $A_{iso} > 0$ MHz), whereas one bound to Ca should not.

The lack of definitive Mn ligand assignments from crystal structure data have led several researchers to model the OEC by saturating the Mn coordination sphere with water molecules. Yet, simulations of the ^2H modulation pattern assuming 12 additional coordinating waters with canonical ^2H HFC coupling parameters ($A_{iso} = 0.34$ MHz; $T = 0.47$ MHz) give unrealistic modulation depths. Thus, it is unlikely that all solvent-derived Mn ligands are rapidly exchanging. Using broken-symmetry DFT, Ames et al. evaluated several models of the OEC on the basis of structural parameters determined from EXAFS spectroscopy (and thus assumed not to suffer from photoreduction), predicted ^{55}Mn HFCs, and other magnetic properties [132]. These models were largely based on that originally defined by Siegbahn [120] and the Umena crystal structure [82]. The best model of the low-spin S_2 state was predicted to have one hydroxide ligand and one water ligand to Mn4 and two waters bound to calcium. However, the HFCs for the water-derived protons were not reported; thus, it remains unknown if this protonation state for S_2 is consistent with the above spectroscopic data.

ESEEM spectroscopic results on ^{17}O-labeled water exchanged PSII by Pace and coworkers showed modulation attributed to an ^{17}O nucleus [254]. Their analysis of the frequency domain spectrum led to an evaluation of $A_{iso}(^{17}O) = 5$ MHz. Additionally, Åhrling et al. derived ^2H HFC parameters significantly different from those published by the Britt group (cf. Table 1.2) [161]. Their results indicate that after 3 h of D_2O incubation, up to four classes of solvent-derived deuterons are directly bound (through oxygen, presumably) to a moiety that carries unpaired spin density (i.e., $A_{iso} > 0$ MHz). Notably, however, in their analysis, they constrain the parameter space for A_{iso} and T based on CW-ENDOR results (Figure 1.9b), which lack resonances from the more strongly hyperfine-coupled protons evident in the Davies pulse ENDOR spectra (cf. Figure 1.9a). By omitting from their analysis the large values of $A_\perp$ evident in the ^1H pulse ENDOR spectrum—but not present in the ^1H CW-ENDOR data—the simulation of the $^2H/^1H$ ESEEM time-domain data is handcuffed as large values of A_{iso} and T cannot be supported (cf. Table 1.2). Therefore, Pace and coworkers have to model the deep ^2H modulation using multiple classes of deuterons with small but nonzero A_{iso} values and increase the overall population of deuterons that are coupled to the OEC. These simulations suggest that two to four exchangeable water molecules are coordinated directly to Mn after 3 h of D_2O incubation, whereas Britt and coworkers observed only one after just 30 min of D_2O incubation.

Owing to the presence of the rather large electric quadrupole constant of the $I = 5/2$ ^{17}O nucleus and the expected large hyperfine anisotropy, ESEEM and ENDOR spectroscopy may not be ideally suited for studying strongly coupled ^{17}O in the

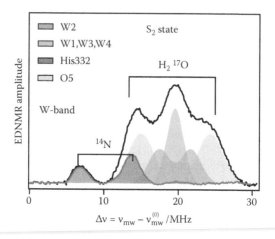

FIGURE 1.10 W-band (94 GHz) EDNMR spectra of PSII exchanged with ^{17}O-water and poised in the S_2 state. (Adapted from L. Rapatskiy et al., *J. Am. Chem. Soc.*, 134, 16619, 2012.)

proximity of the OEC. However, electron–electron double-resonance (ELDOR)-detected nuclear magnetic resonance (EDNMR) has recently proven to be a valuable tool in overcoming these obstacles [264]. In these studies, Cox and Lubitz and coworkers observed three classes of hyperfine-coupled ^{17}O-containing species (Figure 1.10) that washed into solvent-exchangeable sites after <15 s of incubation in ^{17}OH$_2$ [128]. The strongest hyperfine-coupled ^{17}O moiety ($A(^{17}$O) = [10.7, 5.2, 13.1] MHz) was assigned to a μ-oxido bridge that in the low-spin S_2 state is bound between two Mn ions. Analysis of the hyperfine anisotropy suggested that the principal axis of the ^{17}O hyperfine tensor is aligned perpendicularly to that of the D1-His332 nitrogen bound to Mn1. Using this piece of information in conjunction with the x-ray crystal structure coordinates and energetically feasible DFT-computed geometries, the authors determined that this bridging oxido likely resides between Mn3 and Mn4, either O4 or O5 (Figure 1.1). Results from further spectroscopic studies suggest that O5 derives from a substrate water molecule and gets incorporated into the O_2 product [126].

As described previously, low concentrations of small alcohols tend to increase the intensity of the MLS without affecting water oxidation activity. To explore the nature of the alcohol–OEC interaction, ratioed ^{2}H/^{1}H ESEEM data were collected as above using the perdeuterated isotopologue of the alcohol [160,161].* The ^{2}H modulation depth for PSII particles poised in the S_2 state and treated with methanol, ethanol, n-propanol, and isopropanol were compared [160]. Excepting isopropanol, deuterons for each of the additives exhibited nonzero values for T (e.g., specifically for methanol, T = 0.18, 0.20, and 0.45 MHz for the three methyl deuterons) consistent

---

* The study by Åhrling et al. determined slightly larger values of T for the methyl deuterons (T = 0.39, 0.39, and 0.60 MHz) [161]. This inconsistency with the Force et al. data [160] likely results from the fact that neither analysis is constrained by corresponding ^{1}H ENDOR data, as was necessary for the water-counting studies.

with the close proximity (i.e., binding) of the respective alcohol. Intriguingly, while isopropanol does not appear to bind directly to the OEC, its addition to PSII forces the generation of the S_2 MLS at the expense of the $g = 4.1$ signal. The origins of this counterintuitive behavior remain unknown. Given that neither isopropanol nor dimethylsulfoxide seem to bind, it was concluded that the bulkiness of these probes may prevent access to the water-oxidizing active site.

1.3.4   OTHER EFFECTORS OF THE MLS

As described previously, many small molecules can deleteriously affect the oxygen-evolving ability of PSII. Coincident with these changes of enzymatic activity, alterations in the appearance of the MLS can sometimes occur [265]. This phenomenon was first observed for PSII particles treated with 100 mM NH_4Cl, poised in the S_2 state by 190 K illumination, and then annealed at 273 K [155,266]. Such ammonia-treated PSII gives rise to an MLS in which the splitting between adjacent peaks collapses from 85 to 90 G to $\approx$68 G (Figure 1.3e). Examples of other chemical manipulations that similarly distort the MLS include Ca^{2+} depletion (Figure 1.3f), Sr^{2+} substitution (Figure 1.3g), and several site-directed mutants (e.g., D1-H332E; Figure 1.3h) [196]. The advanced EPR spectroscopic properties of the S_2 MLS as a function of each of these effectors will be described in turn.

1.3.4.1   Ammonia Binding

Results from biochemical and spectroscopic studies have shown that ammonia (derived from treatment with NH_4Cl) can occupy two distinct binding sites in the vicinity of the OEC. Binding at one site (defined as "SY II" in older literature) leads to the altered MLS described previously, and binding at the other site (defined as "SY I") inhibits O_2 production and induces formation of the $g = 4.1$ signal (Figure 1.3d) and will be discussed in depth in Section 1.4. Results from ESEEM studies by Britt et al. reveal that a single ^{14}N nucleus from ammonia is coupled to the OEC in spinach PSII ($A_{iso} = 2.29$ MHz, $T = 0.2$ MHz; $e^2Qq_{zz}/h = 1.61$ MHz; $\eta = 0.59$) and binds only following the annealing procedure [267]. These ammonia nitrogen magnetic values are only slightly different in the NH_4Cl-treated PSII from *T. elongatus* ($A_{iso} = 2.36$ MHz, $e^2Qq_{zz}/h = 1.52$ MHz; $\eta = 0.47$), suggesting the ammonia-derived moiety binds similarly to PSII from both organisms [126]. The high value for the asymmetry of the EFG suggests that ammonia is deprotonated upon binding to Mn. This fact and the relatively low magnitude of e^2Qq_{zz}/h is consistent with the ^{14}N nucleus as being part of an amido bridge between two metal centers, one of which must be an Mn ion as $|A_{iso}| > 0$ MHz (Section 1.2).

This replacement of, ostensibly, an oxido bridge with an amido group dramatically alters the exchange coupling between Mn centers, as evidenced by the smaller splitting of the MLS in the corresponding CW-EPR spectra (cf. Figure 1.3a,e). Indeed, the pulse ENDOR spectrum of NH_3-treated PSII [83] shows rather significant changes in the effective ^{55}Mn HFC constants (Table 1.1; Figure 1.5c), although a more recent study performed at higher frequency reports slightly smaller changes [126]. These changes could be accounted for by moving the Mn(III) ion out of the trimer (for untreated or methanol-treated PSII) to the monomer or dangler position (Figure 1.6a).

However, such a change in valance locality was ruled out by a recent ESEEM study of ammonia-treated PSII, which showed that the magnitude of the ^{14}N HFC arising from D1-His332 is unchanged by the addition of ammonia [126]. As D1-His332 was previously established as being bound to the lone Mn(III) ion [204], it is clear that the Mn(III) ion did not change its locality within the OEC. Navarro et al. proposed instead that ammonia must displace the water bound terminally to the dangling Mn(IV) ion. In the high-resolution crystal structure [82], this water is *trans* to O5 whose HFC parameters are dramatically altered by ammonia binding [126]. Furthermore, O5 exchanges on the time scale of the slower-exchanging water [261], disclosing it as one of the locations of substrate binding [126]. Despite this compelling evidence for the site of ammonia binding being terminal to the dangling Mn, the corresponding ^{14}N(ammonia) NQI parameters are not well reproduced by DFT. Perhaps there is a very strong H-bond to an ammonia proton that would give rise to the large observed ^{14}N(ammonia) NQI asymmetry that is not accounted for in the computational model, or perhaps the actual mode of ammonia binding is very different.

1.3.4.2 Ca²⁺ Depletion

Removal of the Ca^{2+} ion from the OEC by any of a variety of means (e.g., high-concentration salt [NaCl] wash followed by treatment with either 10 mM pyrophosphate or ethylene glycol tetraacetic acid [EGTA], sulfate, or citrate at low pH) [268–271] has a similar effect as ammonia incorporation does on the appearance of the CW-EPR spectrum of the treated PSII poised in a pseudo-S_2 state. Low-temperature illumination (195 K) leads to the generation an MLS with reduced interpeak spacings of approximately 55 G (Figure 1.3f) [271]. This altered S_2 state is unusually dark-stable, meaning that once generated, the MLS can be observed even after the sample has been annealed at high temperatures ($\approx$273 K) that would normally allow relaxation of the OEC back to S_1. Nitrogen isotope-sensitive features are present at 4.1, 7.9, and $\approx$11 MHz in the X-band two-pulse ESEEM spectrum of PSII that is depleted of Ca^{2+} then treated with pyrophosphate (features appear at 3.9, 7.3, 8.1, and $\approx$11 MHz in the three-pulse ESEEM spectrum) [272].* This represents a significant reduction in frequency compared with the 4.6 MHz feature first observed for native PSII trapped in the S_2 state [193]. Superficially, this difference may indicate that the nature of N-coordination to the Mn cluster is perturbed by Ca^{2+} depletion. However, the various reported crystal structures show that the Ca^{2+} ion is situated on the opposite side of the OEC from potential N-donor ligands D1-His332 and D1-His337 [218, 273]. Furthermore, results from EXAFS studies show that Mn$\cdots$Mn internuclear distances are largely unaffected by Ca^{2+} depletion precluding any significant structural modification of the Mn cluster [274]. Additionally, the ^{55}Mn HFC are only slightly modified by removal of Ca^{2+} [275]. These modest spectral changes were rationalized in terms of an increase in the magnitude of the site-specific ZFS for the Mn(III) ion from $d = -1.3$ cm^{-1} for Ca-containing PSII to $d = -2.3$ cm^{-1} upon removal of calcium. This change in ZFS was attributed to some unidentified but small structural

---

* Additional features are observed in the ESEEM spectrum of EGTA-treated Ca^{2+}-depleted PSII; however, these peaks are believed to result from the amide nitrogens of EGTA interacting directly with the $S = 1/2$ Mn cluster.

variation around the Mn(III) site, and leads to a diminution of the corresponding equatorial projection factor [130]. Therefore, the origin of the 0.5 MHz downshift of the observed ^{14}N frequency is most likely due to this reduction in the projection factor for the Mn(III) ion that is the bonding partner to the D1-His332 nitrogen [272].

1.3.4.3   Sr^{2+} Substitution

While many mono-, di-, and even trivalent ions are capable of binding in the Ca^{2+} site (e.g., Na$^+$, K$^+$, Cd^{2+}, La^{3+}, and Dy^{3+}(OH$^-$)) [276–278], only Sr^{2+} can functionally substitute for Ca^{2+} in the OEC, although the rate of oxygen evolution (S$_3$ → S$_0$) is slowed by a factor of 4–8 [270,279–283]. Similarly to Ca^{2+} depletion, Sr^{2+} ion substitution causes the MLS to narrow significantly with spacings between peaks shrinking to 55 G (Figure 1.3g) [271]. ESEEM studies carried out by Kim et al. showed new low-frequency (0.64 MHz) modulations present in the spectra of PSII samples enriched with ^{87}Sr (I = 9/2) [284]. These features, which are not present in the spectra of PSII substituted with natural-abundance Sr (7% ^{87}Sr), correspond to the Larmor frequency of the ^{87}Sr nucleus. From the analysis of the experimentally observed modulation depth using formulas from Dikanov and Tsvetkov [285], an upper limit of $T = 0.04$ MHz is inferred. Using the dangler model projection factors established by Peloquin et al. for the MLS [83], this value for the dipolar HFC places the Sr^{2+} nucleus within 4.5 Å of the Mn cluster, a finding consistent with those from Sr-EXAFS studies [281,286] and a recent crystallographic characterization of Sr-substituted PSII [287]. Comparison of the ^{55}Mn ENDOR spectrum of Sr-substituted PSII to that of native, calcium-containing preparations reveals only modest changes in the isotropic part of the HFC of the four manganese ions [177]. Rather, the anisotropic portion of the HFC shrank by ≈18 and 50 MHz for two of the Mn(IV) ions, leading to the narrowing of the CW-EPR signal upon Sr substitution. These changes are attributed to a small (<0.1 cm^{-1}) decrease in the magnitude of the site-specific ZFS of the Mn(III) ion caused by elongation of the Sr···Mn distances compared with the Ca···Mn internuclear spacings [287].

1.3.4.4   Mutation of D1-His332

To further explore the potential role of D1-His332 as a ligand to the Mn cluster, several site-directed mutants were constructed in *Synechocystis* sp. 6803 [182]. Of these, mutation of D1-His332 to glutamate (D1-H332E) abolishes O$_2$-evolving activity; however, short (30 s) white-light illumination at 273 K leads to formation of a narrow modified MLS, indicating that the Mn cluster is assembled (Figure 1.3h) [196]. As previously noted, the 5 MHz modulation attributed to the interaction of the τ-N nucleus of D1-His332 with the Mn cluster is absent from X-band ESEEM spectra of the D1-H332E mutant [197]. Furthermore, the well-resolved features from the OEC-coordinating histidine in WT PSII in the 30 GHz spectrum are completely abolished in the spectrum of D1-H332E PSII [204].

1.4   S$_2$–g = 4.1 (HIGH-SPIN SPECIES)

The derivative-shaped signal centered at g = 4.1 in X-band CW-EPR spectra was first attributed to the OEC in the S$_2$ state as both it and the MLS are generated by illumination—at <140 and 195 K, respectively—of dark-adapted PSII particles (Figure 1.3a). Further evidence that inferred the Mn cluster origin of the $g = 4.1$ signal was found in the fact that the Mn K-edge XAS spectrum edge shifts upon 140 K illumination to a degree that is nearly identical to that achieved by 195 K illumination, which is coincident with generation of the MLS [35]. Definitive proof, however, came when Kim et al. observed [55]Mn hyperfine structure atop a slightly narrower $g = 4.1$ signal in oriented PSII membranes that had been treated with ammonium chloride and illuminated at 195 K (Figure 1.3d) [145,288]. The experimentally determined [55]Mn-induced splittings (3.6 mT) compare well with the 85 MHz (3.1 mT) computed by Pantazis et al. [129]. In contrast to the case described in Section 1.3.4.1, the creation of this NH_3-altered EPR signal is competitive with Cl^-. Thus, the "SY I" ammonia binding site may correspond to the binding site of the essential chloride cofactor. The enhanced spectral resolution encountered for the $g = 4.1$ signal in NH_3-treated samples could be attributed to a reduction in g-anisotropy through alteration of ZFS parameters within the cluster. Alternatively, ammonia binding could induce changes in the exchange-coupling scheme between Mn centers leading to a more resolved spectrum. Regardless, it is very likely that to induce such changes in the spectral properties, the NH_3-derived moiety must bind directly to the Mn cluster, perhaps in the Cl^- site.

The addition of high concentrations of fluoride—also a competitive inhibitor of the Cl^- binding site—followed by 195 K illumination yields near-complete conversion of the MLS to the $g = 4.1$ signal (Figure 1.3c). Through multifrequency CW-EPR studies (S-, X-, P-, and, most important, Q-band), this feature was unambiguously attributed to a resonance between the middle Kramers doublet ($m_s = \pm 3/2$) of an $S_T = 5/2$ species. This set of multifrequency CW-EPR data was best simulated using the ZFS parameters $D = 0.455$ cm^{-1} and $E/D = 0.25$, which are dominated by the contribution of the ZFS of the sole Mn(III) ion [92,178]. Analogously, these multifrequency studies were able to rationalize the observed narrowing of the X-band resonance at $g = 4.1$ upon either ammonia or fluoride treatment in terms of a slight increase in E/D. For NH_3-treated PSII, $D = 0.44$ cm^{-1} and $E/D = 0.30$, whereas the spectra of F^--treated PSII were best fit using $D = 0.53$ cm^{-1} and $E/D = 0.27$.

Owing to the high-spin nature and fast relaxation properties of the species responsible for the $g = 4.1$ signal [289], pulse EPR studies are extremely difficult to perform. Nonetheless, Astashkin and Kawamori observed a Hahn echo signal at about $g = 4.1$ in the field-swept ESE-EPR spectrum at 4.1 K of untreated PSII membranes from spinach [290]. The dependence of the signal intensity on the power of the incident pulse microwave radiation is consistent with the formulation of this "$g = 4.1$" species as originating from an almost completely rhombic $S = 5/2$ system. Importantly, with respect to future pulse EPR experiments, the authors noted that the echo intensity was completely lost when using microwave pulses separated by 200 ns or more. Thus, simple Hahn echo detection sequences will be ineffective and researchers will need to look to use stimulated echo sequences (i.e., three-pulse ESEEM with very short τ) or remote echo detection methods for data collection.

NIR irradiation of PSII particles poised in the S_2 state via 195 K white-light illumination substantially increases the intensity of the signal at $g = 4.1$ relative to the MLS. This behavior has been rationalized in terms of an Mn(III) → Mn(IV) intervalence charge transfer transition. Support for such a mechanism is gained from the speculations of Peloquin and Charlot and their respective coworkers [83,166]. Both authors promote an exchange-coupling scheme for the OEC in the state that gives rise to the $g = 4.1$ signal that differs from that for the MLS in the position of the Mn(III) ion within the trimer and the values of some of the J-couplings (see Figure 1.6). Recent DFT results strongly support this formulation, namely that the Mn_3CaO_4 core contains only Mn(IV) ions that are ferromagnetically coupled to give rise to an $S_{cube} = 9/2$ ground state that is then antiferromagnetically coupled to the dangling Mn(III) ion to yield an $S_T = 5/2$ system [129]. The computed ^{55}Mn HFCs for the Mn(IV) ions are consistent with those measured for an $Mn(IV)_3CaO_4$ model complex that also has an $S_T = 9/2$ ground state [291]. The facile interconversion from the low-spin form to high-spin form of S_2 is rationalized by the movement of O5 from the bridging position between Mn4 and Mn3 to a μ3 position bridging Mn1, Mn3, and calcium. Coincident with this atomic movement is reduction of Mn4 to the 3+ oxidation state leading to a lengthening of the Mn1···Mn4 distance by 0.5 Å, consistent with EXAFS results [292].

However, the NIR-irradiated PSII S_2–S_1 difference FTIR spectrum is virtually identical to that obtained via white-light illumination. These results are interpreted as indicating that the geometric arrangement of the three Mn(IV) ions and single Mn(III) ion is the same in both the $S_T = 1/2$ and $S_T = 5/2$ forms of the OEC in the S_2 state [293]. The authors suggest that a spin transition within the Mn(III) ion leads to the observed change in the spin state of the cluster. These findings remain incongruous with the magnetic and computational characterizations of the high-spin form of PSII in the S_2 state described previously.

1.5 $S_2Y_Z^{\bullet}$

An EPR signal known as the split signal (Figure 1.11) is achieved via short (5 s) 0°C illumination of PSII particles for which the Ca^{2+} or Cl^- binding site has been disrupted [294]. Originally attributed to the S_3 state of the OEC [244,295–298], this split signal is so-called owing to unusually broad (sometimes in excess of 160 G) Gaussian-shaped features that flank either side of the resonance from $Y_D^{\bullet}$ centered at

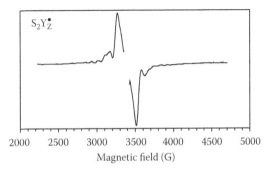

FIGURE 1.11   X-band CW-EPR spectrum (illuminated minus dark-adapted/annealed) of acetate-treated PSII from spinach trapped in the $S_2Y_Z^{\bullet}$ state via 5 s illumination at 273 K.

$g = 2.00$. It has since been determined that these features arise from the exchange and dipolar coupling of the OEC poised in the S_2 state ($S_T = 1/2$) with the photooxidized organic radical, Y_Z^* ($S = 1/2$) [27,257,299]. The advanced EPR spectral properties of this species generated by chloride depletion, acetate treatment, and calcium depletion will be discussed in turn.

Split signals of varying widths (40–200 G) have also been reported for a number of site-directed mutants of PSII, including D1-H332E, D1-E189Q, D1-E189D, D1-E189N, D1-E189S, and D1-E189H [196,300]. Of these, only the D1-E189Q mutant was able to significantly oxidize water (67% of wild-type activity) as well as yield an unaltered S_2 MLS upon 200 K illumination [300]. However, the authors caution that the appearance of a split signal could result from the interaction of Y_Z^* with only partially assembled Mn clusters.

1.5.1   Cl⁻ Depleted

Any of a series of small molecules (amines, acetate, F⁻, I⁻, NO_2^-, NO_3^-, and OH⁻) [301,302] are known to inhibit oxygen evolution activity by interfering with binding of the native Cl⁻ cofactor. Short room-temperature illumination of sulfate-treated Cl⁻-depleted PSII gives rise to a very narrow (60 G) split signal—although SO_4^{2-} is not thought to interact directly with the OEC [157,244]. Likewise, narrow split signals (splitting measured from $g = 2.00$ in Gauss) are generated from Cl⁻-depleted samples following reconstitution with F⁻ (160 G) [303] or NH_3 (100 G) [297] substitution.* The differences in spectral breadth of these various inhibited forms of PSII likely stem from alteration of the exchange pathway between the two paramagnets, Y_Z^* and the $S = 1/2$ form of the Mn cluster. It is attractive to postulate that the native exchange pathway involves one or more H-bonds, and the fact that Cl⁻-depleted/-inhibited PSII cannot advance to the S_3 state could allude to the role of the Cl⁻ cofactor in proton release and/or substrate binding [157].

1.5.2   Acetate Inhibited

The addition of high concentrations (>100 mM) of acetate to PSII greatly retards water oxidation; however, full activity is recovered upon treatment of chloride [304,305]. These findings suggest that acetate displaces the Cl⁻ ion thought to be incorporated into the OEC and, in this form, PSII cannot advance to the S_3 state. Similarly, 200 K continuous illumination does not achieve the familiar S_2 MLS. Instead, illumination of acetate-treated PSII at ≈0°C causes an EPR signal to develop that is now known to represent the interaction of the S_2 form of the Mn cluster with Y_Z^* [29,298,299,306]. In addition to the two large features flanking $g = 2.00$ and separated by 230 G, the CW-EPR spectrum of the $S_2Y_Z^*$ state of acetate-treated PSII possesses a number of small resonances reminiscent of the ^{55}Mn hyperfine structure present in the native S_2 MLS (Figure 1.11). In fact, these features are shifted by ≈20–40 G compared with their S_2 counterparts, further evincing the dipolar coupled

---

* Here, "NH_3 treatment" refers to that described in Section 1.2.1.3, which enhances the intensity of the $g = 4.1$ signal in the S_2 state.

nature of this state [306]. Simulations of the CW-EPR data for this triplet system have led to a wide range of reported interspin distances spanning from 3.5 to 20 Å [27–29,256,257,297–299,306,307]. However, consensus seems to be centering on a value between 7.5 and 9.5 Å. A CW-EPR study of acetate-treated membranes oriented on Mylar strips resulted in single-crystal-like spectra in one dimension and found that this interspin vector is canted 75° away from the membrane normal [29]. This parameter can be used to help organize models of the OEC and, in particular, to determine the relative orientation of phenyl ring of $Y_Z^{\cdot}$ to the Mn cluster.

Addition of nitric oxide to PSII in the $S_2Y_Z^{\cdot}$ state generated by acetate inhibition quenches the tyrosine radical and yields an EPR spectrum that is extremely similar to the S_2 MLS [255,308]. This finding strongly suggests that the exchange-coupling scheme within the Mn cluster is unaffected by either acetate binding or the oxidation of Y_Z. Indeed, X-band ^{55}Mn pulse ENDOR data acquired for acetate-treated PSII poised in the $S_2Y_Z^{\cdot}$ state showed that the effective ^{55}Mn HFC constants were only modestly perturbed compared with those determined for S_2 (cf. Figure 1.5b,d) [28]. However, a new prominent feature is apparent at 66 MHz in the pulse ENDOR spectrum. Its presence is understood in terms of a system of four electronic states that results from the weak coupling of the two doublet wave functions of the S_2 state of the OEC and the oxidized tyrosine [257,307,309]. The effect of this coupling leaves the HFC interactions unchanged in two of the new electronic states. Therefore, ENDOR transitions involving these unperturbed coupled states should have identical frequencies to those expected for the isolated paramagnets. The remaining two electronic states are mixed according to the degree of isotropic exchange (J) and dipolar coupling (D) between the two radical species. Consequently, the frequencies for nuclear transitions within these states are reduced by a factor f that is dependent on J and D.

$$f = \nu_{Mn} \pm \frac{3J - D}{6} \left[\sqrt{1 + \sum_i^4 \frac{3A_i\left(m_i + 1\right)^2}{3J - D}} - \sqrt{1 + \sum_i^4 \frac{\left(3A_i m_i\right)^2}{3J - D}} \right] \quad (1.15)$$

Thus, the position of the low-frequency feature in the pulse ENDOR spectrum for the $S_2Y_Z^{\cdot}$ state of acetate-treated PSII serves as a sensitive measure of the exchange ($J = -850$ MHz) and dipolar ($D = 150$ MHz) couplings between the Mn cluster and $Y_Z^{\cdot}$ [28]. The magnitudes for J and D were later confirmed by simulation of high-frequency CW-EPR spectra (95–285 GHz) of acetate-treated PSII membranes ($J = +820$ MHz; $D = -120$ MHz) [307]. However, Dorlet et al. point out that the high-field spectra can only be properly simulated if the sign of J is positive, indicating that the two spins are ferromagnetically coupled [299]. This fact has important implications for the pathway of exchange, especially as the same study found that the sign of J was negative for the $S_2Y_Z^{\cdot}$ state of Ca^{2+}-depleted PSII (see Section 1.5.3). Notably, J is significantly reduced to +680 MHz in acetate-treated reaction center cores, possibly reflecting structural changes in the exchange pathway induced by the removal of the 17 (PsbQ) and 23 (PsbP) kDa extrinsic proteins. Regardless, the value for D indicates that the phenoxyl O atom on $Y_Z^{\cdot}$ and the Mn cluster are separated by approximately 8.6–11.5 Å [28], a distance that is consistent with all currently available crystal structure data [1,218,273]. This long interspin distance precludes the direct involvement of $Y_Z^\bullet$ as an H atom abstractor from an Mn-bound substrate water.

While often assumed to be closely associated with the Mn cluster of the OEC, the location of the Cl$^-$ ion is not actually known (see Section 1.3.2). However, given the competitive inhibition behavior of Cl$^-$ with acetate, this molecule can be used to probe the proximity of the Cl$^-$ binding site to Mn. Results from early ^{2}H/^{1}H ratioed ESEEM spectroscopic studies of the $S_2Y_Z^\bullet$ state using acetate deuterated at the methyl position show very shallow modulations that correspond to $T = 0.12$ MHz for a single deuteron situated 3.1 Å from $Y_Z^\bullet$ or three equivalent deuterons that are 3.7 Å away [303]. It is noteworthy that these interspin distances are calculated using a number of approximations, including (i) the point-dipole approximation (see Equation 1.4); (ii) for the phenoxyl O atom, a value for $\rho = 0.28$ derived from EPR studies of ^{17}O-labeled $Y_D^\bullet$ [310–313], not $Y_Z^\bullet$ in the $S_2Y_Z^\bullet$ state; and (iii) complete neglect of the contribution to the effective ^{2}H HFC from the S_2 ($S = 1/2$) form of the Mn cluster (see Equation 1.9). A more reliable interpretation is gained in the simplified spin landscape provided by NO$^\bullet$ treatment of acetate-inhibited PSII membranes (i.e., acetate-treated PSII poised in the S_2 state). These studies directly probed the strength of coupling between the Mn cluster and the methyl deuterons on acetate [255]. The ^{2}H/^{1}H ratioed ESEEM data were well simulated using two deuterons with $T = 0.10$ and 0.19 MHz. Using the projection factors determined from pulse ENDOR experiments on the S_2 state (see Section 1.3.1 and Equation 1.9) [83], an isosurface map can be constructed, which helps visualize the location of the acetate molecule with respect to a given geometric and oxidation state configuration for the Mn ions of the OEC. It is easily seen from this plot that a range of Mn$\cdots^2$H distances are possible: 3.9–6.1 and 3.5–5.6 Å for the weak and more strongly coupled deuterons, respectively, depending on the nature of the coordinating Mn ion.

Results from ESEEM spectroscopic studies also indicated that methanol binding is unaffected by the presence of acetate, suggesting that acetate ligates to the Mn cluster at a site different from that for methanol [255]. It does appear, however, that acetate displaces a number of water molecules from the OEC. A comparison of the ^{2}H/^{1}H ratioed ESEEM spectra of untreated and acetate/NO$^\bullet$-treated PSII in D_2O reveals a significant loss of modulation depth upon addition of acetate (and NO$^\bullet$) [255]. The quotient of these two modulation patterns is simulated using six deuterons: two strongly isotropically coupled ($A_{iso} = 0.80$ MHz, $T = 0.62$ MHz) and four weakly coupled ($A_{iso} = 0$ MHz, $T = 0.45$ MHz). This loss of the strongly coupled deuterons (presumably from the same water molecule) coincident with the loss of activity is rationalized as a displacement of an Mn-bound water by acetate binding and disruption of the electron (and maybe proton) transfer pathway between the cluster and $Y_Z^\bullet$. Unfortunately, neither of these classes of deuterons exhibits HFC parameters similar to those measured for exchangeable waters in the native S_2 state (see Section 1.3.3 and Table 1.2). This anomaly could suggest that the water ejected from the OEC by acetate binding is not exchanged in its absence, or that acetate perturbs the hydrogen bonding network in the active site as to affect the ^{2}H HFC constants. Alternatively, the accuracy of the ^{2}H HFC parameters derived in this study could suffer from the lack of constraints provided by complementary ^{1}H ENDOR data.

ESEEM spectroscopy was also employed to assess the number of exchangeable water molecules coupled to the OEC in the acetate-inhibited $S_2Y_Z^\bullet$ state. Force et al. modeled the ratioed $^2H/^1H$ data using modulations from two deuterons with a coupling of $T = 0.57$ MHz [256], while Dorlet et al. simulated their data with an average dipolar coupling of 0.42 MHz from four deuterons [257]. This rather striking difference in conclusion from qualitatively quite similar data can be explained somewhat by the differences in analytical approach applied by the two groups. Dorlet et al. required that the number and strength of coupled deuterons explain the relative damping behavior and intensity of the fundamental and sum-combination ^2H modulations. Alternatively, Force et al. observed that the width of the ^2H pulse ENDOR spectrum of $Y_Z^\bullet$ in acetate-inhibited PSII was quite similar to that measured for $Y_D^\bullet$, and therefore, the anisotropic HFC determined for $Y_D^\bullet$ ($T = 0.57$ MHz) should be held constant for deuterons participating in hydrogen bonds to $Y_Z^\bullet$ [67]. Using these spin Hamiltonian parameters ($A_{iso} = 0$ MHz, $T = 0.57$ MHz, $e^2q_{zz}Q = 0.20$ MHz, $\eta = 0$), the number of equivalently coupled deuterons was varied until the experimentally observed ^2H modulation depth was achieved. In this case, two deuterons were necessary.

1.5.3 CA²⁺ DEPLETED

Like the acetate-inhibited form, Ca^{2+}-depleted PSII also exhibits the so-called split signal. However, the flanking lobes are separated by only 164 G compared with the nearly 230 G observed for acetate-treated PSII in the $S_2Y_Z^\bullet$ state (see Section 1.5.2) [271]. The similarity of the pulse ENDOR spectrum of Ca^{2+}-depleted PSII to that of $Y_Z^\bullet$ alone (obtained using Mn-depleted PSII) confirmed the participation of a photo-oxidized tyrosine residue in the spin-coupled species that gives rise to the split signal [27,313]. CW-EPR and pulse EPR studies on oriented Ca^{2+}-depleted membranes showed that the vector connecting these two paramagnets is at an angle of 65° with respect to the membrane normal, almost identical to that found for acetate-treated PSII [314].

High-field CW-EPR spectra of Ca^{2+}-depleted PSII in the $S_2Y_Z^\bullet$ state are well simulated using a dipolar exchange-coupling constant ($D = -100$ MHz) on the order of that found for the acetate-treated sample ($D = -110$ to -120 MHz) [307]. Strikingly, however, the isotropic exchange-coupling parameter J must be negative ($J = -420$ MHz) to achieve a satisfactory simulation of the 285 GHz data. The authors propose that this sign change in J compared with the positive value measured for acetate-treated PSII could result from (i) a dramatic change in the inter-Mn couplings within the tetramer leading or (ii), more simply, the Mn ion nearest to $Y_Z^\bullet$ changes its oxidation state. The latter mechanism is attractive as Ca^{2+} depletion leads to an altered S_2 MLS that can be understood in terms of a formal relocation of the Mn(III) ion to a new position within the cluster (see Section 1.3.4.2). Contrastingly, acetate-inhibited PSII poised in the $S_2Y_Z^\bullet$ state and then treated with nitric oxide gives rise to a normal S_2-like MLS, suggesting that the distribution of oxidation states within the OEC differs from that for Ca^{2+}-depleted membranes.

The number of exchangeable water molecules was again probed by ratioing the ESEEM spectra obtained after 30–60 min of incubation in either natural-abundance water or D_2O [258]. Tommos et al. were able to best simulate the ^2H modulation pattern using two deuterons, each with $T = 0.48$ MHz. Assuming, as before, that the phenoxyl O has an unpaired spin density of $\rho = 0.28$, this value for the anisotropic hyperfine corresponds to a $^2H-O(Y_Z^{\bullet})$ spin-only dipolar distance of 1.92 Å.

1.6 $Y_Z^{\bullet}$ COUPLED TO OTHER S-STATES

Akin to the split signal described previously that is achieved by $Y_Z^{\bullet}$ coupling to the S_2 form of the Mn cluster (see Section 1.5), photooxidation of Y_Z at low temperature (5 K) generates a relatively long-lived tyrosine radical that can couple to the OEC in any state of the Kok cycle excluding S_4 [315–320]. These spectra are generated without the addition of inhibitor (e.g., acetate) and thus may provide a more realistic picture of the electronic structure of the reaction intermediate. The properties of these CW-EPR spectra and the method of their generation is the subject of a recent review [321].

1.7 S_3

Matsukawa and coworkers [109] observed that by exposing dark-adapted PSII membranes to 10 min of continuous white-light illumination or to two saturating laser flashes ($\lambda = 532$ nm) at 235 K, new weak features emerged at $g = 8$ and 12 in the ‖-mode X-band EPR spectrum and at $g = 6.7$ in the ⊥-mode spectrum (Figure 1.12). Both ‖- and ⊥-mode spectra were well simulated, assuming an $S = 1$ spin system with ZFS parameters $D = 0.435$ cm^{-1} and $E = -0.137$ cm^{-1}. The temperature-dependent behavior of the EPR spectrum led to the conclusion that S_3 possesses an $S_T = 0$ ground state with at least one low-lying excited state ($\Delta < 1$ cm^{-1}) [109,321]. Unlike

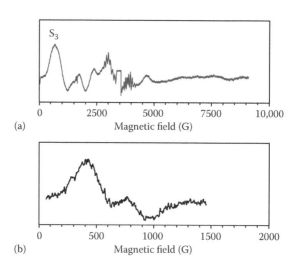

(a)

(b)

FIGURE 1.12   X-band CW-EPR spectra of the S_3 form of PSII observed in (a) perpendicular mode and (b) parallel mode. These data were obtained by subtracting the dark-adapted (S_1) spectrum from that of samples that had been illuminated for 12 min at 235 K.

for S_1 (Section 1.9), no [55]Mn hyperfine structure was found within these features. Recently, Boussac and coworkers detected additional resonances at higher applied magnetic fields with effective g-values of 2.94, 2.20, 1.4, and 0.85 [38]. These features could only be simulated using a value of $S_T = 3$ and the following magnetic parameters: $g_{iso} = 2$, $|D| = 0.175$ cm^{-1}, and $E/D = 0.275$. The authors point out that the strong magnetic coupling that must be present to achieve an $S_T = 3$ spin state strongly implies that oxidation of the OEC (either Mn centered or centered on a ligand of the first coordination sphere) is occurring on the $S_2 \rightarrow S_3$ transition.

1.8 S_0

As the S_2 form of PSII exhibits a ⊥-mode CW-EPR spectrum characteristic of a Kramers-type system, it is not surprising that the S_0 state, twice reduced from S_2, does as well (Figure 1.13) [322]. S_0 can be prepared in a variety of methods that yield qualitatively similar CW-EPR spectra. Messinger et al. found that hydrazine reduces dark-adapted PSII by two electrons, effectively yielding an S_{-1} form of the OEC [323,324]. This treatment in the presence of methanol (1–3% v/v) followed by illumination at 273 K gives rise to an MLS attributed to S_0. Additionally, hydroxylamine reduction of PSII in the S_1 state yields an almost identical EPR signature. A more physiologically relevant S_0 MLS achieved by laser flashing dark-adapted PSII was later observed separately by Åhrling et al. [325]. This method involves, first, illuminating dark-adapted PSII with a saturating laser flash at room temperature to oxidize all Y_D to $Y_D^\bullet$ and thus synchronizing all centers in the S_1 state [325]. Following the addition of 3% methanol, three laser flashes deliver S_0. The resulting CW-EPR spectrum is broader (≈2500 G) than that for the S_2 form (≈1800 G) and shows at least 20 [55]Mn-induced hyperfine lines with spacings ≈82 G [326]. Temperature-dependent EPR studies confirm that the S_0 state possesses an $S_T = 1/2$ ground state [325,326], allowing for a straightforward comparison between the EPR properties associated with the S_0 and S_2 forms of PSII. The S_0 MLS relaxes much faster than the MLS for S_2 [327]. Such behavior could be due to the presence of a relaxation enhancer near

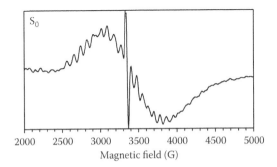

FIGURE 1.13 X-band CW-EPR spectrum of 1% (v/v) methanol-treated spinach PSII trapped in the S_0 state using three flash illumination at 277 K in the presence of carbonyl cyanide 4-trifluoromethoxy phenylhydrazone (FCCP) to enhance relaxation of centers trapped in the S_2 and S_3 states back to S_1.

the OEC in the S_0 state that is absent or distant in the S_2 state. Alternatively, owing to the differences in Mn oxidation states, the exchange-coupling scheme for S_0 will certainly be different from that for S_2 and could lead to a diminution of the energy gap between the ground and first excited spin manifold. Such low-lying excited states often quicken the relaxation kinetics of spin systems (i.e., Orbach process). Indeed, results from pulse EPR inversion recovery experiments indicate that there is an excited spin manifold just 21.7 cm^{-1} above the $S_T = 1/2$ ground state [328], a smaller energy gap than that found for S_2 (30–40 cm^{-1}).

Measurement of the g-values for S_0 has not yet been achieved via the high-frequency EPR single-crystal experiments that yielded precise values for the S_2 state (see Section 1.3.1 and Table 1.1). However, a very recent report required a considerably anisotropic g-matrix ($g = [2.009, 1.855, 1.974]$) to satisfactorily simulate both Q-band ESE-EPR and ^{55}Mn pulse ENDOR spectra of methanol-treated PSII trapped in the S_0 state [143,144]. These ENDOR data yield ^{55}Mn HFC matrices for each Mn ion and provide the basis for evaluating possible J-coupling schemes as above for S_2. Following the approach described in Charlot et al. [166], Kulik and coworkers explored a wide exchange-coupling parameter space for three classes of Mn cluster models (the most relevant of which are summarized in Figure 1.6). Importantly, they could not find a solution that included an Mn(II) ion in its description of the Mn cluster in the S_0 state. In their preferred model, it was noted that only one J-coupling— that between the trimer of Mn ions and the dangling monomer (J_{AB})—changed significantly upon oxidation to S_2 (from −38 cm^{-1} to −180 cm^{-1}). This change is coincident with a shortening of an Mn$\cdots$Mn internuclear distance observed by EXAFS spectroscopy [329]. It has been proposed that this spectral and structural evolution can be understood in terms of deprotonation of a μ-hydroxido bridge between Mn$_A$ and Mn$_B$ upon S-state advancement from S_0 [330], a hypothesis that is testable via ENDOR and ESEEM spectroscopies.

Analogous to the studies described previously for S_2, the presence of exchangeable, coupled protons present in the S_0 form of the OEC was detected by both CW-ENDOR [251] and ESEEM [91,161] spectroscopies. Results from the ENDOR studies revealed similar classes of exchangeable ^{1}H to those observed for the S_2 state (cf. Table 1.2). However, both features at 4.01 and 2.33 MHz disappeared after only 3 h of D$_2$O incubation, whereas in the S_2 state, the ^{1}H with $A_{obs} \approx 4$ MHz required 24 h to exchange. Assuming that the origin of this resonance is unchanged upon S-state advancement, this modestly strongly coupled proton is either bound to an amino acid residue that is directly coordinated to the Mn cluster or to a component of a protonated μ-oxido bridge, as described in Section 1.3.3. The variation in exchange rate as a function of S-state could imply that the coordinating Mn center is oxidized at some point between S_0 and S_2, lowering the pK_a of the amino acid or hydroxy bridge.

The ratioed ^{2}H/^{1}H ESEEM spectra of PSII in the S_0 state show significant modulation from deuterium following D$_2$O incubation [91,161]. The ^{2}H HFC parameters extracted from these data are presented in Table 1.2. Britt and coworkers distinguish four distinct classes of deuterons in the vicinity of the Mn cluster. The first two sets of deuterons, each with a population of two, have large values of T (0.85 and 0.70 MHz) and seem to correspond to the waters bound directly to Mn and Ca ($r = 2.43$ Å and 2.59 Å, respectively) observed in water-binding studies of S_2. The members of the next class are more distant ($r = 3.12$ Å) from the nearest Mn center and lack any isotropic HFC. In contrast to the results on S_2, an additional, strongly coupled ($a_{iso} = 0.30$ MHz; $T = 0.75$ MHz) deuteron is required to fit the experimental modulation depth. This single H atom could correspond to the proton that is postulated to be lost sometime during the $S_0 \rightarrow S_2$ transitions.

Much as small alcohols increase the intensity of the MLS for the S_2 state of PSII, the presence of methanol is required to resolve the ^{55}Mn hyperfine features in the S_0 MLS [323,324,331]. Using methanol perdeuterated at the methyl position in their ESEEM spectroscopic studies, Åhrling et al. were able to show that the alcohol binds directly to the Mn cluster with a calculated Mn–O bond distance of 1.96 Å (assuming the methanol molecule is bound to Mn ion with a projection factor $c = 1$, i.e., Mn(III)) [161]. This value is in close agreement with that determined for the S_2 state ($r_{Mn-O} = 1.98$ Å, also determined using $c = 1$); hence, it is likely that methanol binds early in the Kok cycle and remains bound at least through to S_2. In addition, there is a feature present at 4.8 MHz, which the authors attribute to the same ^{14}N nucleus (i.e., from D1-His332) as is bound to the S_2 form of the OEC described in Section 1.3.2. The modest change in the frequency of this feature upon S-state advancement (4.8 MHz in S_0 to 5.1 MHz in S_2) could reflect small differences in exchange-coupling parameters between both forms of the Mn cluster. Were the oxidation state of the N-coordinating Mn ion increasing from Mn(III) in the S_0 state to Mn(IV) in the S_2 state, one would expect a much more dramatic shift in the hyperfine frequency based on the large difference in projection factors ($c_{Mn(III)} \approx 1.5$; $c_{Mn(IV)} \approx 1$–1.2). Thus, it seems unlikely that the Mn center bound by D1-His332 undergoes metal-centered oxidation during the $S_0 \rightarrow S_2$ transitions. Nonetheless, a more accurate determination of the HFC parameters for this ^{14}N nucleus is required before the corresponding bonding description in the S_0 state can be inferred.

1.9 S_1

Dark-adapted PSII membrane particles exhibit no signal when using $\perp$-mode CW-EPR spectroscopy at X-band frequency. This is consistent with the formulation of the S_1 state as an integer spin or $S_T = 0$ system. For PSII isolated from spinach, Dexheimer and Klein observed a wide (≈ 600 G peak-to-peak; ≈ 1800 G overall width), nearly symmetric derivative-shaped resonance at $g = 4.8$ in the ||-mode spectrum at X-band. This feature exhibited no clear hyperfine-induced structure and was attributed to an $S = 1$ species that was simulated with ZFS parameters $D = -0.125$ cm^{-1} and $E = 0.025$ cm^{-1} [332]. Through analysis of the temperature-dependent behavior, Yamauchi et al. ascribed the $g = 4.8$ signal to an $S = 1$ excited state approximately 2.5 K above the $S_T = 0$ ground state with nearly identical ZFS parameters as above ($D = -0.140$ cm^{-1} and $E = 0.035$ cm^{-1}) [333]. Notably, the addition of methanol inhibits the formation of this signal. It was also observed that this ||-mode signal could not be observed in the spectra of PSII particles from which Ca^{2+} or Cl^- had been removed. Thus, the loss of either of these cofactors appears to induce a dramatic change in the exchange-coupling parameters, leading to an increase in the magnitude of the corresponding ZFS parameters and splitting the spin levels responsible for the ||-mode signal to a degree beyond that accessible by the X-band microwave photon.

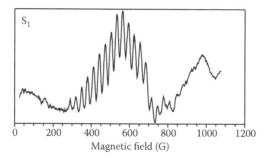

FIGURE 1.14 Parallel-mode X-band (9.41 GHz) CW-EPR spectrum of dark-adapted PSII particles from *Synechocystis* sp. 6803. The EPR spectrum from centers illuminated for 6 min at 195 K (S_2) is subtracted.

Later ‖-mode studies by Campbell et al. on PSII particles from the cyanobacterium *Synechocystis* sp. PCC 6803 revealed an altogether different spectrum with 18 lines and centered at $g \approx 12$ (Figure 1.14) [146,334]. The number and spacing ($\approx$32 G on average) of hyperfine lines visible in the spectrum suggest that all four of the Mn centers are strongly spin coupled. Indeed, the S_1 MLS is well simulated using $S_T = 2$ (based on temperature-dependent behavior), assuming four equivalent [55]Mn HFC matrices ($A_{x,y} = 196$ MHz, $A_z = 166$ MHz) and the following spin Hamiltonian parameters: $g_{x,y} = 2.0$, $g_z = 1.942$, $D = 0.805$ cm^{-1}, and $E = -0.26$ cm^{-1}. These findings are consistent with an Mn(III)$_2$Mn(IV)$_2$ oxidation state formulation for the S_1 form, as was determined from recent XANES (x-ray absorption near edge structure) studies and redox titrations [335]. It should be noted, however, that the appearance of this S_1 MLS in spinach is dependent on the presence or absence of certain extrinsic proteins [146]. While no direct connection has yet been made between the original S_1 ‖-polarization signal observed at $g = 4.8$ and the S_1 MLS at $g \approx 12$, a mechanism of interconversion should not be ruled out given the existence of two distinct but interconverting spin states in the S_2 form of PSII (cf. Section 1.3).

1.10 INSIGHTS FROM EPR SPECTROSCOPY INTO THE ELECTRONIC STRUCTURE OF THE OEC

Results from a combination of multifrequency CW-EPR and pulse EPR spectroscopic studies have afforded unique insights into the electronic structure of the OEC during the S_0, S_1, S_2, and S_3 states of the Kok cycle. Interpreted within the context of findings from EXAFS and x-ray crystallographic experiments, precise pictures of the dynamic architecture of the tetranuclear Mn cluster are beginning to take form. Nonetheless, there remain many essential questions concerning (i) the assignment of Mn oxidation states within the OEC at each stage of the Kok cycle; (ii) the nature of the protein-derived set of ligands to Mn; (iii) and, most important, the locality and electronic structure of the bound substrate water molecules. The most salient and recent progress on these three fronts is summarized in the following.

The [55]Mn HFC constants determined using pulse ENDOR in conjunction with results from CW-EPR and EXAFS spectroscopies have helped formulate the most modern models of the Mn cluster topography [28,83,143,144]. The dangler Mn model was first proposed on the basis of such results. Recent findings from Q-band ENDOR studies of the S_0 state have shown that the Mn cluster cannot include an Mn(II) center. Thus, the S_0 state is most likely composed of one Mn(IV) and three Mn(III) ions.

Multifrequency ESEEM studies on native preparations of PSII poised in the S_2 state have, for the first time, afforded a complete electronic structure description of the nitrogenous Mn ligand first observed via EPR studies carried out in 1988 [202,204,336]. These data strongly suggest that the τ-nitrogen on the imidazole ring of D1-His332 is the lone nitrogen-based ligand to the OEC, at least in its S_2 form that gives rise to the MLS found in Figure 1.3a. Analysis of this nitrogen HFC shows that D1-His332 is bound to the sole Mn(III) ion in the low-spin S_2 state of PSII [204]. In addition, there is some indication of an ^{14}N coupling to the S_0 state of the OEC; yet, an imperforate electronic structure description is thus far lacking [161]. ESEEM spectroscopic results also showed that a nitrogen atom is coupled to the OEC in the S_2 state of Ca^{2+}-depleted [272] and ammonia-treated PSII [126,267]. Both of these species yield a so-called altered MLS (see Section 1.3.4). A change in the locality of the Mn(III) ion was long thought to be responsible for the dramatic changes in the CW-EPR spectrum. However, using the ^{14}N HFC of D1-His332 as a reporter, recent ESEEM spectroscopic results of ammonia-treated PSII show that manganic ion is unmoved [126].

Despite the high-resolution crystal structure coordinates now available for PSII, given the likelihood of radiation-induced damage/reduction to the OEC, the identity of proteinaceous ligands to the fully assembled Mn cluster in catalytically relevant oxidation states remains largely unknown. As stated previously, results from ESEEM spectroscopic studies indicate that D1-His332 is bound at least in the S_2 state [194]. However, this leaves open several coordination sites that are likely occupied by amino acid-derived carboxylate ligands. While numerous candidates for such carboxylates have been proposed (e.g., D1-Asp170, D1-Glu189, D1-Glu333, D1-Asp342, and CP43-Glu354,), only the carboxy terminus of the D1 polypeptide, D1-Ala344, has been shown by EPR to bind to Mn of the OEC in the S_2 state [215].

Some of the open coordination sites on the OEC must allow for binding of the substrate, water. Indeed, in the S_2 state, exchangeable protons and oxygens attributed to water molecules have been detected by EPR spectroscopy (see Section 1.3.3). Taken collectively, these results suggest to us that there exist two substrate water molecules that are in rapid exchange with bulk solvent. Analysis of the corresponding ^{1}H and [17]O HFC parameters indicates that one class of EPR-detected water is bound directly to an Mn ion, while the other is perhaps coordinated to the Ca^{2+} ion. A further more strongly coupled [17]O nucleus was detected using EDNMR methods. This site is also fast exchanging with bulk solvent and, on the basis of DFT results, is assigned to O5 of the OEC [126,264].

ACKNOWLEDGMENTS

This work was funded by the Division of Chemical Sciences, Geosciences, and Biosciences, Office of Basic Energy Sciences of the U.S. Department of Energy through grant DE- FG02-12ER16282. The authors also appreciate the careful and critical reading of the manuscript by Robert M. McCarrick (Miami University, Oxford, OH), Jamie A. Stull, Gregory Yeagle, Marcin Brynda, and Stefan Stoll (University of Washington, Seattle, WA).

REFERENCES

1. A. Zouni, H.-T. Witt, J. Kern, P. Fromme, N. Krauss, W. Saenger, P. Orth, *Nature*, 409 (2001) 739–743.
2. J. Biesiadka, B. Loll, J. Kern, K.-D. Irrgang, A. Zouni, *Phys. Chem. Chem. Phys.*, 6 (2004) 4733–4736.
3. I. Virgin, T. Hundal, S. Styring, B. Andersson, *NATO ASI Ser., Ser. A*, 168 (1989) 535–538.
4. E.J. Boekema, J. Nield, B. Hankamer, J. Barber, *Eur. J. Biochem.*, 252 (1998) 268–276.
5. S.D. Betts, N. Lydakis-Simantiris, J.R. Ross, C.F. Yocum, *Biochemistry*, 37 (1998) 14230–14236.
6. K.A. Vander Meulen, A. Hobson, C.F. Yocum, *Biochim. Biophys. Acta Bioenerget.*, 1655 (2004) 179–183.
7. F. Mamedov, S. Styring, *Physiol. Plant.*, 119 (2003) 328–336.
8. T.M. Bricker, J.L. Roose, R.D. Fagerlund, L.K. Frankel, J.J. Eaton-Rye, *Biochim. Biophys. Acta Bioenerget.*, 1817 (2012) 121–142.
9. G. Renger, T. Renger, *Photosynth. Res.*, 98 (2008) 53–80.
10. B. Loll, J. Kern, A. Zouni, W. Saenger, J. Biesiadka, K.-D. Irrgang, *Photosynth. Res.*, 86 (2005) 175–184.
11. B.A. Diner, E. Schlodder, P.J. Nixon, W.J. Coleman, F. Rappaport, J. Lavergne, W.F. Vermaas, D.A. Chisholm, *Biochemistry*, 40 (2001) 9265–9281.
12. R. Takahashi, K. Hasegawa, T. Noguchi, *Biochemistry*, 47 (2008) 6289–6291.
13. F. Rappaport, B.A. Diner, *Coord. Chem. Rev.*, 252 (2008) 259–272.
14. G.T. Babcock, B.A. Barry, R.J. Debus, C.W. Hoganson, M. Atamian, L. McIntosh, I. Sithole, C.F. Yocum, *Biochemistry*, 28 (1989) 9557–9565.
15. J.P. McEvoy, G.W. Brudvig, *Biochemistry*, 47 (2008) 13394–13403.
16. A.R. Crofts, C.A. Wraight, *Biochim. Biophys. Acta*, 726 (1983) 149–185.
17. R. Krivanek, J. Kern, A. Zouni, H. Dau, M. Haumann, *Biochim. Biophys. Acta Bioenerget.*, 1767 (2007) 520–527.
18. N. Nelson, C.F. Yocum, *Annu. Rev. Plant Biol.*, 57 (2006) 521–565.
19. R.J. Debus, B.A. Barry, I. Sithole, G.T. Babcock, L. McIntosh, *Biochemistry*, 27 (1988) 9071–9074.
20. P. Faller, R.J. Debus, K. Brettel, M. Sugiura, A.W. Rutherford, A. Boussac, *Proc. Natl. Acad. Sci. U.S.A.*, 98 (2001) 14368–14373.
21. M. Haumann, A. Mulkidjanian, W. Junge, *Biochemistry*, 38 (1999) 1258–1267.
22. C. Berthomieu, R. Hienerwadel, A. Boussac, J. Breton, B.A. Diner, *Biochemistry*, 37 (1998) 10547–10554.
23. W. Junge, M. Haumann, R. Ahlbrink, A. Mulkidjanian, J. Clausen, *Philos. Trans. R. Soc. Lond., Ser. B Biol. Sci.*, 357 (2002) 1407–1418.
24. F. Mamedov, R.T. Sayre, S. Styring, *Biochemistry*, 37 (1998) 14245–14256.
25. A.-M.A. Hays, I.R. Vassiliev, J.H. Golbeck, R.J. Debus, *Biochemistry*, 37 (1998) 11352–11365.
26. A.-M.A. Hays, I.R. Vassiliev, J.H. Golbeck, R.J. Debus, *Biochemistry*, 38 (1999) 11851–11865.
27. M.L. Gilchrist, Jr., J.A. Ball, D.W. Randall, R.D. Britt, *Proc. Natl. Acad. Sci. U.S.A.*, 92 (1995) 9545–9549.
28. J.M. Peloquin, K.A. Campbell, R.D. Britt, *J. Am. Chem. Soc.*, 120 (1998) 6840–6841.

29. K.V. Lakshmi, S.S. Eaton, G.R. Eaton, G.W. Brudvig, *Biochemistry*, 38 (1999) 12758–12767.
30. B. Kok, B. Forbush, M. McGloin, *Photochem. Photobiol.*, 11 (1970) 457–475.
31. P. Joliot, B. Kok, *Bioenergetics of Photosynthesis*, Academic Press, New York, 1975 387–412.
32. J.P. McEvoy, G.W. Brudvig, *Chem. Rev. (Washington, DC)*, 106 (2006) 4455–4483.
33. K. Sauer, J. Yano, V.K. Yachandra, *Photosynth. Res.*, 85 (2005) 73–86.
34. Y. Pushkar, J. Yano, P. Glatzel, J. Messinger, A. Lewis, K. Sauer, U. Bergmann, V. Yachandra, *J. Biol. Chem.*, 282 (2007) 7198–7208.
35. J. Cole, V.K. Yachandra, R.D. Guiles, A.E. McDermott, R.D. Britt, S.L. Dexheimer, K. Sauer, M.P. Klein, *Biochim. Biophys. Acta Bioenerget.*, 890 (1987) 395–398.
36. W. Liang, T.A. Roelofs, R.M. Cinco, A. Rompel, M.J. Latimer, W.O. Yu, K. Sauer, M.P. Klein, V.K. Yachandra, *J. Am. Chem. Soc.*, 122 (2000) 3399–3412.
37. M. Haumann, C. Muller, P. Liebisch, L. Iuzzolino, J. Dittmer, M. Grabolle, T. Neisius, W. Meyer-Klaucke, H. Dau, *Biochemistry*, 44 (2005) 1894–1908.
38. A. Boussac, M. Sugiura, A.W. Rutherford, P. Dorlet, *J. Am. Chem. Soc.*, 131 (2009) 5050–5051.
39. P. Joliot, G. Barbieri, R. Chabaud, *Photochem. Photobiol.*, 10 (1969) 309–329.
40. M. Haumann, P. Liebisch, C. Mueller, M. Barra, M. Grabolle, H. Dau, *Science*, 310 (2005) 1019–1021.
41. H. Dau, M. Haumann, *Photosynth. Res.*, 92 (2007) 327–343.
42. H. Dau, M. Haumann, *Science*, 312 (2006) 1471–1472.
43. J.E. Penner-Hahn, C.F. Yocum, *Science*, 310 (2005) 982–983.
44. H. Dau, M. Haumann, *Biochim. Biophys. Acta Bioenerget.*, 1767 (2007) 472–483.
45. L. Gerencser, H. Dau, *Biochemistry*, 49 (2010) 10098–10106.
46. A.F. Miller, G.W. Brudvig, *Biochim. Biophys. Acta Bioenerget.*, 1056 (1991) 1–18.
47. W. Lubitz, *Electron Paramagn. Reson.*, 19 (2004) 174–242.
48. W. Lubitz, *Phys. Chem. Chem. Phys.*, 4 (2002) 5539–5545.
49. K.V. Lakshmi, O.G. Poluektov, M.J. Reifler, A.M. Wagner, M.C. Thurnauer, G.W. Brudvig, *J. Am. Chem. Soc.*, 125 (2003) 5005–5014.
50. S.G. Zech, J. Kurreck, H.-J. Eckert, G. Renger, W. Lubitz, R. Bittl, *FEBS Lett.*, 414 (1997) 454–456.
51. F. Lendzian, R. Bittl, A. Telfer, W. Lubitz, *Biochim. Biophys. Acta Bioenerget.*, 1605 (2003) 35–46.
52. S.E.J. Rigby, J.H.A. Nugent, P.J. O'Malley, *Biochemistry*, 33 (1994) 10043–10050.
53. P. Dorlet, L. Xiong, R.T. Sayre, S. Un, *J. Biol. Chem.*, 276 (2001) 22313–22316.
54. Y. Deligiannakis, A.W. Rutherford, *J. Am. Chem. Soc.*, 119 (1997) 4471–4480.
55. I. García-Rubio, J.I. Martínez, R. Picorel, I.L. Yruela, P.J. Alonso, *J. Am. Chem. Soc.*, 125 (2003) 15846–15854.
56. A. Kawamori, N. Katsuta, H. Mino, A. Ishii, J. Minagawa, T.A. Ono, *J. Biol. Phys.*, 28 (2002) 413–426.
57. Y. Deligiannakis, A. Boussac, A.W. Rutherford, *Biochemistry*, 34 (1995) 16030–16038.
58. T. Yoshii, H. Hara, A. Kawamori, K. Akabori, M. Iwaki, S. Itoh, *Appl. Magn. Reson.*, 16 (1999) 565–580.
59. A. Schnegg, M. Fuhs, M. Rohrer, W. Lubitz, T.F. Prisner, K. Moebius, *J. Phys. Chem. B*, 106 (2002) 9454–9462.
60. M. Rohrer, P. Gast, K. Moebius, T.F. Prisner, *Chem. Phys. Lett.*, 259 (1996) 523–530.
61. R. Aasa, L.E. Andreasson, S. Styring, T. Vaenngaard, *FEBS Lett.*, 243 (1989) 156–160.
62. C. Goussias, Y. Deligianakis, Y. Sanakis, N. Ioannidis, V. Petrouleas, *Biochemistry*, 41 (2002) 15212–15223.
63. J.M. Peloquin, X.S. Tang, B.A. Diner, R.D. Britt, *Biochemistry*, 38 (1999) 2057–2067.

64. A. Sedoud, N. Cox, M. Sugiura, W. Lubitz, A. Boussac, A.W. Rutherford, *Biochemistry*, 50 (2011) 6012–6021.
65. H. Mino, J.-I. Satoh, A. Kawamori, K. Toriyama, J.-L. Zimmermann, *Biochim. Biophys. Acta*, 1144 (1993) 426–433.
66. C. Tommos, C. Madsen, S. Styring, W. Vermaas, *Biochemistry*, 33 (1994) 11805–11813.
67. D.A. Force, D.W. Randall, R.D. Britt, X.-S. Tang, B.A. Diner, *J. Am. Chem. Soc.*, 117 (1995) 12643–12644.
68. C. Teutloff, S. Pudollek, S. Kessen, M. Broser, A. Zouni, R. Bittl, *Phys. Chem. Chem. Phys.*, 11 (2009) 6715–6726.
69. G.T. Babcock, K. Sauer, *Biochim. Biophys. Acta*, 376 (1975) 315–328.
70. G.T. Babcock, R.E. Blankenship, K. Sauer, *FEBS Lett.*, 61 (1976) 286–289.
71. W. Hofbauer, A. Zouni, R. Bittl, J. Kern, P. Orth, F. Lendzian, P. Fromme, H.T. Witt, W. Lubitz, *Proc. Natl. Acad. Sci. U.S.A.*, 98 (2001) 6623–6628.
72. K. Warncke, G.T. Babcock, J. McCracken, *J. Am. Chem. Soc.*, 116 (1994) 7332–7340.
73. R.G. Evelo, A.J. Hoff, S.A. Dikanov, A.M. Tyryshkin, *Chem. Phys. Lett.*, 161 (1989) 479–484.
74. A.W. Rutherford, A. Boussac, P. Faller, *Biochim. Biophys. Acta Bioenerget.*, 1655 (2004) 222–230.
75. M. Szczepaniak, M. Sugiura, A.R. Holzwarth, *Biochim. Biophys. Acta Bioenerget.*, 1777 (2008) 1510–1517.
76. J. Yano, J. Kern, K.-D. Irrgang, M.J. Latimer, U. Bergmann, P. Glatzel, Y. Pushkar, J. Biesiadka, B. Loll, K. Sauer, J. Messinger, A. Zouni, V.K. Yachandra, *Proc. Natl. Acad. Sci. U.S.A.*, 102 (2005) 12047–12052.
77. R.J. Debus, *Biochim. Biophys. Acta*, 1102 (1992) 269–352.
78. H. Popelkova, C.F. Yocum, *Photosynth. Res.*, 93 (2007) 111–121.
79. J. Yano, V. Yachandra, *Photosynth. Res.*, 92 (2007) 289–303.
80. A. Grundmeier, H. Dau, *Biochim. Biophys. Acta Bioenerget.*, 1817 (2012) 88–105.
81. J. Yano, V. Yachandra, *Chem. Rev.*, 114 (2014) 4175–4205.
82. Y. Umena, K. Kawakami, J.-R. Shen, N. Kamiya, *Nature*, 473 (2011) 55–60.
83. J.M. Peloquin, K.A. Campbell, D.W. Randall, M.A. Evanchik, V.L. Pecoraro, W.H. Armstrong, R.D. Britt, *J. Am. Chem. Soc.*, 122 (2000) 10926–10942.
84. G.C. Dismukes, K. Ferris, P. Watnick, *Photobiochem. Photobiophys.*, 3 (1982) 243–256.
85. G.C. Dismukes, Y. Siderer, *Proc. Natl. Acad. Sci. U.S.A.*, 78 (1981) 274–278.
86. G.C. Dismukes, Y. Siderer, *FEBS Lett.*, 121 (1980) 78–80.
87. A.W. Rutherford, A. Boussac, J.L. Zimmermann, *New J. Chem.*, 15 (1991) 491–500.
88. R.D. Britt, J.M. Peloquin, K.A. Campbell, *Annu. Rev. Biophys. Biomol. Struct.*, 29 (2000) 463–495.
89. J.M. Peloquin, R.D. Britt, *Biochim. Biophys. Acta*, 1503 (2001) 96–111.
90. A. Kawamori, *Prog. Theor. Chem. Phys.*, 10 (2003) 529–563.
91. R.D. Britt, K.A. Campbell, J.M. Peloquin, M.L. Gilchrist, C.P. Aznar, M.M. Dicus, J. Robblee, J. Messinger, *Biochim. Biophys. Acta*, 1655 (2004) 158–171.
92. A. Haddy, *Photosynth. Res.*, 92 (2007) 357–368.
93. L.E. Andreasson, *Biochim. Biophys. Acta Bioenerget.*, 973 (1989) 465–467.
94. A. Abragam, B. Bleany, *Electron Paramagnetic Resonance of Transition Ions*, 2nd ed., Clarendon Press, Oxford, UK, 1970.
95. A. Schweiger, G. Jeschke, *Principles of Pulse Electron Paramagnetic Resonance*, Oxford University Press, Oxford, UK, 2001.
96. M. Zheng, G.C. Dismukes, *Inorg. Chem.*, 35 (1996) 3307–3319.
97. C.I.H. Ashby, C.P. Cheng, T.L. Brown, *J. Am. Chem. Soc.*, 100 (1978) 6057–6063.
98. F. Jiang, J. McCracken, J. Peisach, *J. Am. Chem. Soc.*, 112 (1990) 9035–9044.
99. A. Bencini, D. Gatteschi, *EPR of Exchange Coupled Systems*, Verlag, Springer-Verlag, Berlin, 1990.

100. S.R. Cooper, G.C. Dismukes, M.P. Klein, M. Calvin, *J. Am. Chem. Soc.*, 100 (1978) 7248–7252.

101. K. Wieghardt, U. Bossek, W. Gebert, *Angew. Chem.*, 95 (1983) 320.

102. V.K. Yachandra, R.D. Guiles, A.E. McDermott, J.L. Cole, R.D. Britt, S.L. Dexheimer, K. Sauer, M.P. Klein, *Biochemistry*, 26 (1987) 5974–5981.

103. J. Yano, Y. Pushkar, P. Glatzel, A. Lewis, K. Sauer, J. Messinger, U. Bergmann, V. Yachandra, *J. Am. Chem. Soc.*, 127 (2005) 14974–14975.

104. T.C. Brunold, D.R. Gamelin, E.I. Solomon, *J. Am. Chem. Soc.*, 122 (2000) 8511–8523.

105. H. Rupp, R. Cammack, K.K. Rao, D.O. Hall, *Biochim. Biophys. Acta*, 537 (1978) 255–269.

106. L. Banci, I. Bertini, C. Luchinat, *Nuclear and Electron Relaxation*, VCH, New York, 1991.

107. J.H. van Vleck, *Phys. Rev.*, 57 (1940) 426–447.

108. R. Orbach, *Proc. R. Soc. (Lond.)*, 264 (1960) 485–495.

109. T. Matsukawa, H. Mino, D. Yoneda, A. Kawamori, *Biochemistry*, 38 (1999) 4072–4077.

110. P.E. Doan, B.M. Hoffman, *Inorg. Chim. Acta*, 297 (2000) 400–403.

111. R. Song, P.E. Doan, R.J. Gurbiel, B.E. Sturgeon, B.M. Hoffman, *J. Magn. Reson.*, 141 (1999) 291–300.

112. B.E. Sturgeon, P.E. Doan, K.E. Liu, D. Burdi, W.H. Tong, J.M. Nocek, N. Gupta, J. Stubbe, D.M. Kurtz, S.J. Lippard, B.M. Hoffman, *J. Am. Chem. Soc.*, 119 (1997) 375–386.

113. B.M. Hoffman, *J. Phys. Chem.*, 98 (1994) 11657–11665.

114. B.M. Hoffman, B.E. Sturgeon, P.E. Doan, V.J. DeRose, K.E. Liu, S.J. Lippard, *J. Am. Chem. Soc.*, 116 (1994) 6023–6024.

115. P.E.M. Siegbahn, *Inorg. Chem.*, 39 (2000) 2923–2935.

116. P.E.M. Siegbahn, M. Lundberg, *Photochem. Photobiol. Sci.*, 4 (2005) 1035–1043.

117. M. Lundberg, M.R.A. Blomberg, P.E.M. Siegbahn, *Theor. Chem. Acc.*, 110 (2003) 130–143.

118. E.M. Sproviero, J.A. Gascon, J.P. McEvoy, G.W. Brudvig, V.S. Batista, *J. Chem. Theory Comput.*, 2 (2006) 1119–1134.

119. J.P. McEvoy, J.A. Gascon, V.S. Batista, G.W. Brudvig, *Photochem. Photobiol. Sci.*, 4 (2005) 940–949.

120. P.E.M. Siegbahn, *Acc. Chem. Res.*, 42 (2009) 1871–1880.

121. L. Noodleman, E.J. Baerends, *J. Am. Chem. Soc.*, 106 (1984) 2316–2327.

122. L. Noodleman, D. Post, E.J. Baerends, *Chem. Phys.*, 64 (1982) 159–166.

123. L. Noodleman, *J. Chem. Phys.*, 74 (1981) 5737–5743.

124. J.M. Mouesca, L. Noodleman, D.A. Case, B. Lamotte, *Inorg. Chem.*, 34 (1995) 4347–4359.

125. F. Neese, *Coord. Chem. Rev.*, 253 (2009) 526–563.

126. M.P. Navarro, W.M. Ames, H. Nilsson, T. Lohmiller, D.A. Pantazis, L. Rapatskiy, M.M. Nowaczyk, F. Neese, A. Boussac, J. Messinger, W. Lubitz, N. Cox, *Proc. Natl. Acad. Sci. U.S.A.*, 110 (2013) 15561–15566.

127. T. Lohmiller, W. Ames, W. Lubitz, N. Cox, S.K. Misra, *Appl. Magn. Reson.*, 44 (2013) 691–720.

128. L. Rapatskiy, N. Cox, A. Savitsky, W.M. Ames, J. Sander, M.M. Nowaczyk, M. Roegner, A. Boussac, F. Neese, J. Messinger, W. Lubitz, *J. Am. Chem. Soc.*, 134 (2012) 16619–16634.

129. D.A. Pantazis, W. Ames, N. Cox, W. Lubitz, F. Neese, *Angew. Chem. Int. Ed.*, 51 (2012) 9935–9940.

130. J.-H. Su, N. Cox, W. Ames, D.A. Pantazis, L. Rapatskiy, T. Lohmiller, L.V. Kulik, P. Dorlet, A.W. Rutherford, F. Neese, A. Boussac, W. Lubitz, J. Messinger, *Biochim. Biophys. Acta Bioenerget.*, 1807 (2011) 829–840.

131. N. Cox, W. Ames, B. Epel, L.V. Kulik, L. Rapatskiy, F. Neese, J. Messinger, K. Wieghardt, W. Lubitz, *Inorg. Chem.*, 50 (2011) 8238–8251.
132. W. Ames, D.A. Pantazis, V. Krewald, N. Cox, J. Messinger, W. Lubitz, F. Neese, *J. Am. Chem. Soc.*, 133 (2011) 19743–19757.
133. J. Schraut, A.V. Arbuznikov, S. Schinzel, M. Kaupp, *ChemPhysChem*, 12 (2011) 3170–3179.
134. S. Schinzel, J. Schraut, A.V. Arbuznikov, P.E.M. Siegbahn, M. Kaupp, *Chem. Eur. J.*, 16 (2010) 10424–10438.
135. G.W. Brudvig, J.L. Casey, K. Sauer, *Biochim. Biophys. Acta*, 723 (1983) 366–371.
136. G. Bernat, F. Morvaridi, Y. Feyziyev, S. Styring, *Biochemistry*, 41 (2002) 5830–5843.
137. R.G. Evelo, S. Styring, A.W. Rutherford, A.J. Hoff, *Biochim. Biophys. Acta Bioenerget.*, 973 (1989) 428–442.
138. A. Sedoud, L. Kastner, N. Cox, S. El-Alaoui, D. Kirilovsky, A.W. Rutherford, *Biochim. Biophys. Acta Bioenerget.*, 1807 (2011) 216–226.
139. J.-H. Su, W. Lubitz, J. Messinger, *J. Am. Chem. Soc.*, 130 (2007) 786–787.
140. J.-H. Su, W. Lubitz, J. Messinger, *J. Am. Chem. Soc.*, 133 (2011) 12317.
141. H. Matsuoka, K. Furukawa, T. Kato, H. Mino, J.-R. Shen, A. Kawamori, *J. Phys. Chem. B*, 110 (2006) 13242–13247.
142. C. Teutloff, S. Kessen, J. Kern, A. Zouni, R. Bittl, *FEBS Lett.*, 580 (2006) 3605–3609.
143. L.V. Kulik, B. Epel, W. Lubitz, J. Messinger, *J. Am. Chem. Soc.*, 127 (2005) 2392–2393.
144. L.V. Kulik, B. Epel, W. Lubitz, J. Messinger, *J. Am. Chem. Soc.*, 129 (2007) 13421–13435.
145. D.H. Kim, R.D. Britt, M.P. Klein, K. Sauer, *Biochemistry*, 31 (1992) 541–547.
146. K.A. Campbell, W. Gregor, D.P. Pham, J.M. Peloquin, R.J. Debus, R.D. Britt, *Biochemistry*, 37 (1998) 5039–5045.
147. C. Teutloff, K.-O. Schaefer, S. Sinnecker, V. Barynin, R. Bittl, K. Wieghardt, F. Lendzian, W. Lubitz, *Magn. Reson. Chem.*, 43 (2005) S51–S64.
148. K. Hasegawa, M. Kusunoki, Y. Inoue, T.-A. Ono, *Biochemistry*, 37 (1998) 9457–9465.
149. K. Hasegawa, T.-A. Ono, Y. Inoue, M. Kusunoki, *Chem. Phys. Lett.*, 300 (1999) 9–19.
150. K.A. Åhrling, R.J. Pace, *Biophys. J.*, 68 (1995) 2081–2090.
151. K.A. Åhrling, P.J. Smith, R.J. Pace, *J. Am. Chem. Soc.*, 120 (1998) 13202–13214.
152. R.M. McCarrick, R.D. Britt, in: P. Fromme (ed.) *Photosynthetic Protein Complexes: A Structural Approach*, Wiley, Weinheim, 2008.
153. J.L. Zimmermann, A.W. Rutherford, *Biochim. Biophys. Acta*, 767 (1984) 160–167.
154. J.L. Zimmermann, A.W. Rutherford, *Biochemistry*, 25 (1986) 4609–4615.
155. W.F. Beck, G.W. Brudvig, *Biochemistry*, 25 (1986) 6479–6486.
156. V.J. DeRose, M.J. Latimer, J.-L. Zimmermann, I. Mukerji, V.K. Yachandra, K. Sauer, M.P. Klein, *Chem. Phys.*, 194 (1995) 443–459.
157. P. van Vliet, A. Boussac, A.W. Rutherford, *Biochemistry*, 33 (1994) 12998–13004.
158. A. Haddy, W.R. Dunham, R.H. Sands, R. Aasa, *Biochim. Biophys. Acta Bioenerget.*, 1099 (1992) 25–34.
159. N. Mizusawa, T. Yamanari, Y. Kimura, A. Ishii, S. Nakazawa, T.-A. Ono, *Biochemistry*, 43 (2004) 14644–14652.
160. D.A. Force, D.W. Randall, G.A. Lorigan, K.L. Clemens, R.D. Britt, *J. Am. Chem. Soc.*, 120 (1998) 13321–13333.
161. K.A. Åhrling, M.C.W. Evans, J.H.A. Nugent, R.J. Ball, R.J. Pace, *Biochemistry*, 45 (2006) 7069–7082.
162. A. Boussac, H. Kuhl, S. Un, M. Roegner, A.W. Rutherford, *Biochemistry*, 37 (1998) 8995–9000.
163. A. Boussac, *J. Biol. Inorg. Chem.*, 2 (1997) 580–585.
164. A. Boussac, J.-J. Girerd, A.W. Rutherford, *Biochemistry*, 35 (1996) 6984–6989.
165. K.A. Åhrling, M.C.W. Evans, J.H.A. Nugent, R.J. Pace, *Biochim. Biophys. Acta Bioenerget.*, 1656 (2004) 66–77.

166. M.-F. Charlot, A. Boussac, G. Blondin, *Biochim. Biophys. Acta Bioenerget.*, 1708 (2005) 120–132.
167. R.D. Britt, G.A. Lorigan, K. Sauer, M.P. Klein, J.-L. Zimmermann, *Biochim. Biophys. Acta Bioenerget.*, 1140 (1992) 95–101.
168. Ö. Hannson, R. Aasa, T. Vänngård, *Biophys. Chem.*, 51 (1987) 825–832.
169. R.J. Pace, P. Smith, R. Bramley, D. Stehlik, *Biochim. Biophys. Acta Bioenerget.*, 1058 (1991) 161–170.
170. G.A. Lorigan, R.D. Britt, *Biochemistry*, 33 (1994) 12072–12076.
171. D. Koulougliotis, R.H. Schweitzer, G.W. Brudvig, *Biochemistry*, 36 (1997) 9735–9746.
172. G.A. Lorigan, R.D. Britt, *Photosynth. Res.*, 66 (2001) 189–198.
173. M.I. Belinskii, *Chem. Phys.*, 189 (1994) 451–465.
174. D.W. Randall, B.E. Sturgeon, J.A. Ball, G.A. Lorigan, M.K. Chan, M.P. Klein, W.H. Armstrong, R.D. Britt, *J. Am. Chem. Soc.*, 117 (1995) 11780–11789.
175. H.J. Gerritsen, E.S. Sabisky, *Phys. Rev.*, 132 (1963) 1507–1512.
176. H.G. Andresen, *Phys. Rev.*, 120 (1960) 1606–1611.
177. N. Cox, L. Rapatskiy, J.-H. Su, D.A. Pantazis, M. Sugiura, L. Kulik, P. Dorlet, A.W. Rutherford, F. Neese, A. Boussac, W. Lubitz, J. Messinger, *J. Am. Chem. Soc.*, 133 (2011) 3635–3648.
178. A. Haddy, K.V. Lakshmi, G.W. Brudvig, H.A. Frank, *Biophys. J.*, 87 (2004) 2885–2896.
179. A. Boussac, S. Un, O. Horner, A.W. Rutherford, *Biochemistry*, 37 (1998) 4001–4007.
180. Y. Sanakis, N. Ioannidis, G. Sioros, V. Petrouleas, *J. Am. Chem. Soc.*, 123 (2001) 10766–10767.
181. P.J. Nixon, B.A. Diner, *Biochemistry*, 31 (1992) 942–948.
182. H.-A. Chu, A.P. Nguyen, R.J. Debus, *Biochemistry*, 34 (1995) 5839–5858.
183. Y. Allahverdiyeva, Z. Deak, A. Szilard, B.A. Diner, P.J. Nixon, I. Vass, *Eur. J. Biochem.*, 271 (2004) 3523–3532.
184. B.A. Diner, D.F. Ries, B.N. Cohen, J.G. Metz, *J. Biol. Chem.*, 263 (1988) 8972–8980.
185. P.J. Nixon, J.T. Trost, B.A. Diner, *Biochemistry*, 31 (1992) 10859–10871.
186. N. Mizusawa, Y. Kimura, A. Ishii, T. Yamanari, S. Nakazawa, H. Teramoto, T.-A. Ono, *J. Biol. Chem.*, 279 (2004) 29622–29627.
187. R.O. Cohen, P.J. Nixon, B.A. Diner, *J. Biol. Chem.*, 282 (2007) 7209–7218.
188. M.A. Strickler, H.J. Hwang, R.L. Burnap, J. Yano, L.M. Walker, R.J. Service, R.D. Britt, W. Hillier, R.J. Debus, *Philos. Trans. R. Soc. B Biol. Sci.*, 363 (2008) 1179–1187.
189. R.J. Debus, *Metal Ions Biol. Syst.*, 37 (2000) 657–711.
190. B.A. Diner, *Biochim. Biophys. Acta Bioenerget.*, 1503 (2001) 147–163.
191. X.S. Tang, M. Sivaraja, G.C. Dismukes, *J. Am. Chem. Soc.*, 115 (1993) 2382–2389.
192. R.D. Britt, X.S. Tang, M.L. Gilchrist, G.A. Lorigan, B.S. Larsen, B.A. Diner, *Biochem. Soc. Trans.*, 22 (1994) 343–347.
193. V.J. DeRose, V.K. Yachandra, A.E. McDermott, R.D. Britt, K. Sauer, M.P. Klein, *Biochemistry*, 30 (1991) 1335–1341.
194. X.S. Tang, B.A. Diner, B.S. Larsen, M.L. Gilchrist, Jr., G.A. Lorigan, R.D. Britt, *Proc. Natl. Acad. Sci. U.S.A.*, 91 (1994) 704–708.
195. H.J. Hwang, P. Dilbeck, R.J. Debus, R.L. Burnap, *Biochemistry*, 46 (2007) 11987–11997.
196. R.J. Debus, K.A. Campbell, J.M. Peloquin, D.P. Pham, R.D. Britt, *Biochemistry*, 39 (2000) 470–478.
197. R.J. Debus, K.A. Campbell, W. Gregor, Z.-L. Li, R.L. Burnap, R.D. Britt, *Biochemistry*, 40 (2001) 3690–3699.
198. H.L. Flanagan, G.J. Gerfen, D.J. Singel, *J. Chem. Phys.*, 88 (1988) 20–24.
199. H.L. Flanagan, D.J. Singel, *Chem. Phys. Lett.*, 137 (1987) 391–397.
200. H.L. Flanagan, D.J. Singel, *J. Chem. Phys.*, 87 (1987) 5606–5616.
201. I.L. McConnell, V.M. Grigoryants, C.P. Scholes, W.K. Myers, P.-Y. Chen, J.W. Whittaker, G.W. Brudvig, *J. Am. Chem. Soc.*, 134 (2011) 1504–1512.

202. G.J. Yeagle, M.L. Gilchrist, R.M. McCarrick, R.D. Britt, *Inorg. Chem.*, 47 (2008) 1803–1814.
203. G.J. Yeagle, M.L. Gilchrist, Jr., L.M. Walker, R.J. Debus, R.D. Britt, *Philos. Trans. R. Soc. B*, 363 (2008) 1157–1166.
204. T.A. Stich, G.J. Yeagle, R.J. Service, R.J. Debus, R.D. Britt, *Biochemistry*, 50 (2011) 7390–7404.
205. M. Sugiura, F. Rappaport, W. Hillier, P. Dorlet, Y. Ohno, H. Hayashi, A. Boussac, *Biochemistry*, 48 (2009) 7856–7866.
206. M. Sugiura, A. Boussac, T. Noguchi, F. Rappaport, *Biochim. Biophys. Acta Bioenerget.*, 1777 (2008) 331–342.
207. S.A.P. Merry, P.J. Nixon, L.M.C. Barter, M. Schilstra, G. Porter, J. Barber, J.R. Durrant, D.R. Klug, *Biochemistry*, 37 (1998) 17439–17447.
208. K. Cser, I. Vass, *Biochim. Biophys. Acta Bioenerget.*, 1767 (2007) 233–243.
209. S. Milikisiyants, R. Chatterjee, A. Weyers, A. Meenaghan, C. Coates, K.V. Lakshmi, *J. Phys. Chem. B*, 114 (2010) 10905–10911.
210. P. Hofer, A. Grupp, H. Nebenfuhr, M. Mehring, *Chem. Phys. Lett.*, 132 (1986) 279–282.
211. H.-A. Chu, W. Hillier, R.J. Debus, *Biochemistry*, 43 (2004) 3152–3166.
212. M.A. Strickler, L.M. Walker, W. Hillier, R.J. Debus, *Biochemistry*, 44 (2005) 8571–8577.
213. K. Cser, B.A. Diner, P.J. Nixon, I. Vass, *Photochem. Photobiol. Sci.*, 4 (2005) 1049–1054.
214. S. Kuvikova, M. Tichy, J. Komenda, *Photochem. Photobiol. Sci.*, 4 (2005) 1044–1048.
215. J.A. Stull, T.A. Stich, R.J. Service, R.J. Debus, S.K. Mandal, W.H. Armstrong, R.D. Britt, *J. Am. Chem. Soc.*, 132 (2010) 446–447.
216. K.A. Campbell, D.A. Force, P.J. Nixon, F. Dole, B.A. Diner, R.D. Britt, *J. Am. Chem. Soc.*, 122 (2000) 3754–3761.
217. A.F. Miller, G.W. Brudvig, *Biochemistry*, 28 (1989) 8181–8190.
218. K.N. Ferreira, T.M. Iverson, K. Maghlaoui, J. Barber, S. Iwata, *Science*, 303 (2004) 1831–1838.
219. H.-A. Chu, R.J. Debus, G.T. Babcock, *Biochemistry*, 40 (2001) 2312–2316.
220. R.J. Debus, M.A. Strickler, L.M. Walker, W. Hillier, *Biochemistry*, 44 (2005) 1367–1374.
221. R.J. Debus, C. Aznar, K.A. Campbell, W. Gregor, B.A. Diner, R.D. Britt, *Biochemistry*, 42 (2003) 10600–10608.
222. M.L. Ghirardi, T.W. Lutton, M. Seibert, *Biochemistry*, 37 (1998) 13559–13566.
223. V.V. Klimov, S.I. Allakhverdiev, Y.M. Feyziev, S.V. Baranov, *FEBS Lett.*, 363 (1995) 251–255.
224. V.V. Klimov, S.V. Baranov, *Biochim. Biophys. Acta*, 1503 (2001) 187–196.
225. V.V. Klimov, R.J. Hulsebosch, S.I. Allakhverdiev, H. Wincencjusz, H.J. van Gorkom, A.J. Hoff, *Biochemistry*, 36 (1997) 16277–16281.
226. S.V. Baranov, A.M. Tyryshkin, D. Katz, G.C. Dismukes, G.M. Ananyev, V.V. Klimov, *Biochemistry*, 43 (2004) 2070–2079.
227. G.M. Ananyev, L. Zaltsman, C. Vasko, G.C. Dismukes, *Biochim. Biophys. Acta*, 1503 (2001) 52–68.
228. G.C. Dismukes, V.V. Klimov, S.V. Baranov, Y.N. Kozlov, J. DasGupta, A. Tyryshkin, *Proc. Natl. Acad. Sci. U.S.A.*, 98 (2001) 2170–2175.
229. Y.N. Kozlov, S.K. Zharmukhamedov, K.G. Tikhonov, J. Dasgupta, A.A. Kazakova, G.C. Dismukes, V.V. Klimov, *Phys. Chem. Chem. Phys.*, 6 (2004) 4905–4911.
230. J. Dasgupta, A.M. Tyryshkin, Y.N. Kozlov, V.V. Klimov, G.C. Dismukes, *J. Phys. Chem. B*, 110 (2006) 5099–5111.
231. J. Dasgupta, R.T. van Willigen, G.C. Dismukes, *Phys. Chem. Chem. Phys.*, 6 (2004) 4793–4802.
232. A. Boussac, *Chem. Phys.*, 194 (1995) 409–418.
233. M. Haumann, M. Barra, P. Loja, S. Loescher, R. Krivanek, A. Grundmeier, L.-E. Andreasson, H. Dau, *Biochemistry*, 45 (2006) 13101–13107.

234. H. Wincencjusz, H.J. van Gorkom, C.F. Yocum, *Biochemistry*, 36 (1997) 3663–3670.
235. K. Kawakami, Y. Umena, N. Kamiya, J.R. Shen, *Proc. Natl. Acad. Sci. U.S.A.*, 106 (2009) 8567–8572.
236. P. van Vliet, A.W. Rutherford, *Biochemistry*, 35 (1996) 1829–1839.
237. D.I. Bryson, N. Doctor, R. Johnson, S. Baranov, A. Haddy, *Biochemistry*, 44 (2005) 7354–7360.
238. T. Ono, H. Nakayama, H. Gleiter, Y. Inoue, A. Kawamori, *Arch. Biochem. Biophys.*, 256 (1987) 618–624.
239. H. Wincencjusz, C.F. Yocum, H.J. van Gorkom, *Biochemistry*, 38 (1999) 3719–3725.
240. A. Jajoo, S. Bharti, A. Kawamori, *Photochem. Photobiol. Sci.*, 4 (2005) 459–462.
241. A. Haddy, J.A. Hatchell, R.A. Kimel, R. Thomas, *Biochemistry*, 38 (1999) 6104–6110.
242. H. Yu, C.P. Aznar, X. Xu, R.D. Britt, *Biochemistry*, 44 (2005) 12022–12029.
243. A. Boussac, A.W. Rutherford, *FEBS Lett.*, 236 (1988) 432–436.
244. A. Boussac, P. Setif, A.W. Rutherford, *Biochemistry*, 31 (1992) 1224–1234.
245. K. Luebbers, W. Drevenstedt, W. Junge, *FEBS Lett.*, 336 (1993) 304–308.
246. P.L. Dilbeck, H.J. Hwang, I. Zaharieva, L. Gerencser, H. Dau, R.L. Burnap, *Biochemistry*, 51 (2011) 1079–1091.
247. I. Rivalta, M. Amin, S. Luber, S. Vassiliev, R. Pokhrel, Y. Umena, K. Kawakami, J.-R. Shen, N. Kamiya, D. Bruce, G.W. Brudvig, M.R. Gunner, V.S. Batista, *Biochemistry*, 50 (2011) 6312–6315.
248. A. Boussac, N. Ishida, M. Sugiura, F. Rappaport, *Biochim. Biophys. Acta Bioenerg.*, 1817 (2012) 802–810.
249. A. Kawamori, T. Inui, T. Ono, Y. Inoue, *FEBS Lett.*, 254 (1989) 219–224.
250. R. Fiege, W. Zweygart, R. Bittl, N. Adir, G. Renger, W. Lubitz, *Photosynth. Res.*, 48 (1996) 227–237.
251. H. Yamada, H. Mino, S. Itoh, *Biochim. Biophys. Acta Bioenerget.*, 1767 (2007) 197–203.
252. C.P. Aznar, R.D. Britt, *Philos. Trans. R. Soc. Lond., Ser. B*, 357 (2002) 1359–1366.
253. C.P. Aznar, M.L. Gilchrist, J.M. Peloquin, D.A. Force, J. Sarrou, X. Xu, R.D. Britt, unpublished (2007).
254. M.C.W. Evans, J.H.A. Nugent, R.J. Ball, I. Muhiuddin, R.J. Pace, *Biochemistry*, 43 (2004) 989–994.
255. K.L. Clemens, D.A. Force, R.D. Britt, *J. Am. Chem. Soc.*, 124 (2002) 10921–10933.
256. D.A. Force, D.W. Randall, R.D. Britt, *Biochemistry*, 36 (1997) 12062–12070.
257. P. Dorlet, M. Di Valentin, G.T. Babcock, J.L. McCracken, *J. Phys. Chem. B*, 102 (1998) 8239–8247.
258. C. Tommos, J. McCracken, S. Styring, G.T. Babcock, *J. Am. Chem. Soc.*, 120 (1998) 10441–10452.
259. W. Hillier, T. Wydrzynski, *Phys. Chem. Chem. Phys.*, 6 (2004) 4882–4889.
260. W. Hillier, G. Hendry, R.L. Burnap, T. Wydrzynski, *J. Biol. Chem.*, 276 (2001) 46917–46924.
261. W. Hillier, T. Wydrzynski, *Biochemistry*, 39 (2000) 4399–4405.
262. E.R. Davies, *Phys. Lett. A*, 47 (1974) 1–2.
263. M.C.W. Evans, A.M. Rich, J.H.A. Nugent, *FEBS Lett.*, 477 (2000) 113–117.
264. N. Cox, W. Lubitz, A. Savitsky, *Mol. Phys.*, 111 (2013) 2788–2808.
265. W.F. Beck, G.W. Brudvig, *Chem. Scr.*, 28A (1988) 93–98.
266. W.F. Beck, J.C. De Paula, G.W. Brudvig, *J. Am. Chem. Soc.*, 108 (1986) 4018–4022.
267. R.D. Britt, J.L. Zimmermann, K. Sauer, M.P. Klein, *J. Am. Chem. Soc.*, 111 (1989) 3522–3532.
268. T. Ono, Y. Inoue, *FEBS Lett.*, 227 (1988) 147–152.
269. K.A. Vander Meulen, A. Hobson, C.F. Yocum, *Biochemistry*, 41 (2002) 958–966.
270. A. Boussac, A.W. Rutherford, *Biochemistry*, 27 (1988) 3476–3483.
271. A. Boussac, J.L. Zimmermann, A.W. Rutherford, *Biochemistry*, 28 (1989) 8984–8989.
272. J.L. Zimmermann, A. Boussac, A.W. Rutherford, *Biochemistry*, 32 (1993) 4831–4841.

273. B. Loll, J. Kern, W. Saenger, A. Zouni, J. Biesiadka, *Nature*, 438 (2005) 1040–1044.
274. M.J. Latimer, V.J. DeRose, V.K. Yachandra, K. Sauer, M.P. Klein, *J. Phys. Chem. B*, 102 (1998) 8257–8265.
275. T. Lohmiller, N. Cox, J.-H. Su, J. Messinger, W. Lubitz, *J. Biol. Chem.*, 287 (2012) 24721–24733.
276. C.F. Yocum, *Biochim. Biophys. Acta Bioenerget.*, 1059 (1991) 1–15.
277. K. Burda, K. Strzalka, G.H. Schmid, *Z. Naturforsch. C Biosci.*, 50 (1995) 220–230.
278. C.-I. Lee, K.V. Lakshmi, G.W. Brudvig, *Biochemistry*, 46 (2007) 3211–3223.
279. A. Boussac, F. Rappaport, P. Carrier, J.-M. Verbavatz, R. Gobin, D. Kirilovsky, A.W. Rutherford, M. Sugiura, *J. Biol. Chem.*, 279 (2004) 22809–22819.
280. K.L. Westphal, N. Lydakis-Simantiris, R.I. Cukier, G.T. Babcock, *Biochemistry*, 39 (2000) 16220–16229.
281. R.M. Cinco, J.H. Robblee, A. Rompel, C. Fernandez, V.K. Yachandra, K. Sauer, M.P. Klein, *J. Phys. Chem. B*, 102 (1998) 8248–8256.
282. M.J. Latimer, V.J. DeRose, I. Mukerji, V.K. Yachandra, K. Sauer, M.P. Klein, *Biochemistry*, 34 (1995) 10898–10909.
283. A. Boussac, A.W. Rutherford, *Chem. Scr.*, 28A (1988) 123–126.
284. S.H. Kim, W. Gregor, J.M. Peloquin, M. Brynda, R.D. Britt, *J. Am. Chem. Soc.*, 126 (2004) 7228–7237.
285. S.A. Dikanov, Y.D. Tsvetkov, *Electron Spin-Echo Envelope Modulation Spectroscopy*, CRC Press, Boca Raton, FL, 1992.
286. R.M. Cinco, J.H. Robblee, A. Rompel, C. Fernandez, K. Sauer, V.K. Yachandra, M.P. Klein, *J. Synchrotron Radiat.*, 6 (1999) 419–420.
287. F.H.M. Koua, Y. Umena, K. Kawakami, J.-R. Shen, *Proc. Natl. Acad. Sci. U.S.A.*, 110 (2013) 3889–3894.
288. D.H. Kim, R.D. Britt, M.P. Klein, K. Sauer, *J. Am. Chem. Soc.*, 112 (1990) 9389–9391.
289. J.C. De Paula, W.F. Beck, G.W. Brudvig, *J. Am. Chem. Soc.*, 108 (1986) 4002–4009.
290. A.V. Astashkin, Y. Kodera, A. Kawamori, *J. Magn. Reson.*, 105 (1994) 113–119.
291. S. Mukherjee, J.A. Stull, J. Yano, T.C. Stamatatos, K. Pringouri, T.A. Stich, K.A. Abboud, R.D. Britt, V.K. Yachandra, G. Christou, *Proc. Natl. Acad. Sci. U.S.A.*, 109 (2012) 2257–2262.
292. W. Liang, M.J. Latimer, H. Dau, T.A. Roelofs, V.K. Yachandra, K. Sauer, M.P. Klein, *Biochemistry*, 33 (1994) 4923–4932.
293. K. Onoda, H. Mino, Y. Inoue, T. Noguchi, *Photosynth. Res.*, 63 (2000) 47–57.
294. N. Lydakis-Simantiris, P. Dorlet, D.F. Ghanotakis, G.T. Babcock, *Biochemistry*, 37 (1998) 6427–6435.
295. A. Boussac, J.L. Zimmermann, A.W. Rutherford, *FEBS Lett.*, 277 (1990) 69–74.
296. J. Messinger, U. Wacker, G. Renger, *Biochemistry*, 30 (1991) 7852–7862.
297. B.J. Hallahan, J.H.A. Nugent, J.T. Warden, M.C.W. Evans, *Biochemistry*, 31 (1992) 4562–4573.
298. D.J. MacLachlan, J.H.A. Nugent, *Biochemistry*, 32 (1993) 9772–9780.
299. K.V. Lakshmi, S.S. Eaton, G.R. Eaton, H.A. Frank, G.W. Brudvig, *J. Phys. Chem. B*, 102 (1998) 8327–8335.
300. R.J. Debus, K.A. Campbell, D.P. Pham, A.-M.A. Hays, R.D. Britt, *Biochemistry*, 39 (2000) 6275–6287.
301. P.O. Sandusky, C.F. Yocum, *Biochim. Biophys. Acta Bioenerget.*, 766 (1984) 603–611.
302. P.O. Sandusky, C.F. Yocum, *FEBS Lett.*, 162 (1983) 339–343.
303. M. Baumgarten, J.S. Philo, G.C. Dismukes, *Biochemistry*, 29 (1990) 10814–10822.
304. O. Saygin, S. Gerken, B. Meyer, H.T. Witt, *Photosynth. Res.*, 9 (1986) 71–78.
305. H. Kuehne, V.A. Szalai, G.W. Brudvig, *Biochemistry*, 38 (1999) 6604–6613.
306. V.A. Szalai, H. Kuehne, K.V. Lakshmi, G.W. Brudvig, *Biochemistry*, 37 (1998) 13594–13603.

307. P. Dorlet, A. Boussac, A.W. Rutherford, S. Un, *J. Phys. Chem. B*, 103 (1999) 10945–10954.
308. V.A. Szalai, G.W. Brudvig, *Biochemistry*, 35 (1996) 15080–15087.
309. T.D. Smith, J.R. Pilbrow, *Coord. Chem. Rev.*, 13 (1974) 173.
310. F. Dole, B.A. Diner, C.W. Hoganson, G.T. Babcock, R.D. Britt, *J. Am. Chem. Soc.*, 119 (1997) 11540–11541.
311. R.J. Hulsebosch, J.S. van den Brink, S.A.M. Nieuwenhuis, P. Gast, J. Raap, J. Lugtenburg, A.J. Hoff, *J. Am. Chem. Soc.*, 119 (1997) 8685–8694.
312. C.T. Farrar, G.J. Gerfen, R.G. Griffin, D.A. Force, R.D. Britt, *J. Phys. Chem. B*, 101 (1997) 6634–6641.
313. A.V. Astashkin, H. Mino, A. Kawamori, T.-A. Ono, *Chem. Phys. Lett.*, 272 (1997) 506–516.
314. H. Mino, A. Kawamori, T. Ono, *Biochemistry*, 39 (2000) 11034–11040.
315. N. Ioannidis, V. Petrouleas, *Biochemistry*, 41 (2002) 9580–9588.
316. J.H.A. Nugent, I.P. Muhiuddin, M.C.W. Evans, *Biochemistry*, 41 (2002) 4117–4126.
317. D. Koulougliotis, C. Teutloff, Y. Sanakis, W. Lubitz, V. Petrouleas, *Phys. Chem. Chem. Phys.*, 6 (2004) 4859–4863.
318. D. Koulougliotis, J.-R. Shen, N. Ioannidis, V. Petrouleas, *Biochemistry*, 42 (2003) 3045–3053.
319. N. Ioannidis, J.H.A. Nugent, V. Petrouleas, *Biochemistry*, 41 (2002) 9589–9600.
320. C. Zhang, S. Styring, *Biochemistry*, 42 (2003) 8066–8076.
321. V. Petrouleas, D. Koulougliotis, N. Ioannidis, *Biochemistry*, 44 (2005) 6723–6728.
322. P. Geijer, S. Peterson, K.A. Åhrling, Z. Deak, S. Styring, *Biochim. Biophys. Acta Bioenerget.*, 1503 (2001) 83–95.
323. J. Messinger, J. Robblee, W.O. Yu, K. Sauer, V.K. Yachandra, M.P. Klein, *J. Am. Chem. Soc.*, 119 (1997) 11349–11350.
324. J. Messinger, J.H.A. Nugent, M.C.W. Evans, *Biochemistry*, 36 (1997) 11055–11060.
325. K.A. Åhrling, S. Peterson, S. Styring, *Biochemistry*, 36 (1997) 13148–13152.
326. K.A. Åhrling, S. Peterson, S. Styring, *Biochemistry*, 37 (1998) 8115–8120.
327. S. Peterson, K.A. Åhrling, S. Styring, *Biochemistry*, 38 (1999) 15223–15230.
328. L.V. Kulik, W. Lubitz, J. Messinger, *Biochemistry*, 44 (2005) 9368–9374.
329. J.H. Robblee, J. Messinger, R.M. Cinco, K.L. McFarlane, C. Fernandez, S.A. Pizarro, K. Sauer, V.K. Yachandra, *J. Am. Chem. Soc.*, 124 (2002) 7459–7471.
330. M.J. Baldwin, T.L. Stemmler, P.J. Riggs-Gelasco, M.L. Kirk, J.E. Penner-Hahn, V.L. Pecoraro, *J. Am. Chem. Soc.*, 116 (1994) 11349–11356.
331. Z. Deak, S. Peterson, P. Geijer, K.A. Åhrling, S. Styring, *Biochim. Biophys. Acta Bioenerget.*, 1412 (1999) 240–249.
332. S.L. Dexheimer, M.P. Klein, *J. Am. Chem. Soc.*, 114 (1992) 2821–2826.
333. T. Yamauchi, H. Mino, T. Matsukawa, A. Kawamori, T.-A. Ono, *Biochemistry*, 36 (1997) 7520–7526.
334. K.A. Campbell, J.M. Peloquin, D.P. Pham, R.J. Debus, R.D. Britt, *J. Am. Chem. Soc.*, 120 (1998) 447–448.
335. T. Kuntzleman, C.F. Yocum, *Biochemistry*, 44 (2005) 2129–2142.
336. R.D. Britt, in: *Electron Spin Echo Spectroscopy of Photosynthesis*, Lawrence Berkeley Laboratory, Berkeley, CA, 1988, pp. 214.

2 Radical SAM Enzymes and Their Roles in Complex Cluster Assembly

Jeremiah N. Betz, Eric M. Shepard, and Joan B. Broderick

CONTENTS

2.1 INTRODUCTION

The abstraction or transfer of a hydrogen atom (H·) from an unactivated bond *in vivo* is often carried out by transition metal-containing enzymes. Owing to its multiple biologically accessible oxidation states, iron is more often than not called upon to aid in hydrogen atom abstraction mechanisms. Under aerobic conditions, cytochrome P450 enzymes may utilize an oxyferryl heme group to abstract an H-atom from a C–H bond, eventually producing an alcohol [1]. This oxygen insertion reaction is used in solubilizing foreign hydrophobic chemicals and synthesizing biologically important chemicals. Under anaerobic conditions, the radical *S*-adenosyl-L-methionine (SAM) superfamily [2] is primarily responsible for H-atom abstraction events that initiate a wide array of biochemical transformations. Radical SAM enzymes all contain a [4Fe–4S] cubane cluster, whose irons are coordinated by sulfurs from three cysteine residues and in a bidentate fashion by the carboxyl and amine functional groups of SAM (Figure 2.1). More than 5000 anaerobic and aerobic organisms from all three domains of life possess radical SAM enzymes. Roughly 50,000 radical SAM enzymes have been putatively identified through bioinformatics analyses of sequences in the National Center for Biotechnology Information sequence database.

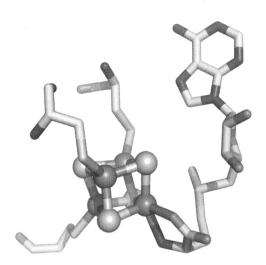

FIGURE 2.1   Site-differentiated [4Fe–4S] cluster coordinated by the Cys residues of the CX_3CX_2C motif and SAM (PDB ID: 3IIZ).

2.1.1 Early Characterization and Superfamily Status of Radical SAM Enzymes

As one can only see the forest when the number of trees is high and the vantage point is sufficiently far, the description of the radical SAM superfamily arose only after the discovery and biochemical and spectroscopic characterization of several key enzymes. New bioinformatics tools and low-cost sequencing further catalyzed the superfamily designation [2,3]. The classification scheme is largely built around the highly conserved CX_3CX_2C motif whose thiolate moieties bind three irons of a site-differentiated [4Fe–4S] cluster. Early members of the radical SAM family, which are still currently used as model enzymes, include lysine amino mutase (LAM), pyruvate formate lyase activating enzyme (PFL-AE), and biotin synthase (BioB). While these enzymes catalyze very different overall chemistries, it is held that they have conserved early mechanistic steps as described later in this chapter.

2.1.2 Catalytic Reaction Examples

A small sample set of the breadth of reactions that radical SAM enzymes are capable of catalyzing is represented in Figure 2.2 by isomerizations (a and e), enzyme activation (b), sulfur insertion (c), small-molecule production (d), and complex rearrangements (f). Four examples of some of the better-characterized radical SAM enzymes are highlighted in Sections 2.1.2.1 to 2.1.2.4.

2.1.2.1 Lysine 2,3-Aminomutase (LAM)

Radical SAM enzymes are often employed to catalyze reactions involving amino acid metabolism. LAM stereoselectively converts L-lysine to L-β-lysine in the first step of the lysine fermentation pathway. LAM was shown to require pyridoxal phosphate (PLP) as a cofactor in addition to SAM for full mutase activity, and to exhibit enhanced activity with the addition of exogenous iron (Fe^{2+}) [4]. Subsequent investigations showed that four irons bound per subunit, and these were assigned to a [4Fe–4S] cluster [5]. Electron nuclear double resonance (ENDOR) studies demonstrated that SAM coordinated to this [4Fe–4S] cluster via the amino and carboxylate moieties, and suggested that upon reductive cleavage the methionine would be bound as a tridentate chelate with the added coordination of the sulfur [6]. The LAM isomerization reaction is completely reversible, and it is largely held that SAM is used as a cofactor and is therefore regenerated after each catalytic cycle. PLP forms a Schiff base with the α or β amine nitrogen of lysine immobilizing the substrate. Upon reductive cleavage of SAM, the 5′-deoxyadenosyl radical (dAdo•) intermediate is produced, which abstracts a hydrogen atom from lysine to initiate rearrangement. The x-ray crystal structure of LAM from *Clostridium subterminale* revealed a dimer of dimers where each structural unit contains a splayed $(\beta\alpha)_6$ motif (Figure 2.3a,b) similar to a triosephosphate isomerase (TIM) $(\beta\alpha)_8$ barrel [7]. LAM is rather unusual among radical SAM enzymes as its quaternary structure is stabilized by both zinc ion binding and domain swapping. The active site is located within the partial TIM barrel, and the distance between C5′ of SAM and the β carbon of lysine was determined to be 3.6 Å; at this distance, the radical can swing

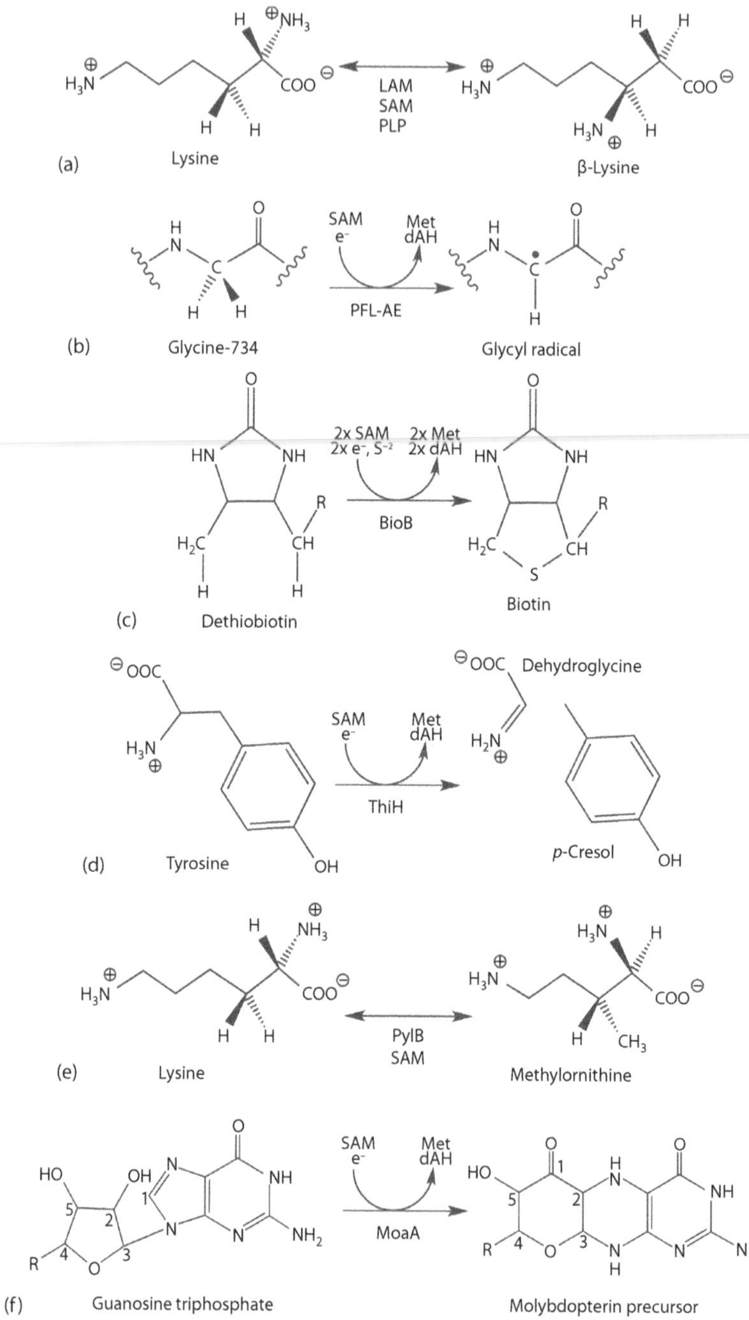

FIGURE 2.2 Overall reactions of select radical SAM enzymes. (a) Lysine 2,3-aminomutase (LAM). (b) Pyruvate formate lyase activating enzyme (PFL-AE). (c) Biotin synthase (BioB). (d) Thiamine biosynthesis enzyme H (ThiH). (e) Pyrrolysine biosynthesis enzyme B (PylB). (f) Molybdopterin biosynthetic enzyme A (MoaA). R = triphosphate group.

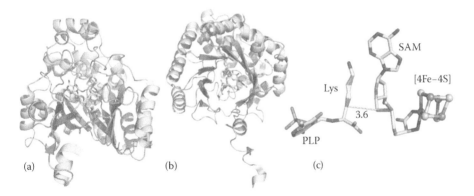

FIGURE 2.3 Structure of the LAM subunit as viewed from side (a) and bottom (b) of the partial TIM barrel. (c) The active site highlighting the [4Fe–4S] cluster SAM, PLP, and Lys. Distance in Å between lysine's carbon with abstractable hydrogen and 5'C of SAM is marked with a dash (PDB ID: 2A5H).

over and abstract a hydrogen from either the β carbon of lysine or the α carbon of β-lysine [8]. The generally accepted isomerization mechanism has a cyclic intermediate that includes the α carbon and nitrogen and the β carbon of lysine (Figure 2.4) in a manner that is likely similar to the analogous enzyme glutamate 2,3-aminomutase [9,10].

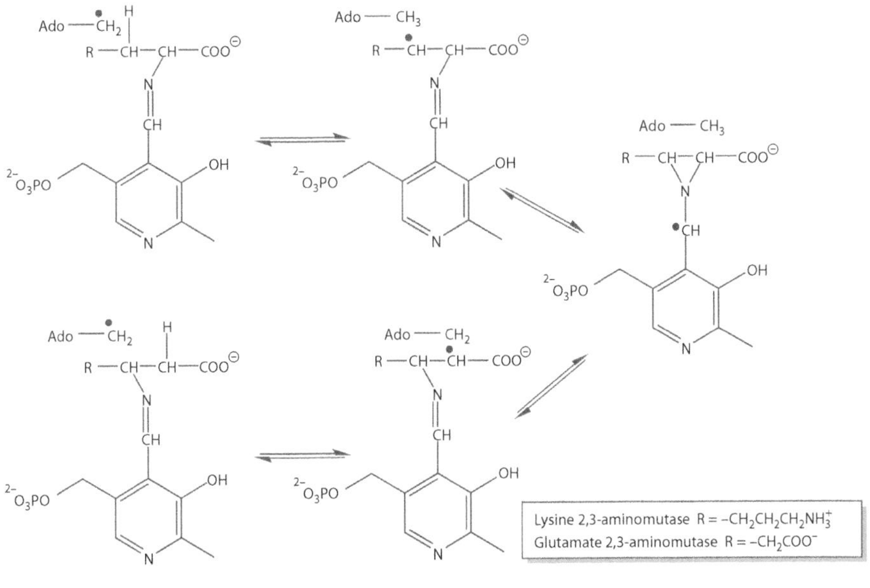

FIGURE 2.4 Proposed equilibrium-driven mechanism for LAM after reductive cleavage of SAM.

2.1.2.2 Pyruvate Formate Lyase Activating Enzyme (PFL-AE)

While LAM has a relatively small substrate, PFL-AE abstracts a hydrogen atom from a conserved glycine residue on the enzyme pyruvate formate lyase (PFL), a 170-kDa dimer (Figures 2.2b and 2.5b). The proposed mechanism begins by coupling the oxidation of a $[4Fe–4S]^{1+}$ cluster to the reductive cleavage of SAM producing methionine and the reactive dAdo• [11,12]. This radical directly abstracts the pro-*S* hydrogen from Gly-734 of PFL [13]. Activated PFL can subsequently reversibly catalyze multiple conversions of coenzyme A and acetate to acetyl-CoA and formate. Upon exposure to O_2, activated PFL is cleaved into two catalytically inactive peptides with the masses of 3 and 82 kDa [14]. Some organisms produce a "spare part" to repair the inactivated PFL enzyme; in *Escherichia coli*, this spare part is the protein YfiD, and it can restore activity after binding to the larger PFL degradation product [15]. YfiD contains a catalytic glycine residue in a highly similar 50 amino acid sequence to that in PFL. While full activity is only observed with these natural substrates, lower-level glycyl radical formation can be observed for as little as a heptamer peptide fragment of the glycyl radical loop shared by PFL and YfiD [13]. PFL-AE induces a large conformational shift in PFL, causing it to expose Gly-734, allowing for the generation of a relatively stable radical on PFL [16,17]. Unlike LAM, PFL-AE employs SAM as a cosubstrate and, thus, SAM must be continually resupplied for catalytic activity. PFL-AE is the enzyme for which the SAM coordination to the site-differentiated [4Fe–4S] cluster was first demonstrated; this same coordination mode has now been observed in all radical SAM enzymes for which structural information is available, suggesting that the SAM–cluster interaction is an essential aspect of the radical SAM mechanism [6,18].

2.1.2.3 Biotin Synthase (BioB)

The biologically essential vitamin biotin is synthesized from dethiobiotin *in vivo* by BioB. This enzyme is the first documented and most studied member of the subfamily of radical SAM enzymes that insert sulfur into unactivated C–H bonds. The disputed role or roles of auxiliary [2Fe–2S] and [4Fe–4S] (Figure 2.6b) clusters in sulfur-inserting enzymes has led to lively discussions between research groups. The primary point of contention is whether auxiliary clusters are immediately sacrificed to provide the substrate sulfur, or if exogenous sulfur supplies are tapped during or

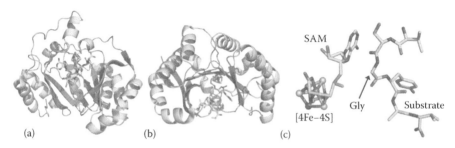

(a)   (b)   (c)

FIGURE 2.5 Structure of PFL-AE as viewed from side (a) and bottom (b). The active site highlighting the [4Fe–4S] cluster, SAM, and model substrate peptide (c). The black line denotes the location of the abstractable H-atom from the peptidyl Gly residue (PDB ID: 3CB8).

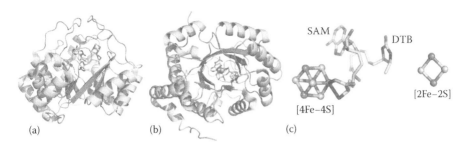

FIGURE 2.6   Structure of BioB as viewed from side (a) and bottom (b). The active site highlighting the [4Fe–4S] and [2Fe–2S] clusters, SAM, and DTB (c) (PDB ID: 1R30).

even before catalysis. In addition to the canonical [4Fe–4S]$^{2+}$ cluster observed in all other radical SAM enzymes, BioB was found to contain an auxiliary [2Fe–2S]$^{2+}$ cluster demonstrated by Mössbauer spectroscopy with differential isotopic iron labeling [19]. While multiple lines of evidence indicated that SAM was not the sulfur source for biotin biosynthesis [20,21], other experiments demonstrated that when ^{35}S-reconstituted BioB was assayed, ^{35}S was inserted into dethiobiotin to produce labeled biotin [22,23]. Further spectroscopic studies indicated that the [2Fe–2S] cluster was degraded during turnover, suggesting that this cluster served as the sacrificial sulfur donor during biotin synthesis [24]. Such a role for the [2Fe–2S] cluster was supported by the x-ray crystal structure of BioB, which revealed this cluster optimally positioned to react with the dethiobiotin subsequent to H-atom abstraction [25]. The current mechanistic proposal involves the cannibalistic abstraction of a sulfide from the [2Fe2S] cluster by two subsequent rounds of radical-mediated hydrogen atom abstraction from the methyl groups of dethiobiotin by way of a 9-mercaptodethiobiotin intermediate producing biotin [26]. In addition to supplying the sulfur for biotin production, it has been proposed that the [2Fe–2S] cluster may also act as a radical-quenching Lewis acid after both hydrogen atom abstraction events [27]. The eventual replenishment of the incorporated sulfur *in vivo* is likely supplied by cysteine desulfurase activity. It has been contended that as BioB is partially degraded during turnover, it should be referred to as a substrate and not an enzyme [22]. The source of sulfur in other sulfur-insertion enzymes such as lipoyl synthase (LipA) and MiaB is still contended [28–30].

2.1.2.4   Molybdenum Cofactor Biosynthetic Enzyme (MoaA)

In 1998, investigators shed significant insight into the complex intramolecular rearrangements that yield the precursor to the molybdopterin cofactor (Figure 2.2f) [31]. Cells containing a *moaABC* cassette were supplied with isotopically labeled ribulose 5-phosphate (pentose) or guanine produced labeled molybdopterin precursors identified through nuclear magnetic resonance spectroscopy. The pentose and guanine moieties in GTP have been subsequently confirmed to be the source of the molybdopterin skeleton [32,33]. Hänzelmann and Schindelin solved the third crystal structure of a radical SAM enzyme in 2004 and confirmed the presence of an additional [4Fe–4S] cluster located ~17 Å from the canonical N-terminal [4Fe–4S] cluster [33]. The same authors solved an MoaA crystal structure with the substrate 5′-GTP and

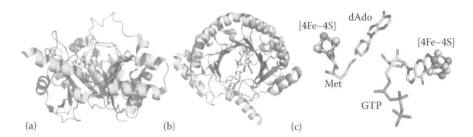

FIGURE 2.7   Structure of MoaA as viewed from side (a) and bottom (b). The active site highlighting the two [4Fe–4S] clusters, 5′-deoxyadenosine (dAdo), methionine, and GTP (c) (PDB ID: 2FB3).

the SAM degradation products methionine and 5′-deoxyadenosine bound, which surprisingly revealed GTP coordination to the accessory cluster (Figure 2.7) [32]. ENDOR studies carried out in collaboration with Brian Hoffman clarified GTP coordination to the second cluster in an unusual bidentate fashion to the enol tautomer of guanine [34]. With structural information from the active site and deuterium-labeled GTP, researchers in 2013 identified the likely hydrogen abstraction site to be the 3′C of the ribose moiety and subsequently proposed a mechanism for GTP rearrangement [35].

2.1.3   HYDROGENASE/NITROGENASE MATURASE RADICAL SAM ENZYMES (HMDB, HYDE, HYDG, AND NIFB)

Interestingly, a number of radical SAM enzymes have been identified that are involved in complex metallocofactor biosynthesis. These include the HmdB protein involved in the synthesis of the FeGP cofactor of the [Fe-only]-hydrogenase, HydE and HydG that play key roles in the synthesis of the H-cluster of the [FeFe]-hydrogenase, and NifB that catalyzes the production and/or insertion of carbide at the center of the FeMo-cofactor of nitrogenase. These radical SAM enzymes will be discussed in further detail in Section 2.3.

2.2   ARCHITECTURE AND MECHANISM OF RADICAL SAM ENZYMES

In addition to similarities in the primary structure, radical SAM enzymes share elements of the same tertiary fold [36]. Two years after the formal classification of the superfamily, the first crystal structure of a radical SAM enzyme (HemN) was solved [2,37]. The HemN active-site structure was consistent with earlier spectroscopic investigations of PFL-AE that had demonstrated coordination of the amino and carboxy groups of the methionine moiety of SAM to the site-differentiated [4Fe–4S] cluster [6,18]. HemN was shown to consist of two domains, one of which was a $(\beta\alpha)_6$ barrel. To date, 15 radical SAM crystal structures have been published, and all contain modestly modified TIM barrels, generally either a complete $(\beta\alpha)_8$ or an incomplete $(\beta\alpha)_6$ TIM barrel with the exceptions of BtrN and QueE [38,39]. Table 2.1 highlights some of the details of these structures. In addition to the core

TABLE 2.1

Consolidated Details of Radical SAM Enzymes with Published Crystal Structure Solutions

Radical SAM Enzyme (Abbreviation)	Gene	Genome	PDB Deposition	Protein Data Bank ID	Resolution (Å)	References
Coproporphyrinogen III oxidase (HemN)	hemN	E. coli	Aug-13-2003	1OLT (SAM, [4Fe–4S])	2.07	[37]
Biotin synthase (BioB)	bioB	E. coli	Sep-30-2003	1R30 (SAM, [4Fe–4S], [2Fe–2S], Sub)	3.40	[25]
Molybdenum cofactor biosynthesis protein A (MoaA)	moaA	S. aureus	Jun-28-2004	1TV7 ([4Fe–4S]), 1TV8 (SAM, [4Fe–4S])	2.80, 2.20	[33]
L-lysine 2,3-aminomutase (LAM)	kamA	C. subterminale	Jun-30-2005	2A5H (SAM, [4Fe–4S], PLP, Sub)	2.10	[7]
Molybdenum cofactor biosynthesis protein A (MoaA)	moaA	S. aureus	Dec-8-2005	2FB2 (Var, SAM, 2x[4Fe–4S]), 2FB3 (Var, 2x[4Fe–4S], Met, dAdo, PP, Sub)	2.25, 2.35	[32]
N(1)-methylguanosine cyclizing enzyme (TWY1)	taw1	P. horikoshii	Apr-23-2007	2YX0	2.21	[40]
N(1)-methylguanosine cyclizing enzyme (TWY1)	taw1	M. jannaschii	May-28-2007	2Z2U	2.40	[41]
Ribosomal protein S12 methylthiotransferase (RimO)	rimO	T. maritima	Jun-29-2007	2QGQ	2.00	[42]
Pyruvate formate-lyase 1-activating enzyme (PFL–AE)	pflA	E. coli	Feb-11-2008	3C8F (disSAM, [4Fe–4S]), 3CB8 (SAM, [4Fe–4S])	2.25, 2.77	[16]
FeFe-hydrogenase maturase (HydE)	hydE	T. maritima	Mar-12-2008	3CIW (SAH, [4Fe–4S]), 3CIX (SAH, [4Fe–4S], [2Fe–2S], SCN)	1.35, 1.70	[43]

(Continued)

TABLE 2.1 (CONTINUED)

Consolidated Details of Radical SAM Enzymes with Published Crystal Structure Solutions

Radical SAM Enzyme (Abbreviation)	Gene	Genome	PDB Deposition	Protein Data Bank ID	Resolution (Å)	References
HMP-P synthase (ThiC)	thiC	C. crescentus	Sep-29-2008	3EPM, 3EPN (SubA), 3EPO (Prod)	2.79, 2.11, 2.10	[44]
FeFe-hydrogenase maturase (HydE)	hydE	T. maritima	Aug-3-2009	3IIX ([4Fe–4S], Met, dAdo), 3IIZ (SAM, [4Fe–4S], [2Fe–2S])	1.25, 1.62	[45]
Ribosomal RNA large subunit methyltransferase N (RlmN)	rlmN	E. coli	Apr-5-2011	3RF9 ([4Fe–4S]), 3RFA(SAM, [4Fe–4S])	2.20, 2.05	[46]
Methylornithine synthase (PylB)	pylB	M. barkeri	Jul-31-2011	3T7V (SAM, [4Fe–4S], Prod)	1.50	[47]
Spore photoproduct lyase (SPL)	GTNG_2348	G. thermodenitrificans	Jun-6-2012	4FHC (SeSAM, [4Fe–4S]), 4FHD (SeSAM, [4Fe–4S], PP, SubA), 4FHE (Var, SeSAM, [4Fe–4S]), 4FHF (Var, SeSAM, [4Fe–4S], PP, SubA), 4FHG (Var, SeSAM, [4Fe–4S])	2.20, 2.00, 2.30, 2.00, 2.00	[48]
Ribosomal protein S12 methylthiotransferase (RimO)	rimO	T. maritima	Feb-20-2013	4JCO (2x[4Fe–4s])	3.30	[30]

Enzyme	Gene	Organism	Date	PDB codes	Resolution	Ref.
FeFe-hydrogenase maturase (HydE)	hydE	T. maritima	Mar-28-2013	4JXC, 4JY8, 4JY9, 4JYD, 4JYE, 4JYF (see reference)	1.50, 2.90, 1.60, 1.71, 1.65, 1.45	[49]
Anaerobic sulfatase-maturating enzyme (anSMEcpe)	CPF_0616	C. perfringens	Apr-10-2013	4K36 (SAM, 3x[4Fe–4S]), 4K37 (SAM, 3x[4Fe–4S]), 4K38 (SAM, 3x[4Fe–4S], SubA), 4K39 (SAM, 3x[4Fe–4S], SubA)	1.62, 1.62, 1.83, 1.78	[50]
Spore photoproduct lyase (SPL)	GTNG_2348	G. thermodenitrificans	Apr-20-2011	4K9R (Var, SeSAM, [4Fe–4S])	2.30	[51]
Butirosin biosynthetic enzyme (BtrN)	btrN	B. circulans	Aug-12-2013	4M7S (2x[4Fe–4S]) 4M7T (SAM, 2x[4Fe–4S], Sub)	2.02, 1.56	[38]
HMP-P synthase (ThiC)	thiC	A. thaliana	Oct-16-2013	4N7Q	1.60	[52]
7-carboxy-7-deazauanine synthase (QueE)	queE	B. multivorans	Nov-10-2013	4NJG (SAM, [4Fe–4S], SubA), 4NJH (SAM, [4Fe–4S], Sub), 4NJI (SAM, [4Fe–4S], Sub, Mg2+), 4NJJ (SAM, [4Fe–4S], Sub, Mn2+), 4NJK (SAM, [4Fe–4S], Prod, Mg2+)	2.60, 1.90, 2.20, 2.70, 1.91	[39]
Ribosomal RNA large subunit methyltransferase N(RlmN)	rlmN	E. coli	May-16-2014	4PL1 (Var, SAM, [4Fe–4S]), 4PL2 (Var, [4Fe–4S])	2.58, 2.20	[53]

Note: dAdo, 5′-deoxyadenosine; disSAM, disordered SAM; Met, methionine; PLP, pyridoxal-L-phosphate; PP, pyrophosphate; Prod, product; SAH, S-adenosyl-L-homocysteine; SAM, S-adenosyl-L-methionine; SeSAM, selenoSAM; SCN, thiocyanate ion; Sub, substrate; SubA, substrate analogue; Var, variant protein.

TIM barrel motif, many of the solved structures contain smaller auxiliary domains that are believed to aid in substrate recognition and help seal the active site from bulk solvent. The TIM barrel serves as a convenient container for the unique initiating mechanism common to all radical SAM enzymes.

2.2.1 Tertiary Structure Similarities

The smallest radical SAM enzyme characterized is the anaerobic ribonucleotide reductase activating enzyme (aRNR-AE), which has been predicted to have a $(\beta\alpha)_4$ barrel although its structure has not yet been solved [36]. Recently, the crystal structure solutions of two radical SAM enzymes, BtrN and QueE, have been solved that exhibit significant modifications of the core $(\beta\alpha)_6$ barrel [38,39]. BtrN, which catalyzes the anaerobic dehydrogenation of the antibiotic precursor 2-deoxy-*scyllo*-inosamine, was found to have a unique $(\beta_5\alpha_4)$ fold less reminiscent of a typical TIM barrel [36]. However, like other superfamily members, the CX_3CX_2C motif is located in a loop between the first β strand and α helix in the primary sequence (Figure 2.8). 7-carboxy-7-deazaguanine synthase (QueE) crystalized as a dimer with each monomer containing a $(\beta_6\alpha_3)$ core [39]. The three helices missing from the core $(\beta\alpha)_6$ TIM barrel have been largely replaced by loops. Alignments of radical SAM enzymes

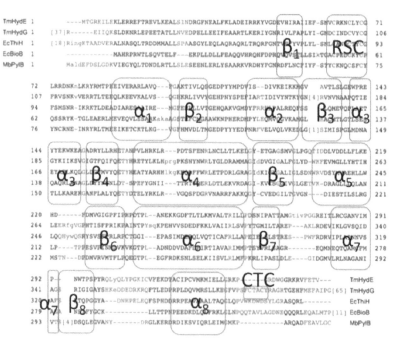

FIGURE 2.8 Sequence alignment of select radical SAM enzymes with complete TIM barrels. The CX_3CX_2C motif that binds the radical SAM cluster is labeled RSC. HydG's C-terminal domain contains a $CX_2CX_{22}C$ motif of which the first two Cys are in a box labeled CTC.

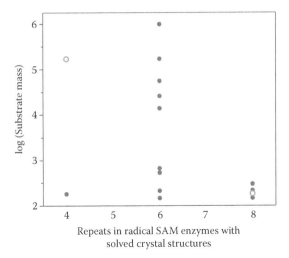

FIGURE 2.9 Plots of substrate mass (Da) compared with the completeness of the core domain. Closed circles: radical SAM enzymes with crystal structures. Open circles: aRNR-AE $(\beta\alpha)_4$ and HydG/ThiH $(\beta\alpha)_8$ predicted completeness. SPL is not plotted as the substrate size is variable; however, it would follow the general trend of larger substrates with less complete TIM barrels. Owing to the highly modified architectures of QueE and BtrN, they have been plotted as having six and four repeats, respectively.

revealed that sequence similarities between different radical SAM subfamilies end at around β_5, and the differences that extend beyond this region might play a role in substrate diversity [36,54].

The completeness of the barrel appears to inversely correlate with the size of the substrate, with the less complete barrels generally accommodating larger substrates (PFL-AE/PFL) and full barrels inherently providing more shielding for radical-based chemistry on smaller substrates (BS/dethiobiotin) (Figure 2.5a) [55]. Four exceptions to the overall trend include BtrN, LAM, MoaA, and HemN, which have smaller substrates (<750 Da) but have only a partial TIM barrel. A stronger correlation occurs with large substrates, which appear to be completely incompatible with full $(\beta\alpha)_8$ barrels (Figure 2.9).

2.2.2 Conserved Motifs

Despite the conserved canonical CX_3CX_2C motif and a common super-secondary structure, radical SAM enzymes have low sequence homology between subfamilies. Not surprisingly, glycine and proline residues are semiconserved, often punctuating the α helices and β strands. These two residues can relieve stress in transitioning between secondary structural elements by adopting unusual dihedral angles. Structurally based sequence alignments have been the most successful at aligning primary structures of radical SAM enzymes in different subgroups (Figure 2.8) [36,54]. These alignments have found two additional semiconserved glycine-rich motifs as discussed further in the following sections.

2.2.2.1 Core Cysteine Motif

In 2001, Sofia et al. identified the radical-SAM superfamily, adding >600 putative members to the five "defining" homologous members [2]. Through a bioinformatics approach, researchers mined the available gene databases for DNA sequences coding for the cysteine-rich CX_3CX_2C motif where X can be any amino acid (Figure 2.10). This motif along with other similarities that include cofactors and iron and sulfide requirements had been previously described in the literature; however, no one had yet applied these as discriminating factors to search for other members. Exceptions to the CX_3CX_2C motif have been reported, including in the thiamin pyrimidine biosynthetic enzyme ThiC (CX_2CX_4C) [56] and in an enzyme involved in phosphonate metabolism, PhnJ ($CX_2CX_{27}C$) [57,58]. Other noncanonical motifs have been described; however, all contain at least three cysteine residues requisite for ligation of an FeS cluster and subsequent catalysis [39,59]. While ~20% of QueE sequences contain additional (11 or 17) residues between the first two cysteines of the three-Cys motif, the insert does not significantly affect the geometry of the cysteines that coordinate the [4Fe–4S] cluster [39].

The three-cysteine motif in radical SAM enzymes coordinates a redox-active [4Fe–4S] cluster or "radical SAM cluster" whose resting state is 2+; the three-cysteine coordination renders the radical SAM [4Fe–4S] cluster site differentiated, with one iron of the four not coordinated by cysteine. Cuboidal [4Fe–4S] clusters are not unique to radical SAM enzymes, and similarly composed clusters are used in electron shuttling, redox sensing, substrate binding, and protein structure [60]. These [4Fe–4S] clusters typically have biologically accessible reduction potentials that single electron-transporting systems, such as flavodoxin, flavodoxin reductase, and NADPH, can reduce the cluster to the 1+ state, which is the catalytically active state for radical SAM enzymes [12].

2.2.2.2 Penultimate Aromatic Residue in Cysteine-Rich Motif

The radical SAM canonical cysteine-rich motif is often written as $CX_3CX\phi C$, with ϕ representing an aromatic residue. The graphical representation of the contribution at the penultimate position in Figure 2.10 shows the relative abundance

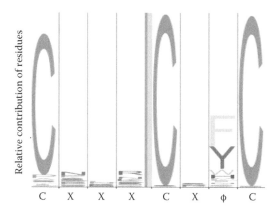

FIGURE 2.10 Sequence alignment logo generated using the Pfam 27.0 database. Heights of single-letter amino acid abbreviations are proportional to their relative abundances in multiple sequences of radical SAM superfamily members.

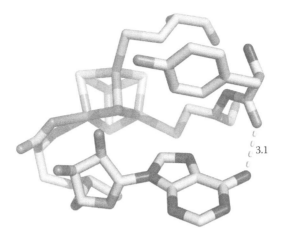

FIGURE 2.11 anSMEcpe's hydrogen bond between the N6 hydrogen of SAM's adenine moiety and the backbone carbonyl on the ϕ residue depicted by the dashed line (distance in Å; PDB ID: 4K38).

to be F, Y, W, H, and M. Interestingly, the only conserved interaction of the residue is a hydrogen bond (average distance of 3.0 Å) between the N6 hydrogen of the SAM adenine moiety and the backbone carbonyl on the ϕ residue (Figure 2.11). In a recent spectroscopic and functional characterization of BlsE, a radical SAM enzyme required for blasticidin S biosynthesis, which contains a penultimate methionine residue, M37F, M37Y, and M37W mutants had near wild-type activities [61]. Often, radical SAM enzymes without aromatic residues at the ϕ position have such amino acids located within two positions of the C-terminal Cys. The aromatic residue may serve as a conduit for electron transfer to the [4Fe–4S] cluster and/or block solvent access to the active site. A crystal structure of an enzyme with an aliphatic residue would perhaps shed light on the nature of this residue and its role in enzyme function.

2.2.2.3 Glycine-Rich Motifs

The second most prominent radical SAM motif is the conserved GGE sequence starting at the end of the β_4 strand. The first Gly of this motif aids in orienting the second Gly and the following Glu residues for hydrogen bonding with the amino group on the methionine moiety of SAM (Figure 2.12). While there are exceptions to these residues in wild-type proteins, mutating these residues may lead to inactive proteins [61]. Second, the "GXIXGXXE" motif located along the β_4 strand runs roughly parallel with SAM and appears to provide the adenosine moiety with a hydrophobic binding pocket.

2.2.3 Common Initial Mechanism

While the reactions and substrates vary tremendously, radical SAM enzymes appear to initiate catalysis by a common mechanism. Upon *in vitro* or *in vivo* reduction to the 1+ state, the radical SAM cluster can donate an electron through an inner sphere

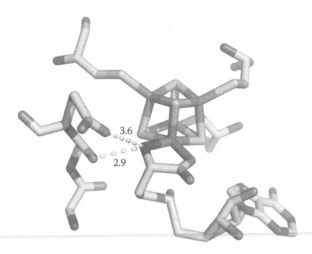

FIGURE 2.12 GGE (PFL-AE) motif interacting with methionine moiety on SAM depicted by dashed lines (distance in Å; PDB ID: 3CB8).

electron transfer to the sulfonium moiety on SAM (Figure 2.13). Computational and experimental evidence suggest the electron transfer is promoted by electron donation from the reduced [4Fe–4S] cluster into the σ* orbital of the C–S bond in SAM's sulfonium moiety [45,62]. Immediately following the electron transfer event, the S–C(5′) is cleaved to produce methionine and the dAdo•. It is expected that the hydrogen atom to be abstracted would be poised very near to the site where the dAdo• is generated; several crystal structures bear this expectation out. The crystal structure for LAM bound to SAM and the substrate L-α-lysine, for example, shows how the 5′-deoxyadenosyl moiety of SAM is poised in close proximity to the substrate H-atom (Figure 2.3c). The dAdo• can rotate or bend while anchored by the adenosine group to abstract a hydrogen atom from the nearby substrate. The substrate-based radicals then embark on a seemingly limitless array of transformations.

2.3 RADICAL SAM METALLOCOFACTOR MATURASES

The utilization of radical SAM enzymes has been recognized in the maturation schemes of at least three complex metallocofactors, with the [Fe-only]-hydrogenase, [FeFe]-hydrogenase, and nitrogenase enzymes all requiring at least one radical SAM enzyme for maturation. The current biosynthetic hypotheses of [FeFe]-hydrogenase and nitrogenase maturation bear striking similarities in their construction of metallocofactor assemblies on scaffold proteins, which are finally delivered to an "apo" enzyme. Other similarities include the involvement of NTPase enzymes, as well as the defined progression from standard iron–sulfur clusters to the modified catalytic precursor units that are then delivered to the structural proteins themselves [63]. While there are significant questions regarding the radical SAM enzymes that are used in synthesizing the active-site cluster precursor units, it is apparent that their chemistries are very diverse even within the radical SAM family.

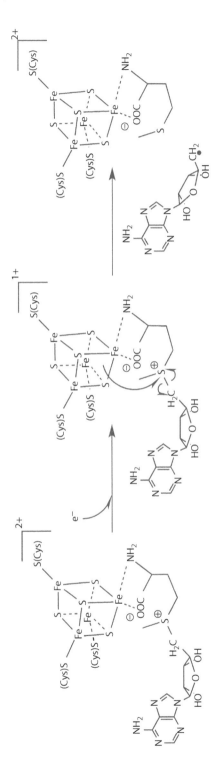

FIGURE 2.13 Reductive cleavage of SAM and the generation of the 5′-deoxyadenosyl radical (dAdo•).

2.3.1 HMDB IN THE MATURATION OF [FE-ONLY]-HYDROGENASE

Three classes of hydrogenases are differentiated on the basis of their metal content, and are referred to as the [NiFe]-, [FeFe]-, and [Fe-only]-hydrogenases (Figure 2.14a through c). While their catalyzed reactions are similar and all contain at least one redox-active iron in the catalytic center, the active sites differ significantly in structure. The most simple but enigmatic hydrogenase is the [Fe-only]-hydrogenase, also known as Hmd-hydrogenase or simply HmdA. Under limited nickel availability, the expression of [NiFe]-hydrogenase is downregulated while [Fe-only]-hydrogenase is upregulated in *Methanothermobacter marburgensis* [64]. The [Fe-only]-hydrogenase catalyzes the reversible reaction $CH\equiv H_4MPT^+ + H_2 \leftrightarrows CH_2 = H_4MPT + H^+$ through a hydride intermediate [65,66]. HmdA is employed to oxidize H_2 to provide electrons for the reduction of CO_2 to CH_4. The most recent proposed active-site structure contains a single low-spin Fe(II) bound to two CO ligands, the acyl and nitrogen moieties of a pyridinol ring, a protein-based cysteine thiolate, and perhaps a solvent molecule in nearly octahedral geometry [67]. The iron/protein stoichiometry of nearly 1 to 1 indicates that this protein does not harbor accessory iron–sulfur clusters as do most members of the other two hydrogenase classes. The maturation machinery for [Fe-only]-hydrogenase includes at least six separate gene products HmdB–HmdG [68].

HmdB has been implicated as a radical SAM enzyme by three complementary lines of evidence [69]. First, its noncanonical CX_5CX_2C motif likely binds a requisite [4Fe–4S] cluster, since reconstitution with exogenous Fe^{2+} and S^{2-} under reducing conditions reveals the binding of 3.9 ± 0.4 Fe atoms per protein. Second, the reconstituted protein reduced with NaDT exhibits an axial signal attributed to an $S = 1/2$ $[4Fe–4S]^{1+}$ cluster that is altered upon addition of SAM. Last, HmdB catalyzes the production of 5′-deoxyadenosine (dAdo) from SAM, a reaction typical of radical SAM enzymes. The role of this maturase enzyme in hydrogenase maturation is yet unknown [69]. HmdB was previously speculated to catalyze the formation of CO for

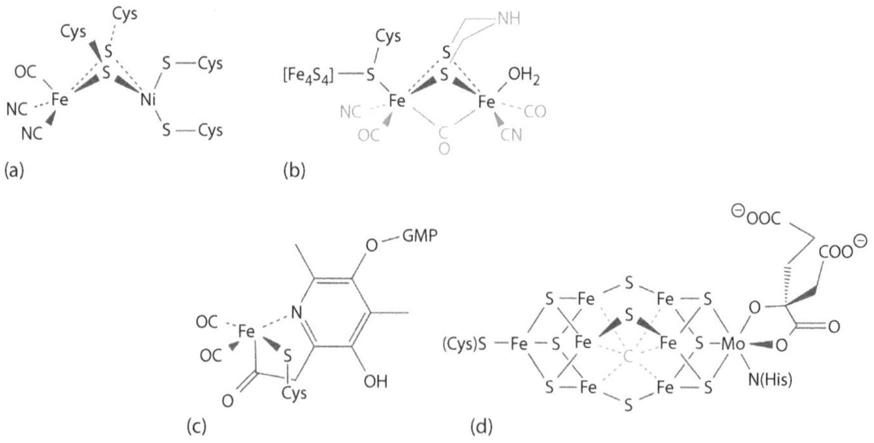

FIGURE 2.14 (a) [NiFe]-, (b) [FeFe]-, and (c) [Fe-only]-hydrogenases, and (d) nitrogenase's active site primary coordination environments. Gray atoms represent those putatively derived by radical SAM chemistry.

the HmdA active site; however, biochemical and spectroscopic studies using stable-isotope-labeled growth substrates revealed the carbon sources for the CO ligands in HmdA to originate from CO_2 [70]. This result indicates that HmdB's radical SAM chemistry does not produce CO. [Fe-only]-hydrogenase complements the more traditional hydrogenase enzymes, and its unique chemistry and active site make it an exciting potential area of research.

2.3.2 TWO RADICAL SAM ENZYMES ARE REQUIRED FOR BIOSYNTHESIS OF [FeFe]-HYDROGENASE ACTIVE-SITE CLUSTER

[FeFe]-hydrogenase is used *in vivo* primarily to catalyze the reduction of protons to hydrogen gas ($2 \text{ H}^+ + 2 \text{ e}^- \leftrightarrows \text{H}_2$). The first look at the structure and active site of the [FeFe]-hydrogenase was provided by Peters et al. in 1998 [71]. The catalytic cluster responsible for H_2 activation is composed of a [4Fe–4S] cluster coordinated by four Cys thiolates, with one of these bridging to a [2Fe–2S] cluster. This 2Fe subcluster (Figure 2.14c) is ornately decorated with several unusual biological ligands. Both iron atoms are singularly coordinated by a carbon monoxide and cyanide ligand, and together are bridged by two thiolates of a 1,3-di-(thiomethyl)amine (DTMA) and a third carbon monoxide [72–74]. During catalysis, the [4Fe–4S] cluster may donate and accept electrons to and from the 2Fe subcluster, in addition to modulating its electronic properties [75].

Sixty years after Gaffron and Rubin's seminal report on algal hydrogen production, Posewitz et al. determined the maturase enzymes required for [FeFe]-hydrogenase (HydA) activation [76]. Isolates from a *Chlamydomonas reinhardtii* insertional mutagenesis library were screened for H_2 production, and a strain unable to produce hydrogen was found to have an insertion in the *hydEF-1* gene, destroying the activity of the gene product HydEF on HydA [76]. HydEF in *C. reinhardtii* is homologous to two separate proteins found in all prokaryotic organisms containing [FeFe]-hydrogenase. Genomic analysis and comparison of organisms with [FeFe]-hydrogenases revealed *hydG* was also strictly conserved in these organisms. In the same report, the investigators noted that HydE and HydG contain the canonical CX_3CX_2C motif and were putatively radical-SAM enzymes. Heterologous coexpression of [FeFe]-hydrogenase, HydE, HydF, and HydG with HydA in *E. coli* was found sufficient and essential to yield hydrogenase activity [77]. The combined findings of several laboratories have been used to construct a proposed [FeFe]-hydrogenase maturation scheme (Figure 2.15) that will be discussed further in Sections 2.3.2.1 and 2.3.2.2 [76–86].

A modified reaction scheme has been recently proposed by Kuchenreuther et al. that differs from that in Figure 2.15 in three key points [87]. First, the modified scheme does not include a [2Fe–2S] precursor on HydF$^{\Delta EG}$ to which the diatomic ligands are eventually affixed. Rather, a one or two iron complex bound with CO and CN$^-$ ligands is constructed on HydG, resulting in a species termed HydG-co. Second, they propose that HydG may also produce the bridging dithiolate moiety from the leftover component of tyrosine degradation. Last, by reassigning the roles of HydG to include the complete synthesis of the decorated 2Fe H-cluster precursor, the role for HydE is relegated to that of a chaperone assisting in the cluster transfer

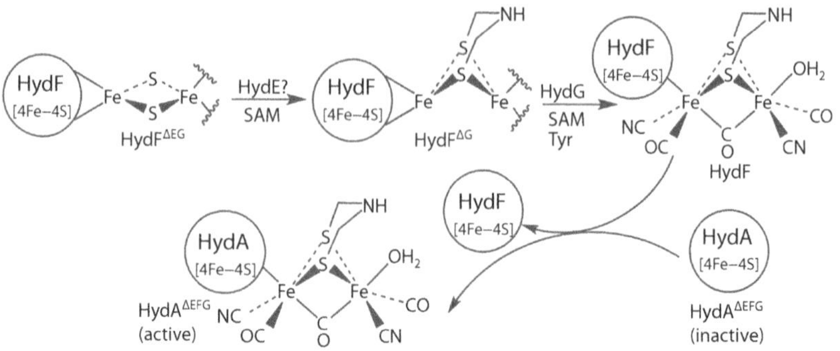

FIGURE 2.15 Proposed composite [FeFe]-hydrogenase maturation scheme.

from HydG to HydF. These proposed alternate roles for HydE and HydG will be briefly discussed in their respective sections in the following.

2.3.2.1 HydG

2.3.2.1.1 HydG Initial Characterization and FeS Clusters

Following the assignment of HydE and HydG as radical SAM enzymes based on their possession of the canonical CX_3CX_2C motif [76], an initial biochemical and spectroscopic characterization served to confirm the place of these maturases in the radical SAM superfamily [88]. Recombinant HydG from *Thermotoga maritima* overexpressed aerobically in *E. coli* could be reconstituted with iron and sulfide to yield enzyme with ~4 Fe/protein. The complex EPR signal of the reconstituted HydG indicated multiple FeS cluster states were present. The temperature relaxation and power saturation profiles suggested that an additional [4Fe–4S] cluster might be located at the $CX_2CX_{22}C$ motif near the C-terminus of HydG. In addition, HydG promoted the reductive cleavage of SAM producing dAdo at a rate of 0.25 mol dAdo/h. Such uncoupled cleavage of SAM in the absence of substrates is characteristic of many radical SAM enzymes, which then typically exhibit an increase in SAM consumption in the presence of substrate.

2.3.2.1.2 HydG Structure and Biochemistry

Despite the limited sequence homology of radical SAM enzymes, which generally pushes the limits of alignment algorithms, HydG was found to share a 27% identity with the thiamine biosynthetic enzyme ThiH [80]. While tyrosine was shown to be a factor in the synthesis of thiamine in the 1970s, it was only recently that this requirement was linked to ThiH [89–91]. Under the hypothesis that the similarities between HydG and ThiH may extend to a common substrate, Pilet et al. found that HydG also demonstrated tyrosine lyase activity, producing *p*-cresol [80]. It was proposed at the time that the dehydroglycine intermediate produced upon cleavage of tyrosine might be a precursor to the DTMA bridge of the H-cluster. Kuchenreuther et al. have recently revisited this hypothesis by investigating HydA activation independent of either HydF (HydE and HydG only) or HydE (HydF and HydG only) [87]. Their results led them to speculate that HydG produces the DTMA bridge (in addition to the diatomic ligands discussed in the following), forming a pre-H-cluster species termed HydG-co; the involvement of HydG in synthesizing the DTMA bridge has not, however, been conclusively demonstrated. The proposed involvement of tyrosine in HydA maturation was further supported by stimulation of H_2 production by cell lysates containing heterologously overexpressed HydAEFG upon addition of exogenous tyrosine [92]. Sequence alignments of $T.m.$HydG with $T.m.$HydE, $M.b.$PylB, and $E.c.$BioB indicate that all have $(\beta\alpha)_8$ complete TIM barrels (Figure 2.8). While HydG does not have a published crystal structure, perhaps the most intriguing feature is a C-terminal domain beginning immediately after the last α helix [80]. This domain contains the $CX_2CX_{22}C$ believed to take part in the chemistry that differentiates HydG from ThiH, as discussed further in the following sections.

Following the demonstration that tyrosine served as a substrate for HydG, a race ensued among several laboratories to determine the ultimate product(s) and how these fit into the HydA maturation process. In addition to dAdo and p-cresol, HydG was also shown to produce the cyanide ligands likely from a dehydroglycine intermediate [83]. HydG incubated with SAM, Tyr, and NaDT was shown to produce CN^- via derivatization with naphthalene-2,3-dicarbaldehyde, which generated a compound detected by fluorescence and liquid chromatography–mass spectrometry. Utilization of U-$[^{13}C,^{15}N]$Tyr produced a mass shift in the derivatized product, confirming Tyr as the origin of the CN^-. Shortly thereafter, it was demonstrated that CO was also produced by HydG [84]. The production of CO was detected by monitoring a shift in the λ_{max} values of deoxyhemoglobin included in the HydG assay as it was converted to carboxyhemoglobin. Fourier transform infrared (FTIR) analysis of the carboxyhemoglobin produced in the assay carried out with U-$[^{13}C,^{15}N]$Tyr revealed the expected shifts in the vibrational bands for HbCO, confirming that the CO formed was derived from Tyr. Furthermore, FTIR studies of HydF expressed in the presence of HydE and HydG showed that this protein bound a CO and CN^- ligated iron species likely similar to the 2Fe subcluster of the H-cluster [82,85]. The Swartz laboratory utilized FTIR analysis to demonstrate that all five diatomic ligands in the H-cluster were derived from tyrosine [93].

2.3.2.1.3 *Analysis of HydG Variants and Proposed Mechanism*

Conclusive spectroscopic and analytical evidence that two separate [4Fe–4S] clusters are required for HydG activity was provided in a paper by Shepard et al. [84]. Anaerobically expressed, purified, and reconstituted protein was found to contain 8.7 ± 0.7 Fe/protein and exhibit EPR spectra similar to those previous reported [88]. Upon addition of SAM to the photoreduced, reconstituted HydG, however, a significant shift in the g-values for only one of the two clusters could be observed. Variants of HydG lacking the C-terminal iron–sulfur cluster, made by either deleting the C-terminal domain (ΔCTD) or by changing the first two cysteine residues of the $CX_2CX_{22}C$ motif to serine or alanine, were compromised in their ability to generate the diatomic ligands [94,95]. While the ΔCTD variant did not produce measurable CO or CN^-, the C→S and C→A variants produced half the CN^- compared with wild-type HydG with no CO production [94,95].

Although the HydG mechanism had been previously proposed to involve homolytic (rather than heterolytic, as in the proposed ThiH mechanism) C_α–C_β bond cleavage [94], recent spectroscopic evidence for the existence of a 4-oxidobenzyl radical (4OB$^\bullet$) intermediate in HydG from *Shewanella oneidensis* (*S.o.*), favors a mechanism where C_α–C_β heterolytic cleavage produces dehydroglycine and 4OB$^\bullet$ that abstracts an H-atom and is released as *p*-cresol [96]. These recent spectroscopic studies also pointed to the possibility of tyrosine binding to the C-terminal [4Fe–4S] cluster, and the authors proposed a mechanism for HydG catalysis with tyrosine coordination to this cluster as a central feature [96]. This proposal, however, is difficult to resolve with the predicted structure for HydG based on similarity to other radical SAM enzymes with complete TIM barrels (similar to HydG) and solved crystal structures (HydE, PylB, BioB) (Figure 2.8). These comparisons suggest that HydG's C-terminal cluster would likely lie at the bottom of the TIM barrel roughly 30 Å from the location of the radical SAM cluster (Figure 2.16). For comparison, in structures where the substrate/analog/product and SAM are cocrystalized, the distances between the 5′C of SAM and the abstractable H-atom are generally only 4.0 ± 0.4 Å (Figure 2.17). This short distance is important in allowing rapid and specific H-atom abstraction. Given the predicted distance between the HydG clusters, it seems unlikely that Tyr could be bound to the C-terminal cluster and be at a sufficient distance for H-atom abstraction, especially given the relative rigidity of the complete TIM barrel. Alternatively, the Tyr may bind near SAM in the typical radical SAM/substrate binding site where

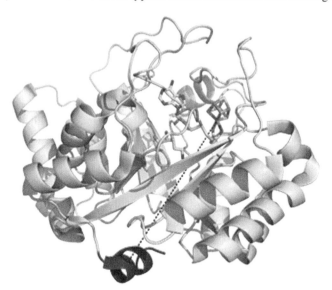

FIGURE 2.16 PylB (PDB ID: 3T7V) crystal structure with complete TIM barrel (light gray) and C-terminal domain of PylB at approximate location of CTC in HydG (dark gray). The CTC most likely lies outside the TIM barrel, at an approximate distance of 30 Å from the radical SAM cluster (Figure 2.8). The limited structural plasticity of the complete TIM barrel in radical SAM enzymes (Figure 2.9), coupled to the presumed location of the CTC in HydG directly after the terminal α_8 helix, would place the distance between the iron sulfur clusters as too large for H-atom abstraction from coordinated tyrosine (Figure 2.8).

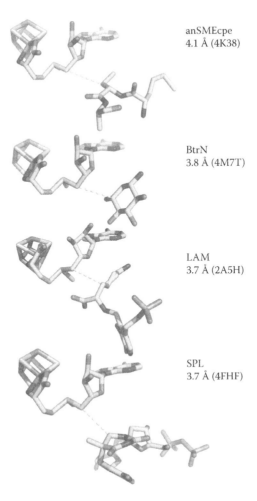

anSMEcpe
4.1 Å (4K38)

BtrN
3.8 Å (4M7T)

LAM
3.7 Å (2A5H)

SPL
3.7 Å (4FHF)

FIGURE 2.17 Representative distances between the 5′C of SAM and abstracted/abstractable H-atom on corresponding product/substrate molecules. PDB ID numbers in parentheses.

upon initial Tyr lyase activity, the fragments travel to the bottom of the barrel where subsequent reactions occur [97].

A recent biochemical and spectroscopic characterization of wild-type *C.a.* HydG determined the K_M for Tyr to be 0.3 mM [95]. This value was increased upon mutation of either the C-terminal cluster (C386S) or in the complete removal of the C-terminal domain (ΔCTD), to 1.6 and 10.6 mM, respectively. Despite the increase in K_M for tyrosine in the variant proteins, *p*-cresol production rates were largely unchanged, thereby supporting the idea that the Tyr lyase activity of HydG occurs in the core TIM barrel, likely near the N-terminal, radical SAM cluster. Insight into the role of the C-terminal [4Fe–4S] cluster was recently inferred from the observation of Fe(CO)$_2$CN FTIR bands associated with HydG during turnover; these results coupled with ENDOR data tracking ^{57}Fe species suggest that the shuttling of the diatomic species from HydG to HydF may involve cannibalization of HydG's accessory [4Fe–4S] cluster [98]. Ultimately, the hypotheses relating to the role of the C-terminal cluster in HydG are very exciting and will certainly incite further investigation.

2.3.2.2   HydE

2.3.2.2.1   Initial Characterization, Crystal Structure,
*   and Interactions with HydF*

The same groups that have elucidated the HydF and HydG roles in [FeFe]-hydrogenase maturation have worked on unraveling the function of HydE. An early biochemical characterization of HydE confirmed its place in the radical SAM superfamily by showing that it produced dAdo in the absence of substrate and that it could be reconstituted to contain [4Fe–4S] clusters, but shed little insight into its function or mechanism [88]. The first of several HydE solved crystal structures confirmed that HydE contains a complete TIM barrel [43]. Crystal soaks and *in silico* docking studies indicated that the substrate-binding site would accommodate a small compound with a carboxylate group and a partial positive charge [43]. Also noted was HydE's high affinity for SCN^-, which led to the erroneous speculation that HydE produced CN^- [43].

More recently, Costantini et al. measured binding constants between HydE, HydF, and HydG by surface plasmon resonance [86]. They determined that HydE and HydG have K_D values of 9.19×10^{-8} and 1.31×10^{-6} M, respectively, for binding to HydF. They also found that HydE could displace HydG bound to HydF but not the other way around. This further supports the composite maturation scheme (Figure 2.15) in that HydE operates independently of and before HydG in the synthesis of the 2Fe H-cluster moiety. Despite the efforts of several groups, however, the substrate, product, and mechanism of HydE have yet to be determined.

2.3.2.2.2   Paradigm Shift for HydE Functionality

Strangely, of the 15 radical SAM enzymes with published crystal structure solutions, HydE stands alone as not having a defined biochemical role, substrate, or product despite intense efforts of several independent groups [43,78,86,88,92]. HydE has been frequently annotated as BioB owing to its structural similarities, presence of an auxiliary FeS cluster, and sequence homology to BioB. This has led to proposals that HydE may also have a role in catalyzing a sulfur insertion reaction. HydE and BioB have significant differences that likely contribute to distinct chemistries, however. The auxiliary [2Fe–2S] cluster of BioB is located close to the active site (7.6 Å from the 5′-carbon on SAM) and is likely directly involved in providing the sulfur atom for insertion into dethiobiotin through a 9-mercaptodethiobiotin intermediate [26]. The auxiliary and radical SAM clusters of HydE are located on opposite sides of the TIM barrel at a distance of >21 Å (PBD ID: 4JY8). At this distance, it is unlikely that the sulfur atoms of the auxiliary cluster of HydE could be involved in direct insertion into substrate in a BioB-like reaction. In addition, the auxiliary cluster in HydE is only semiconserved and not required for HydA activation [43].

Recently, the crystal structure of methylornithine synthase (PylB) was solved [47]. The gene product of *pylB* is a radical SAM enzyme that catalyzes the isomerization of L-lysine to L-methylornithine in the biosynthetic pathway of pyrrolysine, the 22nd naturally encoded amino acid [99]. The tertiary structure of PylB is a $(\beta\alpha)_8$ TIM barrel with one conserved [4Fe–4S] cluster coordinated by the canonical CX_3CX_2C motif. The distance alignment matrix method was used to compare PylB with all other solved crystal structures with HydE and BioB scoring 29.0% and 20.6% similarity, respectively [47,100]. Also, HydE and PylB crystal structures have a root-mean-square deviation (RMSD) of only 1.3 Å compared with BioB and PylB with an RMSD of 1.8 Å. The alignment of PylB and HydE crystal structures show similar SAM and putative substrate binding pockets, and together, these similarities may be pointing to a parallel mechanism for these two enzymes. Interestingly, HydG and PylB both have proposed mechanisms involving glycine-like intermediates, and perhaps HydE will as well [47,94].

2.3.3 NIFB IN THE MATURATION OF NITROGENASE

2.3.3.1 Nitrogenase Active Site and Maturation Scheme

The reduction of elemental nitrogen (N_2) to ammonia (NH_3) is a costly reaction both biologically and industrially. It has been estimated that two-thirds of all nitrogen fixation is catalyzed by nitrogenase with a bulk of the remainder by the Haber–Bosch process [101]. Diazotrophic microbes catalyze the reaction $N_2 + 8\ H^+\ 8\ e^- + 16\ MgATP + 16\ H_2O \rightarrow 2\ NH_3 + H_2 + 16\ MgADP + 16\ P_i$ at ambient temperatures and pressures, whereas industrial processes require much higher temperatures and pressures [102].

The holoenzyme of nitrogenase is composed of two polypeptides, NifD and NifK, with NifD housing the active site. The NifH(γ)/NifD(α)NifK(β) complex (Figure 2.18) with a $(\alpha\beta\gamma_2)_2$ stoichiometry crystallized from *Azotobacter vinelandii* likely

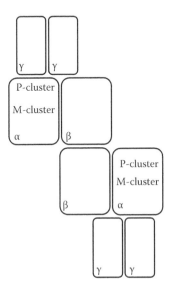

FIGURE 2.18 Nitrogenase complex composed of NifH (γ) and NifD (α) with P- and M-clusters, and NifK (β).

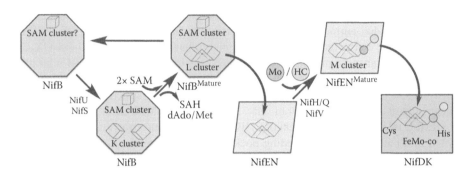

FIGURE 2.19 Proposed *ex situ* construction of the M-cluster. SAH, *S*-adenosyl-L-homocysteine; Mo, molybdenum; HC, homocitrate.

represents a snapshot of active nitrogenase [103]. The most distinctive aspect of nitrogenase is its active-site cluster also known as the M-cluster or FeMo-co (Figure 2.14d) located in NifD. FeMo-co ($Fe_7MoCS_7^{\cdot}$ homocitrate) contains a unique heterometallic cluster whose core may be described as a [4Fe–3S] cubane bridged to a [Mo2Fe3S] cubane by three sulfides and one hexacoordinated carbide [104,105]. The molybdenum at the distal position is coordinated in a bidentate fashion to homocitrate and anchored by a His ligand from the protein. A Cys ligand also from NifD secures the terminal iron on the other side of the cluster. The *ex situ* or external synthesis of the M-cluster before insertion into NifDK has been proposed to be accomplished on two discrete proteins (Figure 2.19) [106]. NifB (described later) constructs an all iron M-cluster precursor (L-cluster) that is transferred to the scaffold protein NifEN [107,108]. On NifEN, NifH or NifQ and NifV likely swap the cluster's distal Fe atom for an Mo atom and aid in its coordination by homocitrate producing the M-cluster [101,106]. This M-cluster can be transferred to the structural protein complex NifDK where it is held in place by Cys and His residues.

2.3.3.2 Initial Characterization of the Radical SAM Enzyme NifB

Nitrogenase produced in the absence of the *nifB* gene was shown to be inactive as it lacked FeMo-co, implicating that NifB was involved in a critical step in cluster maturation [109–112]. The *A. vinelandii nifB* gene product contains a CX_3CX_2C motif and a total of nine Cys and eight His residues highly conserved in NifB proteins, indicating the ligation of a site differentiated [4Fe–4S] cluster and potentially other FeS clusters [109]. Anaerobically purified NifB was shown to contain 12 Fe/dimer and could be reconstituted up to 18 Fe/dimer [112]. UV–visible spectral features of reconstituted NifB that decrease upon addition of dithionite indicated the presence of FeS clusters. Activity dropped significantly for NifB synthesis of FeMo-co if Fe^{2+}, S^{2-}, or SAM was omitted in *in vitro* assays. These findings in conjunction with previously noted sequence analysis strongly indicated that NifB is a radical SAM enzyme.

While NifB is difficult to stabilize in solution, its product NifB-co can be isolated on the small carrier protein, NifX [113,114]. The NifX:NifB-co complex core was found to be very similar to the mature FeMo-co, based on a nuclear resonance vibrational spectroscopy and extended x-ray absorption fine structure spectroscopic investigation, indicating that most of the FeS skeleton including the interstitial atom are assembled before transfer to NifEN [114]. After spectral modeling, the NifX:NifB-co cluster to several models; the best fit was a [6Fe–9SX] cluster. Another group recently proposed that the FeMo-co intermediate produced on NifB is rather similar to the final FeMo-co with an iron number of eight rather than six [114,115].

2.3.3.3   NifEN-B Fusion and Proposed NifB Mechanism

An alternative approach to stabilizing NifB was accomplished by its fusion to the NifN, similar to a natural fusion in *Clostridium pasteurianum* [115]. The NifB stabilization may occur by sequestration of a hydrophobic patch on NifB by the NifEN complex. NifEN-B could be produced with up to 30 Fe/mol, a large increase over NifEN's 16 Fe/mol. This indicated the possibility of multiple FeS clusters on NifB in addition to the radical SAM [4Fe–4S] cluster. EPR and activity assays also indicate the nonradical SAM clusters may undergo structural changes (2× [4Fe–4S] to 8Fe clusters) in a radical SAM-dependent reaction on NifB. It is currently unknown if the K-(two [4Fe–4S] clusters) to L-cluster conversion (Figure 2.19) occurs at the subunit interface of the NifB dimer or is localized on one subunit.

The interstitial species in FeMo-co was identified from a list of light-atom candidates (C, N, O) as C^{4-} in independent x-ray emission spectroscopy and electron spin echo envelope modulation spectroscopic investigations [104,105]. Carbide's origin is likely from SAM's methyl group as the radioactivity from [methyl-^{14}C]-SAM could be traced from NifB (NifEN-B) to NifDK in *in vitro* maturation assays [107]. Also, deuterium atoms from [methyl-d_3]-SAM were observed to be incorporated into dAdo produced enzymatically by NifB (NifEN-B). This indicates that that in a manner similar to RlmN and Cfr, SAM is employed as both a methyl and dAdo• source where two equivalents of SAM are required for each turnover (Figure 2.19).

In light of these experimental results, two NifB mechanisms for K- to L-cluster conversion have been proposed [106,107]. Both mechanisms begin with a SAM-dependent methyl transfer to a [4Fe–4S] component in the K-cluster, differing in whether the methyl is transferred as a methyl cation or a methyl radical equivalent, and in the specific site of initial attachment on the K-cluster. Subsequent H-atom abstraction by the SAM-derived dAdo• then produces a methylene radical that is deprotonated and incorporated in between the two clusters. NifB's proposed chemistry and mechanisms are (like many radical SAM enzymes) unprecedented and leave the field of nitrogenase with many unanswered questions and years of exciting research.

2.4   UNIFYING THEMES IN RADICAL SAM ENZYME CHEMISTRY IN THE MATURATION OF METALLOENZYMES

While relatively simple FeS clusters ([2Fe–2S] and [4Fe–4S]) are generated by a suite of generic housekeeping genes (*Suf/Isc*), complex and/or uniquely decorated active sites often call upon radical SAM enzymes in their multistep maturation processes

[63,116]. The active sites of [Fe-only]- and [FeFe]-hydrogenases and nitrogenase all contain iron coordinated by unusual ligands with limited attachment to the enzyme via cysteine residues. Maturation schemes for [FeFe]-hydrogenase and nitrogenase share remarkable similarity as they involve modifying traditional FeS clusters, and employ carrier/scaffold proteins in an *ex situ* process, while utilizing radical SAM chemistry to catalyze the synthesis of the unique ligands that impart the hydrogenase and nitrogenase functionalities to the H-cluster and FeMo-co, respectively. HydE and HmdB, despite intense spectroscopic, structural, and biochemical efforts, have yet to reveal their specific biochemical roles. There is little doubt that in a manner similar to HydG and NifB, the reactions catalyzed by HydE and HmdB will be exciting and push the fields of metallocofactor maturation and radical SAM chemistry to new heights.

REFERENCES

1. Guallar, V., Baik, M. H., Lippard, S. J., and Friesner, R. A. (2003) Peripheral heme substituents control the hydrogen-atom abstraction chemistry in cytochromes P450, *Proc. Natl. Acad. Sci. U.S.A. 100*, 6998–7002.
2. Sofia, H. J., Chen, G., Hetzler, B. G., Reyes-Spindola, J. F., and Miller, N. E. (2001) Radical SAM, a novel protein superfamily linking unresolved steps in familiar biosynthetic pathways with radical mechanisms: Functional characterization using new analysis and information visualization methods, *Nucleic Acids Res. 29*, 1097–1106.
3. Duin, E. C., Lafferty, M. E., Crouse, B. R., Allen, R. M., Sanyal, I., Flint, D. H., and Johnson, M. K. (1997) [2Fe–2S] to [4Fe–4S] cluster conversion in *Escherichia coli* biotin synthase, *Biochemistry 36*, 11811–11820.
4. Chirpich, T. P., Zappia, V., Costilow, R. N., and Barker, H. A. (1970) Lysine 2,3-aminomutase. Purification and properties of a pyridoxal phosphate and *S*-adenosylmethionine-activated enzyme, *J. Biol. Chem. 245*, 1778–1789.
5. Petrovich, R. M., Ruzicka, F. J., Reed, G. H., and Frey, P. A. (1991) Metal cofactors of lysine-2,3-aminomutase, *J. Biol. Chem. 266*, 7656–7660.
6. Walsby, C. J., Hong, W., Broderick, W. E., Cheek, J., Ortillo, D., Broderick, J. B., and Hoffman, B. M. (2002) Electron-nuclear double resonance spectroscopic evidence that *S*-adenosylmethionine binds in contact with the catalytically active [4Fe–4S](+) cluster of pyruvate formate-lyase activating enzyme, *J. Am. Chem. Soc. 124*, 3143–3151.
7. Lepore, B. W., Ruzicka, F. J., Frey, P. A., and Ringe, D. (2005) The x-ray crystal structure of lysine-2,3-aminomutase from *Clostridium subterminale*, *Proc. Natl. Acad. Sci. U.S.A. 102*, 13819–13824.
8. Moss, M., and Frey, P. A. (1987) The role of *S*-adenosylmethionine in the lysine 2,3-aminomutase reaction, *J. Biol. Chem. 262*, 14859–14862.
9. Ruzicka, F. J., and Frey, P. A. (2007) Glutamate 2,3-aminomutase: A new member of the radical SAM superfamily of enzymes, *Biochim. Biophys. Acta 1774*, 286–296.
10. Baraniak, J., Moss, M. L., and Frey, P. A. (1989) Lysine 2,3-aminomutase. Support for a mechanism of hydrogen transfer involving *S*-adenosylmethionine, *J. Biol. Chem. 264*, 1357–1360.
11. Buis, J. M., and Broderick, J. B. (2005) Pyruvate formate-lyase activating enzyme: Elucidation of a novel mechanism for glycyl radical formation, *Arch. Biochem. Biophys. 433*, 288–296.
12. Henshaw, T. F., Cheek, J., Broderick, J. B. (2000) The [4Fe–4S]1+ cluster of pyruvate formate-lyase activating enzyme generates the glycyl radical on pyruvate formate-lyase: EPR-detected single turnover, *J. Am. Chem. Soc. 122*, 8331–8332.

13. Frey, M., Rothe, M., Wagner, A. F., and Knappe, J. (1994) Adenosylmethionine-dependent synthesis of the glycyl radical in pyruvate formate-lyase by abstraction of the glycine C-2 pro-S hydrogen atom. Studies of [2H]glycine-substituted enzyme and peptides homologous to the glycine 734 site, *J. Biol. Chem. 269*, 12432–12437.

14. Wagner, A. F., Frey, M., Neugebauer, F. A., Schafer, W., and Knappe, J. (1992) The free radical in pyruvate formate-lyase is located on glycine-734, *Proc. Natl. Acad. Sci. U.S.A. 89*, 996–1000.

15. Wagner, A. F., Schultz, S., Bomke, J., Pils, T., Lehmann, W. D., and Knappe, J. (2001) YfiD of *Escherichia coli* and Y06I of bacteriophage T4 as autonomous glycyl radical cofactors reconstituting the catalytic center of oxygen-fragmented pyruvate formate-lyase, *Biochem. Biophys. Res. Commun. 285*, 456–462.

16. Vey, J. L., Yang, J., Li, M., Broderick, W. E., Broderick, J. B., and Drennan, C. L. (2008) Structural basis for glycyl radical formation by pyruvate formate-lyase activating enzyme, *Proc. Natl. Acad. Sci. U.S.A. 105*, 16137–16141.

17. Peng, Y., Veneziano, S. E., Gillispie, G. D., and Broderick, J. B. (2010) Pyruvate formate-lyase, evidence for an open conformation favored in the presence of its activating enzyme, *J. Biol. Chem. 285*, 27224–27231.

18. Walsby, C. J., Ortillo, D., Broderick, W. E., Broderick, J. B., and Hoffman, B. M. (2002) An anchoring role for FeS clusters: Chelation of the amino acid moiety of *S*-adenosylmethionine to the unique iron site of the [4Fe–4S] cluster of pyruvate formate-lyase activating enzyme, *J. Am. Chem. Soc. 124*, 11270–11271.

19. Ugulava, N. B., Surerus, K. K., and Jarrett, J. T. (2002) Evidence from Mossbauer spectroscopy for distinct [2Fe–2S](2+) and [4Fe–4S](2+) cluster binding sites in biotin synthase from *Escherichia coli*, *J. Am. Chem. Soc. 124*, 9050–9051.

20. Florentin, D., Bui, B. T., Marquet, A., Ohshiro, T., and Izumi, Y. (1994) On the mechanism of biotin synthase of *Bacillus sphaericus*, *C. R. Acad. Sci. Ser. III Sci. Vie 317*, 485–488.

21. Sanyal, I., Gibson, K. J., and Flint, D. H. (1996) *Escherichia coli* biotin synthase: An investigation into the factors required for its activity and its sulfur donor, *Arch. Biochem. Biophys. 326*, 48–56.

22. Gibson, K. J., Pelletier, D. A., and Turner, I. M., Sr. (1999) Transfer of sulfur to biotin from biotin synthase (BioB protein), *Biochem. Biophys. Res. Commun. 254*, 632–635.

23. Bui, B. T., Florentin, D., Fournier, F., Ploux, O., Mejean, A., and Marquet, A. (1998) Biotin synthase mechanism: On the origin of sulphur, *FEBS Lett. 440*, 226–230.

24. Ugulava, N. B., Gibney, B. R., and Jarrett, J. T. (2001) Biotin synthase contains two distinct iron–sulfur cluster binding sites: Chemical and spectroelectrochemical analysis of iron–sulfur cluster interconversions, *Biochemistry 40*, 8343–8351.

25. Berkovitch, F., Nicolet, Y., Wan, J. T., Jarrett, J. T., and Drennan, C. L. (2004) Crystal structure of biotin synthase, an *S*-adenosylmethionine-dependent radical enzyme, *Science 303*, 76–79.

26. Fugate, C. J., Stich, T. A., Kim, E. G., Myers, W. K., Britt, R. D., and Jarrett, J. T. (2012) 9-Mercaptodethiobiotin is generated as a ligand to the [2Fe–2S](+) cluster during the reaction catalyzed by biotin synthase from *Escherichia coli*, *J. Am. Chem. Soc. 134*, 9042–9045.

27. Fugate, C. J., and Jarrett, J. T. (2012) Biotin synthase: Insights into radical-mediated carbon–sulfur bond formation, *Biochim. Biophys. Acta 1824*, 1213–1222.

28. Cicchillo, R. M., Tu, L., Stromberg, J. A., Hoffart, L. M., Krebs, C., and Booker, S. J. (2005) *Escherichia coli* quinolinate synthetase does indeed harbor a [4Fe–4S] cluster, *J. Am. Chem. Soc. 127*, 7310–7311.

29. Hernandez, H. L., Pierrel, F., Elleingand, E., Garcia-Serres, R., Huynh, B. H., Johnson, M. K., Fontecave, M., and Atta, M. (2007) MiaB, a bifunctional radical-*S*-adenosyl-methionine enzyme involved in the thiolation and methylation of tRNA, contains two essential [4Fe–4S] clusters, *Biochemistry 46*, 5140–5147.

30. Forouhar, F., Arragain, S., Atta, M., Gambarelli, S., Mouesca, J. M., Hussain, M., Xiao, R., Kieffer-Jaquinod, S., Seetharaman, J., Acton, T. B., Montelione, G. T., Mulliez, E., Hunt, J. F., and Fontecave, M. (2013) Two Fe–S clusters catalyze sulfur insertion by radical-SAM methylthiotransferases, *Nat. Chem. Biol. 9*, 333–338.

31. Rieder, C., Eisenreich, W., O'Brien, J., Richter, G., Gotze, E., Boyle, P., Blanchard, S., Bacher, A., and Simon, H. (1998) Rearrangement reactions in the biosynthesis of molybdopterin—An NMR study with multiply 13C/15N labelled precursors, *Eur. J. Biochem. 255*, 24–36.

32. Hanzelmann, P., and Schindelin, H. (2006) Binding of 5′-GTP to the C-terminal FeS cluster of the radical *S*-adenosylmethionine enzyme MoaA provides insights into its mechanism, *Proc. Natl. Acad. Sci. U.S.A. 103*, 6829–6834.

33. Hänzelmann, P., and Schindelin, H. (2004) Crystal structure of the *S*-adenosylmethionine-dependent enzyme MoaA and its implications for molybdenum cofactor deficiency in humans, *Proc. Natl. Acad. Sci. U.S.A. 101*, 12870–12875.

34. Lees, N. S., Hanzelmann, P., Hernandez, H. L., Subramanian, S., Schindelin, H., Johnson, M. K., and Hoffman, B. M. (2009) ENDOR spectroscopy shows that guanine N1 binds to [4Fe–4S] cluster II of the *S*-adenosylmethionine-dependent enzyme MoaA: Mechanistic implications, *J. Am. Chem. Soc. 131*, 9184–9185.

35. Mehta, A. P., Hanes, J. W., Abdelwahed, S. H., Hilmey, D. G., Hanzelmann, P., and Begley, T. P. (2013) Catalysis of a new ribose carbon-insertion reaction by the molybdenum cofactor biosynthetic enzyme MoaA, *Biochemistry 52*, 1134–1136.

36. Nicolet, Y., and Drennan, C. L. (2004) AdoMet radical proteins—From structure to evolution—Alignment of divergent protein sequences reveals strong secondary structure element conservation, *Nucleic Acids Res. 32*, 4015–4025.

37. Layer, G., Moser, J., Heinz, D. W., Jahn, D., and Schubert, W. D. (2003) Crystal structure of coproporphyrinogen III oxidase reveals cofactor geometry of radical SAM enzymes, *EMBO J. 22*, 6214–6224.

38. Goldman, P. J., Grove, T. L., Booker, S. J., and Drennan, C. L. (2013) X-ray analysis of butirosin biosynthetic enzyme BtrN redefines structural motifs for AdoMet radical chemistry, *Proc. Natl. Acad. Sci. U.S.A. 110*, 15949–15954.

39. Dowling, D. P., Bruender, N. A., Young, A. P., McCarty, R. M., Bandarian, V., and Drennan, C. L. (2014) Radical SAM enzyme QueE defines a new minimal core fold and metal-dependent mechanism, *Nat. Chem. Biol. 10*, 106–112.

40. Goto-Ito, S., Ishii, R., Ito, T., Shibata, R., Fusatomi, E., Sekine, S. I., Bessho, Y., and Yokoyama, S. (2007) Structure of an archaeal TYW1, the enzyme catalyzing the second step of wye-base biosynthesis, *Acta Crystallogr. D Biol. Crystallogr. 63*, 1059–1068.

41. Suzuki, Y., Noma, A., Suzuki, T., Senda, M., Senda, T., Ishitani, R., and Nureki, O. (2007) Crystal structure of the radical SAM enzyme catalyzing tricyclic modified base formation in tRNA, *J. Mol. Biol. 372*, 1204–1214.

42. Arragain, S., Garcia-Serres, R., Blondin, G., Douki, T., Clemancey, M., Latour, J. M., Forouhar, F., Neely, H., Montelione, G. T., Hunt, J. F., Mulliez, E., Fontecave, M., and Atta, M. (2010) Post-translational modification of ribosomal proteins: Structural and functional characterization of RimO from *Thermotoga maritima*, a radical *S*-adenosylmethionine methylthiotransferase, *J. Biol. Chem. 285*, 5792–5801.

43. Nicolet, Y., Rubach, J. K., Posewitz, M. C., Amara, P., Mathevon, C., Atta, M., Fontecave, M., and Fontecilla-Camps, J. C. (2008) X-ray structure of the [FeFe]-hydrogenase maturase HydE from *Thermotoga maritima*, *J. Biol. Chem. 283*, 18861–18872.

44. Chatterjee, A., Li, S., Zhang, Y., Grove, T. L., Lee, M., Krebs, C., Booker, S. J., Begley, T. P., and Ealick, S. E. (2008) Reconstitution of ThiC in thiamine pyrimidine biosynthesis expands the radical SAM superfamily, *Nat. Chem. Biol. 4*, 758–765.

45. Nicolet, Y., Amara, P., Mouesca, J. M., and Fontecilla-Camps, J. C. (2009) Unexpected electron transfer mechanism upon AdoMet cleavage in radical SAM proteins, *Proc. Natl. Acad. Sci. U.S.A. 106*, 14867–14871.

46. Boal, A. K., Grove, T. L., McLaughlin, M. I., Yennawar, N. H., Booker, S. J., and Rosenzweig, A. C. (2011) Structural basis for methyl transfer by a radical SAM enzyme, *Science 332*, 1089–1092.

47. Quitterer, F., List, A., Eisenreich, W., Bacher, A., and Groll, M. (2012) Crystal structure of methylornithine synthase (PylB): Insights into the pyrrolysine biosynthesis, *Angew. Chem. 51*, 1339–1342.

48. Benjdia, A., Heil, K., Barends, T. R. M., Carell, T., and Schlichting, I. (2012) Structural insights into recognition and repair of UV-DNA damage by spore photoproduct lyase, a radical SAM enzyme, *Nucleic Acids Res. 40*, 9308–9318.

49. Nicolet, Y., Rohac, R., Martin, L., and Fontecilla-Camps, J. C. (2013) X-ray snapshots of possible intermediates in the time course of synthesis and degradation of protein-bound Fe4S4 clusters, *Proc. Natl. Acad. Sci. U.S.A. 110*, 7188–7192.

50. Goldman, P. J., Grove, T. L., Sites, L. A., McLaughlin, M. I., Booker, S. J., and Drennan, C. L. (2013) X-ray structure of an AdoMet radical activase reveals an anaerobic solution for formylglycine posttranslational modification, *Proc. Natl. Acad. Sci. U.S.A. 110*, 8519–8524.

51. Yang, L. L., Nelson, R. S., Benjdia, A., Lin, G. J., Telser, J., Stoll, S., Schlichting, I., and Li, L. (2013) A radical transfer pathway in spore photoproduct lyase, *Biochemistry 52*, 3041–3050.

52. Coquille, S., Roux, C., Mehta, A., Begley, T. P., Fitzpatrick, T. B., and Thore, S. (2013) High-resolution crystal structure of the eukaryotic HMP-P synthase (THIC) from *Arabidopsis thaliana*, *J. Struct. Biol. 184*, 438–444.

53. Silakov, A., Grove, T. L., Radle, M. I., Bauerle, M. R., Green, M. T., Rosenzweig, A. C., Boal, A. K., and Booker, S. J. (2014) Characterization of a cross-linked protein-nucleic acid substrate radical in the reaction catalyzed by RlmN, *J. Am. Chem. Soc. 136*, 8221–8228.

54. Vey, J. L., and Drennan, C. L. (2011) Structural insights into radical generation by the radical SAM superfamily, *Chem. Rev. 111*, 2487–2506.

55. Duschene, K. S., Veneziano, S. E., Silver, S. C., and Broderick, J. B. (2009) Control of radical chemistry in the AdoMet radical enzymes, *Curr. Opin. Chem. Biol. 13*, 74–83.

56. Martinez-Gomez, N. C., and Downs, D. M. (2008) ThiC is an [Fe–S] cluster protein that requires AdoMet to generate the 4-amino-5-hydroxymethyl-2-methylpyrimidine moiety in thiamin synthesis, *Biochemistry 47*, 9054–9056.

57. Parker, G. F., Higgins, T. P., Hawkes, T., and Robson, R. L. (1999) *Rhizobium* (*Sinorhizobium*) *meliloti* phn genes: Characterization and identification of their protein products, *J. Bacteriol. 181*, 389–395.

58. Kamat, S. S., Williams, H. J., Dangott, L. J., Chakrabarti, M., and Raushel, F. M. (2013) The catalytic mechanism for aerobic formation of methane by bacteria, *Nature 497*, 132–136.

59. Booker, S. J., and Grove, T. L. (2010) Mechanistic and functional versatility of radical SAM enzymes, *F1000 Biol. Rep. 2*, 52.

60. Beinert, H. (2000) Iron–sulfur proteins: Ancient structures, still full of surprises, *J. Biol. Inorg. Chem. 5*, 2–15.

61. Feng, J., Wu, J., Dai, N., Lin, S., Xu, H. H., Deng, Z., and He, X. (2013) Discovery and characterization of BlsE, a radical *S*-adenosyl-L-methionine decarboxylase involved in the blasticidin S biosynthetic pathway, *PLoS One 8*, e68545.

62. Dey, A., Peng, Y., Broderick, W. E., Hedman, B., Hodgson, K. O., Broderick, J. B., and Solomon, E. I. (2011) S K-edge XAS and DFT calculations on SAM dependent pyruvate formate-lyase activating enzyme: Nature of interaction between the Fe4S4 cluster and SAM and its role in reactivity, *J. Am. Chem. Soc. 133*, 18656–18662.

63. Peters, J. W., and Broderick, J. B. (2012) Emerging paradigms for complex iron–sulfur cofactor assembly and insertion, *Annu. Rev. Biochem. 81*, 429–450.

64. Afting, C., Kremmer, E., Brucker, C., Hochheimer, A., and Thauer, R. K. (2000) Regulation of the synthesis of H2-forming methylenetetrahydromethanopterin dehydrogenase (Hmd) and of *Hmd*II and *Hmd*III in *Methanothermobacter marburgensis*, *Arch. Microbiol. 174*, 225–232.

65. Zirngibl, C., Hedderich, R., and Thauer, R. K. (1990) *N*5,*N*10-methylenetetrahydromethanopterin dehydrogenase from *Methanobacterium thermoautotrophicum* has hydrogenase activity, *FEBS Lett. 261*, 112–116.

66. Shima, S., and Thauer, R. K. (2007) A third type of hydrogenase catalyzing H2 activation, *Chem. Rec. 7*, 37–46.

67. Shima, S., Schick, M., Kahnt, J., Ataka, K., Steinbach, K., and Linne, U. (2012) Evidence for acyl-iron ligation in the active site of [Fe]-hydrogenase provided by mass spectrometry and infrared spectroscopy, *Dalton Trans. 41*, 767–771.

68. Lie, T. J., Costa, K. C., Pak, D., Sakesan, V., and Leigh, J. A. (2013) Phenotypic evidence that the function of the [Fe]-hydrogenase Hmd in *Methanococcus maripaludis* requires seven hcg (hmd co-occurring genes) but not hmdII, *FEMS Microbiol. Lett. 343*, 156–160.

69. McGlynn, S. E., Boyd, E. S., Shepard, E. M., Lange, R. K., Gerlach, R., Broderick, J. B., and Peters, J. W. (2010) Identification and characterization of a novel member of the radical AdoMet enzyme superfamily and implications for the biosynthesis of the Hmd hydrogenase active site cofactor, *J. Bacteriol. 192*, 595–598.

70. Schick, M., Xie, X., Ataka, K., Kahnt, J., Linne, U., and Shima, S. (2012) Biosynthesis of the iron-guanylylpyridinol cofactor of [Fe]-hydrogenase in methanogenic archaea as elucidated by stable-isotope labeling, *J. Am. Chem. Soc. 134*, 3271–3280.

71. Peters, J. W., Lanzilotta, W. N., Lemon, B. J., and Seefeldt, L. C. (1998) X-ray crystal structure of the Fe-only hydrogenase (CpI) from *Clostridium pasteurianum* to 1.8 angstrom resolution, *Science 282*, 1853–1858.

72. Nicolet, Y., Piras, C., Legrand, P., Hatchikian, C. E., and Fontecilla-Camps, J. C. (1999) *Desulfovibrio desulfuricans* iron hydrogenase: The structure shows unusual coordination to an active site Fe binuclear center, *Structure 7*, 13–23.

73. Nicolet, Y., de Lacey, A. L., Vernede, X., Fernandez, V. M., Hatchikian, E. C., and Fontecilla-Camps, J. C. (2001) Crystallographic and FTIR spectroscopic evidence of changes in Fe coordination upon reduction of the active site of the Fe-only hydrogenase from *Desulfovibrio desulfuricans*, *J. Am. Chem. Soc. 123*, 1596–1601.

74. Berggren, G., Adamska, A., Lambertz, C., Simmons, T. R., Esselborn, J., Atta, M., Gambarelli, S., Mouesca, J. M., Reijerse, E., Lubitz, W., Happe, T., Artero, V., and Fontecave, M. (2013) Biomimetic assembly and activation of [FeFe]-hydrogenases, *Nature 499*, 66–69.

75. Mulder, D. W., Ratzloff, M. W., Shepard, E. M., Byer, A. S., Noone, S. M., Peters, J. W., Broderick, J. B., and King, P. W. (2013) EPR and FTIR analysis of the mechanism of H2 activation by [FeFe]-hydrogenase HydA1 from *Chlamydomonas reinhardtii*, *J. Am. Chem. Soc. 135*, 6921–6929.

76. Posewitz, M. C., King, P. W., Smolinski, S. L., Zhang, L., Seibert, M., and Ghirardi, M. L. (2004) Discovery of two novel radical *S*-adenosylmethionine proteins required for the assembly of an active [Fe] hydrogenase, *J. Biol. Chem. 279*, 25711–25720.

77. King, P. W., Posewitz, M. C., Ghirardi, M. L., and Seibert, M. (2006) Functional studies of [FeFe] hydrogenase maturation in an *Escherichia coli* biosynthetic system, *J. Bacteriol. 188*, 2163–2172.

78. McGlynn, S. E., Ruebush, S. S., Naumov, A., Nagy, L. E., Dubini, A., King, P. W., Broderick, J. B., Posewitz, M. C., and Peters, J. W. (2007) *In vitro* activation of [FeFe] hydrogenase: New insights into hydrogenase maturation, *J. Biol. Inorg. Chem. 12*, 443–447.

79. McGlynn, S. E., Shepard, E. M., Winslow, M. A., Naumov, A. V., Duschene, K. S., Posewitz, M. C., Broderick, W. E., Broderick, J. B., and Peters, J. W. (2008) HydF as a scaffold protein in [FeFe] hydrogenase H-cluster biosynthesis, *FEBS Lett. 582*, 2183–2187.

80. Pilet, E., Nicolet, Y., Mathevon, C., Douki, T., Fontecilla-Camps, J. C., and Fontecave, M. (2009) The role of the maturase HydG in [FeFe]-hydrogenase active site synthesis and assembly, *FEBS Lett. 583*, 506–511.

81. Mulder, D. W., Ortillo, D. O., Gardenghi, D. J., Naumov, A. V., Ruebush, S. S., Szilagyi, R. K., Huynh, B., Broderick, J. B., and Peters, J. W. (2009) Activation of HydA(DeltaEFG) requires a preformed [4Fe–4S] cluster, *Biochemistry 48*, 6240–6248.

82. Czech, I., Silakov, A., Lubitz, W., and Happe, T. (2010) The [FeFe]-hydrogenase maturase HydF from *Clostridium acetobutylicum* contains a CO and CN⁻ ligated iron cofactor, *FEBS Lett. 584*, 638–642.

83. Driesener, R. C., Challand, M. R., McGlynn, S. E., Shepard, E. M., Boyd, E. S., Broderick, J. B., Peters, J. W., and Roach, P. L. (2010) [FeFe]-hydrogenase cyanide ligands derived from *S*-adenosylmethionine-dependent cleavage of tyrosine, *Angew. Chem. 49*, 1687–1690.

84. Shepard, E. M., Duffus, B. R., George, S. J., McGlynn, S. E., Challand, M. R., Swanson, K. D., Roach, P. L., Cramer, S. P., Peters, J. W., and Broderick, J. B. (2010) [FeFe]-hydrogenase maturation: HydG-catalyzed synthesis of carbon monoxide, *J. Am. Chem. Soc. 132*, 9247–9249.

85. Shepard, E. M., McGlynn, S. E., Bueling, A. L., Grady-Smith, C. S., George, S. J., Winslow, M. A., Cramer, S. P., Peters, J. W., and Broderick, J. B. (2010) Synthesis of the 2Fe subcluster of the [FeFe]-hydrogenase H cluster on the HydF scaffold, *Proc. Natl. Acad. Sci. U.S.A. 107*, 10448–10453.

86. Vallese, F., Berto, P., Ruzzene, M., Cendron, L., Sarno, S., De Rosa, E., Giacometti, G. M., and Costantini, P. (2012) Biochemical analysis of the interactions between the proteins involved in the [FeFe]-hydrogenase maturation process, *J. Biol. Chem. 287*, 36544–36555.

87. Kuchenreuther, J. M., Britt, R. D., and Swartz, J. R. (2012) New insights into [FeFe] hydrogenase activation and maturase function, *PLoS One 7*, e45850.

88. Rubach, J. K., Brazzolotto, X., Gaillard, J., and Fontecave, M. (2005) Biochemical characterization of the HydE and HydG iron-only hydrogenase maturation enzymes from *Thermatoga maritima*, *FEBS Lett. 579*, 5055–5060.

89. Kriek, M., Martins, F., Leonardi, R., Fairhurst, S. A., Lowe, D. J., and Roach, P. L. (2007) Thiazole synthase from *Escherichia coli*: An investigation of the substrates and purified proteins required for activity *in vitro*, *J. Biol. Chem. 282*, 17413–17423.

90. Estramareix, B., and Therisod, M. (1972) [Tyrosine as a factor of biosynthesis of the thiazole moiety of thiamin in *Escherichia coli*], *Biochim. Biophys. Acta 273*, 275–282.

91. White, R. H. (1978) Stable isotope studies on the biosynthesis of the thiazole moiety of thiamin in *Escherichia coli*, *Biochemistry 17*, 3833–3840.

92. Kuchenreuther, J. M., Stapleton, J. A., and Swartz, J. R. (2009) Tyrosine, cysteine, and *S*-adenosyl methionine stimulate *in vitro* [FeFe] hydrogenase activation, *PLoS One 4*, e7565.

93. Kuchenreuther, J. M., George, S. J., Grady-Smith, C. S., Cramer, S. P., and Swartz, J. R. (2011) Cell-free H-cluster synthesis and [FeFe] hydrogenase activation: All five CO and CN ligands derive from tyrosine, *PLoS One 6*, e20346.

94. Nicolet, Y., Martin, L., Tron, C., and Fontecilla-Camps, J. C. (2010) A glycyl free radical as the precursor in the synthesis of carbon monoxide and cyanide by the [FeFe]-hydrogenase maturase HydG, *FEBS Lett. 584*, 4197–4202.

95. Driesener, R. C., Duffus, B. R., Shepard, E. M., Bruzas, I. R., Duschene, K. S., Coleman, N. J. R., Marrison, A. P. G., Salvadori, E., Kay, C. W. M., Peters, J. W., Broderick, J. B., and Roach, P. L. (2013) Biochemical and kinetic characterization of radical S-adenosyl-L-methionine enzyme HydG, *Biochemistry 52*, 8696–8707.

96. Kuchenreuther, J. M., Myers, W. K., Stich, T. A., George, S. J., Nejatyjahromy, Y., Swartz, J. R., and Britt, R. D. (2013) A radical intermediate in tyrosine scission to the CO and CN–ligands of FeFe hydrogenase, *Science 342*, 472–475.

97. Tron, C., Cherrier, M. V., Amara, P., Martin, L., Fauth, F., Fraga, E., Correard, M., Fontecave, M., Nicolet, Y., and Fontecilla-Camps, J. C. (2011) Further characterization of the [FeFe]-hydrogenase maturase HydG, *Eur. J. Inorg. Chem. 2011*, 1121–1127.

98. Kuchenreuther, J. M., Myers, W. K., Suess, D. L., Stich, T. A., Pelmenschikov, V., Shiigi, S. A., Cramer, S. P., Swartz, J. R., Britt, R. D., and George, S. J. (2014) The HydG enzyme generates an Fe(CO)2(CN) synthon in assembly of the FeFe hydrogenase H-cluster, *Science 343*, 424–427.

99. Gaston, M. A., Zhang, L., Green-Church, K. B., and Krzycki, J. A. (2011) The complete biosynthesis of the genetically encoded amino acid pyrrolysine from lysine, *Nature 471*, 647–650.

100. Holm, L., and Sander, C. (1996) Mapping the protein universe, *Science 273*, 595–603.

101. Rubio, L. M., and Ludden, P. W. (2008) Biosynthesis of the iron–molybdenum cofactor of nitrogenase, *Annu. Rev. Microbiol. 62*, 93–111.

102. Igarashi, R. Y., and Seefeldt, L. C. (2003) Nitrogen fixation: The mechanism of the Mo-dependent nitrogenase, *Crit. Rev. Biochem. Mol. Biol. 38*, 351–384.

103. Schindelin, H., Kisker, C., Schlessman, J. L., Howard, J. B., and Rees, D. C. (1997) Structure of ADP × AIF4(−)-stabilized nitrogenase complex and its implications for signal transduction, *Nature 387*, 370–376.

104. Lancaster, K. M., Roemelt, M., Ettenhuber, P., Hu, Y., Ribbe, M. W., Neese, F., Bergmann, U., and DeBeer, S. (2011) X-ray emission spectroscopy evidences a central carbon in the nitrogenase iron-molybdenum cofactor, *Science 334*, 974–977.

105. Spatzal, T., Aksoyoglu, M., Zhang, L., Andrade, S. L., Schleicher, E., Weber, S., Rees, D. C., and Einsle, O. (2011) Evidence for interstitial carbon in nitrogenase FeMo cofactor, *Science 334*, 940.

106. Hu, Y., and Ribbe, M. W. (2013) Nitrogenase assembly, *Biochim. Biophys. Acta 1827*, 1112–1122.

107. Wiig, J. A., Hu, Y., Lee, C. C., and Ribbe, M. W. (2012) Radical SAM-dependent carbon insertion into the nitrogenase M-cluster, *Science 337*, 1672–1675.

108. Shah, V. K., Allen, J. R., Spangler, N. J., and Ludden, P. W. (1994) *In vitro* synthesis of the iron–molybdenum cofactor of nitrogenase. Purification and characterization of NifB cofactor, the product of NIFB protein, *J. Biol. Chem. 269*, 1154–1158.

109. Dos Santos, P. C., Dean, D. R., Hu, Y., and Ribbe, M. W. (2004) Formation and insertion of the nitrogenase iron–molybdenum cofactor, *Chem. Rev. 104*, 1159–1173.

110. Tal, S., Chun, T. W., Gavini, N., and Burgess, B. K. (1991) The delta nifB (or delta nifE) FeMo cofactor-deficient MoFe protein is different from the delta nifH protein, *J. Biol. Chem. 266*, 10654–10657.

111. Schmid, B., Ribbe, M. W., Einsle, O., Yoshida, M., Thomas, L. M., Dean, D. R., Rees, D. C., and Burgess, B. K. (2002) Structure of a cofactor-deficient nitrogenase MoFe protein, *Science 296*, 352–356.

112. Curatti, L., Ludden, P. W., and Rubio, L. M. (2006) NifB-dependent *in vitro* synthesis of the iron–molybdenum cofactor of nitrogenase, *Proc. Natl. Acad. Sci. U.S.A. 103*, 5297–5301.

113. Hernandez, J. A., Igarashi, R. Y., Soboh, B., Curatti, L., Dean, D. R., Ludden, P. W., and Rubio, L. M. (2007) NifX and NifEN exchange NifB cofactor and the VK-cluster, a newly isolated intermediate of the iron–molybdenum cofactor biosynthetic pathway, *Mol. Microbiol. 63*, 177–192.

114. George, S. J., Igarashi, R. Y., Xiao, Y., Hernandez, J. A., Demuez, M., Zhao, D., Yoda, Y., Ludden, P. W., Rubio, L. M., and Cramer, S. P. (2008) Extended x-ray absorption fine structure and nuclear resonance vibrational spectroscopy reveal that NifB-co, a FeMo-co precursor, comprises a 6Fe core with an interstitial light atom, *J. Am. Chem. Soc. 130*, 5673–5680.

115. Wiig, J. A., Hu, Y., and Ribbe, M. W. (2011) NifEN-B complex of *Azotobacter vinelandii* is fully functional in nitrogenase FeMo cofactor assembly, *Proc. Natl. Acad. Sci. U.S.A. 108*, 8623–8627.

116. Johnson, D. C., Dean, D. R., Smith, A. D., and Johnson, M. K. (2005) Structure, function, and formation of biological iron–sulfur clusters, *Annu. Rev. Biochem. 74*, 247–281.

3 Density Functional Theory-Based Treatments of Metal-Binding Sites in Metalloenzymes
Challenges and Opportunities

Mercedes Alfonso-Prieto and Michael L. Klein

CONTENTS

3.1 INTRODUCTION

Metalloproteins constitute between one-third and one-half of the proteome of living organisms (Thomson and Gray 1998; Harding, Nowicki, and Walkinshaw 2010), and around 40% of the proteins deposited in the Protein Data Bank contain metals (Harding, Nowicki, and Walkinshaw 2010). Among many essential structural and biological functions, they are crucial in respiration and photosynthesis, participate in DNA and RNA processing, and are involved in pharmacologically important reactions, such as metabolism of xenobiotics, antibiotic resistance, or defense against reactive oxygen species. Nevertheless, computational studies on metalloproteins are not as abundant as for other proteins, owing to the difficulties encountered in the theoretical treatment of the metal ions (Banci 2003). Here, we review recent efforts in the field of metalloproteins by our group. The chapter is organized as follows. First, we summarize briefly some of the new approaches and methods developed in our group to tackle the computational problems associated with the study of metalloproteins. Then, we present some selected examples that are chosen to illustrate what is currently possible in the area of applications. The chapter ends with some suggestions for avenues to explore in future works.

3.2   METHODOLOGY

3.2.1   Force Fields for Metalloproteins

Typically, classical molecular dynamics (MD) simulations are based on effective pairwise–additive force fields, which cannot describe properly metal-containing sites owing to the intrinsic difficulty of handling effects such as polarization and ligand–metal charge transfer contributions. Several methods have been proposed to improve the description of active sites (Banci, Gori-Savellini, and Turano 1997; Autenrieth et al. 2004; Peters et al. 2010; Carvalho, Teixeira, and Ramos 2013). Typically, these are based on adding bonded terms between the metal and the coordinated ligands. In contrast, our group has proposed a different strategy, namely a flexible nonbonded model based on reparameterization of the metal site charge distribution (CD) (Dal Peraro et al. 2007a).

In short, in the nonbonded CD model, the interactions between the metal ion and the coordinated ligands are treated using a van der Waals potential plus a Coulomb potential. The nonbonded Lennard–Jones parameters remain unchanged compared with the standard force field (FF); however, the point charges are modified on the basis of a quantum mechanical (QM) calculation to better represent the charge redistribution effect at the metal-containing site. Specifically, a model of the metal and its first shell ligands is extracted from the experimental structure and its geometry is optimized by means of density functional theory (DFT). The resulting electronic density is fitted to atomic charges using the Bader "atoms-in-molecules" (AIM) partitioning scheme (Bader 1994), and these AIM-derived atomic charges (q^{AIM}) are applied as a correction to the standard FF point charges (q^{FF}) of the metal and the coordinated ligands. The charge correction for each ligand L is calculated as the difference between the formal total charge ($Q^{form,L}$, e.g., -1 for Asp and Glu or 0 for His) and the total AIM charge of the ligand ($Q^{AIM,L} = \Sigma_n q^{AIM}$, where n runs over the atoms belonging to ligand L). This difference, $\delta Q^L = Q^{form,L} - Q^{AIM,L}$, represents the mean polarization and charge loss of the ligand upon metal binding and is concentrated on the atoms directly coordinated to the metal. Therefore, only the point charges of the coordinated atoms are corrected; their reparameterized effective charges (q^{eff}) are obtained by subtracting the corresponding ligand correction (weighted by the number of ligand coordinated atoms) from the standard FF atomic charges, i.e., $q^{eff} = q^{FF} - (\delta Q^L/\text{number of ligand coordinated atoms})$. The charge correction for the metal is the sum of all the ligand δQ^L contributions (i.e., the amount of charge transferred from the ligands to the metal), and then the effective charge of the metal ion ($q^{\text{eff,M}}$) is obtained by subtracting this metal correction from the formal charge of the cation ($q^{\text{form,M}}$, e.g., +2 for Mg or Zn), i.e., $q^{\text{eff,M}} = q^{\text{form,M}} - \Sigma_L (\delta Q^L)$. The resulting CD-corrected charges differ only slightly from the standard FF charges of the metal and the ligands.

This reparameterization (Dal Peraro et al. 2007a) is able to describe accurately the active-site structure for several metalloproteins with different number (1 or 2) and type (Mg^{2+}, Mn^{2+}, or Zn^{2+}) of metal ions, as well as different coordination geometries (octahedral, square pyramidal, or tetrahedral). Moreover, the stability of the metal-containing site is maintained during nanosecond-long MD simulations, without disruption of the overall protein fold. Furthermore, in contrast to other models including metal–ligand bonded terms, this nonbonded approach allows the sampling of small structural rearrangements. Hence, this method can be used not only to refine structural data but also to explore the dynamics of the metal-containing site, such as water exchange, ligand rearrangements, or even coordination changes. Nevertheless, the CD model also has limitations. This approach tends to underestimate slightly the metal–ligand distances, indicating that reoptimization of the metal van der Waals parameters is also needed. Besides, our nonbonded model is not able to describe correctly highly directional geometries, such as metal–porphyrin complexes; in this case, bonded terms between the metal ion and the porhyrinic nitrogen atoms and the proximal ligand are required. Furthermore, in its current state, this parameterization is still not transferable; a larger training set of metalloproteins will be needed to obtain a generalized CD model.

3.2.2 HUBBARD–U-CORRECTION TO DENSITY FUNCTIONAL THEORY

Quantum mechanics (QM) and quantum mechanics/molecular mechanics (QM/MM) studies on metalloproteins are commonly performed at the DFT level because it is a reasonable compromise between computational cost and accuracy (Siegbahn and Borowski 2006; Ramos and Fernandes 2008; Rokob, Srnec, and Rulívsek 2012). Nevertheless, one should keep in mind the limitations of DFT and the specific exchange-correlation functionals used, especially when treating the strongly correlated d-electrons of transition metal ions, such as Fe, Mn, or Cu. In particular, the approximations to the unknown exact exchange-correlation functional in the Kohn–Sham equation result in an unphysical repulsive interaction of an electron with its own density, known as the self-interaction error (SIE), which artificially favors delocalized states (Lundberg and Siegbahn 2005). This error specially manifests itself in systems with an odd number of electrons, such as transition metal–containing sites, resulting in an inaccurate description of their electronic configuration and/or an incorrect energy ordering of their spin states. Hybrid functionals can partially correct for the SIE by incorporating a percentage of Hartree–Fock (HF) exact exchange. However, inclusion of HF exchange does not necessarily yield a better structural characterization (Cramer and Truhlar 2009), and the amount of HF exchange needed to eliminate completely the SIE depends on the system (Guidon, Hutter, and VandeVondele 2010). Therefore, in our group, we have adopted an alternative strategy to improve the quantum description of the metal, the Hubbard–U-corrected DFT scheme (Sit et al. 2010).

In a nutshell, in the DFT + U approach, the standard exchange-correlation DFT functional (E_{DFT}) is modified with a correction functional (E_U), dependent on the metal d-orbitals (Hubbard 1963). The magnitude of the correction is controlled by the parameter U, which sets the strength of the on-site screening of the Coulomb repulsion and can be determined for each metal system using first principles calculations. In other words, E_U is a penalty, tuned by the parameter U, disfavoring partial orbital occupations, thus counteracting the SIE artificially stabilizing delocalized states.

Although more applications in the field of metalloproteins are needed, we have shown that this Hubbard–U-corrected DFT approach can improve the description of structural, electronic, and energetic properties of transition metal-containing sites (Sit et al. 2010). Moreover, compared with the use of hybrid functionals with plane-wave basis sets, this increased accuracy comes without any additional computational cost.

3.2.3 ELECTRON TRANSFER PARAMETERS

Besides structural and electronic properties, computer simulations can also be used to study the reactions catalyzed by metalloproteins. A particular case is biological electron transfer (ET) reactions. In the Marcus model of nonadiabatic ET (Marcus and Sutin 1985), the rate of ET is determined by three quantities: the redox potential difference between the electron donor and the acceptor (ΔG), the electronic coupling between them (H) and the reorganization free energy (λ). Whereas redox potentials can be easily measured experimentally, determination of reorganization energies and electronic couplings is more challenging. Computational methods can complement experimental data by calculating these three key ET parameters. Indeed, ET has been successfully studied in different metalloproteins using a variety of theoretical approaches (see, for instance, Olsson, Hong, and Warshel 2003; Ceccarelli and Marchi 2003; Cascella et al. 2006; Sulpizi et al. 2007). In our group, we have applied a protocol (Blumberger and Klein 2006; Blumberger 2008) based on the energy gap method (Warshel 1982; Blumberger, Tateyama, and Sprik 2005), combining classical MD simulations (to sample the time and length scales relevant to biological ET) with QM/MM calculations (to obtain accurate energetics).

In practice, to study the ET reaction between the two redox-active sites in a metalloprotein, RO → OR, two classical MD simulations are performed: one in the RO state (i.e., the reactants of the ET reaction, with site 1 reduced and site 2 oxidized), and the other in the OR state (i.e., the products, with site 1 oxidized and site 2 reduced). The FF parameters describing site 1 are the same in the two simulations, except for the point charges, which are decreased overall by 1 e$^-$ to model the electron being removed; the FF charges in each redox state are estimated from QM calculations (see Section 3.2.1). For site 2, the same parameterization approach is used, increasing the total charge by 1 e$^-$ to model the electron being inserted. These classical MD simulations provide an ensemble of atomistic configurations, where the energetics is calculated using a QM/MM scheme.

Since the distance between the two sites is large (ranging from 4 to 24 Å for biological ET reactions), site 2 is expected to be merely a spectator of the ionization of site 1 and vice versa. Hence, in the QM/MM calculations, only the site where the ionization takes place (either site 1 or site 2) is treated at the quantum level, whereas the other site (site 2 or 1, respectively) and the chemically inert environment (the protein and the solvent) are treated by molecular mechanics. In other words, the nonionized site, along with the protein and the solvent, couple to the ionization of the other site just through electrostatic interactions.

In the nonadiabatic limit (i.e., the overlap between the donor and acceptor orbitals is essentially zero) the ET reaction can be decomposed into two separate processes, oxidation of site 1 and reduction of site 2:

$$RO \rightarrow OO + 1\ e^-$$

$$OO + 1\ e^- \rightarrow OR$$

The ET energy (ΔE) can be calculated as the difference in the ionization energies of the two sites:

$$\Delta E(R_N) = IE_1(R_N) - IE_2(R_N)$$

Here, $IE_1(R_N) = E_{OO}(R_N) - E_{RO}(R_N)$, $IE_2(R_N) = E_{OO}(R_N) - E_{OR}(R_N)$, and N represents either state RO or OR. If the MD is carried out in the RO state, IE_1 is the ionization potential of site 1 and IE_2 is the electron affinity of site 2, whereas, in state OR, IE_1 and IE_2 are the electron affinity of site 1 and the ionization potential of site 2, respectively. Therefore, calculation of the ET energy requires four single-point QM/MM calculations at each atomistic configuration: two with site 1 in the QM region (to obtain IE_1) and two with site 2 in the QM region (to compute IE_2). The oxidation or reduction of the ionizable site is enforced by increasing or decreasing, respectively, the total charge of the QM subsystem by 1 e^-, whereas, for the rest of the system (MM region), the classical point charges, van der Waals parameters, and other FF terms remain unchanged.

These QM/MM calculations render two sets of ET energies, one for each state (i.e., $\{\Delta E_{RO,1}, \Delta E_{RO,2}, ..., \Delta E_{RO,n}\}$ and $\{\Delta E_{OR,1}, \Delta E_{OR,2}, ..., \Delta E_{OR,n}\}$, where n is the number of atomistic configurations). If the distributions of ΔE_{RO} and ΔE_{OR} are Gaussian and their variances are the same ($\sigma^2_{RO} = \sigma^2_{OR}$), the linear response (LR) approximation can be applied, and then the ET parameters ΔG and λ can be calculated as

$$\Delta G^{LR} = (<\Delta E_{RO}> + <\Delta E_{OR}>)/2$$

$$\lambda^{LR} = (<\Delta E_{RO}> - <\Delta E_{OR}>)/2$$

where $<\Delta E_{RO}>$ and $<\Delta E_{OR}>$ are the ensemble averaged ET energies in state RO or OR, respectively.

This protocol has been successfully applied to a variety of metalloproteins, yielding redox potentials and reorganization energies in good agreement with experiments (Blumberger and Klein 2006; Blumberger 2008; Tipmanee et al. 2010). Moreover, the molecular and dynamical picture provided by this method has helped rationalize the factors tuning the parameters of biological ET. This information can be used to tweak natural ET proteins for their use in bioreactors and biosensors or to design artificial ET proteins for use in nanoscale materials applications.

3.3   APPLICATIONS

3.3.1   REFINEMENT OF EXPERIMENTAL STRUCTURES

Structural determination by X-ray crystallography and nuclear magnetic resonance (NMR) spectroscopy relies on the use of an effective pairwise-additive FF that is typically not optimized for metalloproteins. Hence, distance restraints are often imposed on the metal–ligand bonds during the refinement. However, this approach assumes a certain coordination number, metal–ligand distances, and protonation state of the ligands, which can be difficult to ascertain unambiguously from the raw crystallographic or NMR data. The use of classical MD simulations in combination with the nonbonded CD model presented in Section 3.2.1 can overcome this limitation because it allows for small structural rearrangements of the metal-containing site. For instance, this approach has been applied to thermolysin (Dal Peraro et al. 2007a; Blumberger, Lamoureux, and Klein 2007), one of the most studied metalloproteases, to decipher the coordination mode of the Glu166 residue bound to the Zn metal ion (Figure 3.1a). There are several structures available for this enzyme, showing either a monodentate or bidentate glutamate coordination. Moreover, the carboxylate group

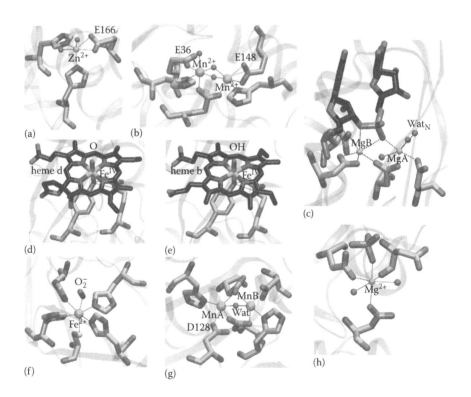

FIGURE 3.1   Some of the metalloproteins mentioned in this chapter: (a) thermolysin, (b) manganese catalase, (c) ribonuclease H, (d) *Penicillium vitale* catalase, (e) *Helicobacter pylori* catalase, (f) superoxide reductase, (g) arginase, and (h) epoxide hydrolase.

exhibits unusually high B-factors, precluding the assignment of the coordination mode with confidence. MD simulations with CD-corrected charges for the metal ion and its ligands show that the glutamate residue may be actually flipping between the two coordination modes. A similar carboxylate shift has been observed for the Glu148 residue bound to Mn in manganese catalase (Spiegel, DeGrado, and Klein 2006) (Figure 3.1b), an enzyme essential for the removal of toxic hydrogen peroxide.

Structural determination of metalloproteins by NMR faces the challenge that metal sites are silent in NMR and, thus, no metal–ligand restraints can be measured experimentally. Hence, NMR refinement uses metal–ligand restraints derived from experimental (the crystal structure of the protein or other proteins of the same family) or computational (typically, QM calculations on gas phase models) data. However, this bonded approach can result in artifacts or distortions in the geometry of the metal-containing site if the guessed coordination number and metal–ligand distances differ from the actual values. Hence, our group has proposed a second refinement step to optimize the metal–ligand geometry by using MD simulations in combination with the nonbonded CD model (Dal Peraro et al. 2007a; Calhoun et al. 2008). This approach allows one to relax the local frustrations present in the NMR structure calculated with standard metal–ligand restraints. Specifically, we have demonstrated that this second refinement step is crucial to obtain a more realistic structure for the *de novo* designed dimetal binding protein di-Zn(II) DFsc (Calhoun et al. 2008). The initial NMR structure obtained using metal–ligand restraints shows a distorted metal site and an occluded solvent access channel. MD simulations with the nonbonded CD model and explicit solvent result in the reorientation of helix 2, changing the metal coordination from 5 to 4 and allowing the entrance of water molecules into the channel. The refined structure obtained is more consistent with the protein NMR restraints and exhibits a metal site geometry in better agreement with similar solution complexes.

In some cases, the low resolution of the X-ray structure precludes the identification of water molecules present in the metal-containing site. Moreover, it is difficult to obtain information about the solvent accessibility and water exchange, even with NMR. Classical MD simulations with explicit solvent and a nonbonded CD description of the metal-containing site can provide further insight into its hydration. For instance, this approach has been applied to ribonuclease H (RNase H), an endonuclease that contains two Mg^{2+} ions (denoted MgA and MgB; see Figure 3.1c) and that is involved in tumorigenesis and HIV infection. Simulations of RNase H (Dal Peraro et al. 2007a; De Vivo, Dal Peraro, and Klein 2008; Ho et al. 2010) show that the catalytic water coordinated to MgA (Wat_N in Figure 3.1c) can be exchanged with solvent molecules several times on a nanosecond time scale, without disrupting the metal coordination geometry. Besides water exchange, this method can also reproduce changes in the number of water molecules coordinated to the metal ion. Indeed, simulations of the *de novo* designed dimanganese DF1 protein (Spiegel, DeGrado, and Klein 2006) reveal how the wetting transition between the two active-site conformations present in the crystal structure occurs (Figure 3.2). During the simulation of the open state, a water molecule is observed to enter the active site and bind to one of the two manganese ions. Water insertion increases the intermetal distance and results in the sliding of helices 1 and 3, yielding the alternative closed state.

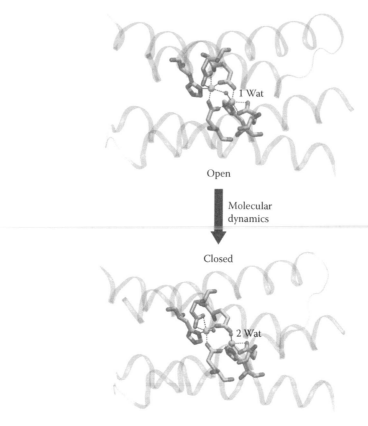

FIGURE 3.2 Open/closed transition in the *de novo* designed dimanganese protein DF1. Molecular dynamics simulations with the nonbonded charge distribution model show a second water molecule entering the active site. (From Spiegel, K. et al., *Proteins Struct. Funct. Bioinform.* 65 (2), 317, 2006.)

As mentioned previously, the nonbonded CD model does not perform as well for metalloproteins with highly directional metal–ligand geometries (Dal Peraro et al. 2007a). In this case, refinement of experimental structures can be accomplished using QM and QM/MM methods (Ryde and Nilsson 2003; Ryde 2007). Calculations varying the oxidation state of the metal, the coordinated ligands, and their protonation state can be used to obtain the reference geometry for each of the coordination complexes likely to be present in the metal-containing site. Comparison of the calculated structures with the experimental crystallographic data allows one to identify unambiguously the actual species trapped in the experiment and to define better metal–ligand restraints to be used in the refinement. For instance, crystallographic analysis of heme proteins (i.e., metalloproteins containing an iron–porphyrin or heme complex) is complicated by photoreduction, which is reduction of the heme by electrons produced by the X-ray radiation used to collect the diffraction data. Moreover, some of these redox intermediates are such powerful oxidants that they can even oxidize Tyr or Trp residues of their own protein scaffold. In other words, the redox species initially trapped when preparing the sample can undergo a transformation during crystallization or irradiation, complicating the assignment of the actual species present. This is the case for heme catalase, an enzyme that dismutates hydrogen peroxide into water and oxygen, thus avoiding the harmful effects of H_2O_2. Crystallographic analysis of two different heme catalases, *Penicillium vitale* catalase (PVC) (containing heme d, see Figure 3.1d) and *Helicobacter pylori* catalase (HPC) (heme b, see Figure 3.1e), exhibit two different Fe–O bond lengths (1.72 and 1.80 Å, respectively). QM and QM/MM calculations (Alfonso-Prieto et al. 2007) show that a short Fe–O distance correspond to an oxo group coordinated to the ferryl ion, whereas a long Fe–O distance represents a hydroxyl ligand. Taken together with the spectroscopic data, the calculated geometries reveal that the species trapped in the PVC crystal is an oxoferryl heme cation radical (i.e., $heme^{+\bullet}-Fe^{IV} = O$), whereas for HPC it is a hydroxoferryl neutral heme (i.e., $heme-Fe^{IV}-OH$).

3.3.2 STRUCTURAL PREDICTION

Besides refinement of experimental structures, computational methods are also useful to make structural predictions. Classical MD simulations with the nonbonded CD model, as well as QM and QM/MM simulations, can provide structural information of reaction intermediates or redox states that are too short lived to be studied by X-ray crystallography or NMR spectroscopy. Besides, they can be used to guide mutagenesis studies, by foreseeing whether the mutation has the expected effect in the protein structure or function. Furthermore, theoretical methods are an invaluable tool in protein design. On the one hand, they can be used to assess the feasibility of the artificial protein; on the other, they can provide reference structural determinants to aid in the refinement of the experimental structure, since the synthetic protein does not have a natural counterpart. For instance, classical MD simulations with the nonbonded CD model of the synthetic dimanganese DF1 protein (Spiegel, DeGrado, and Klein 2006) have been used to determine the structure of the oxidized Mn^{3+}– Mn^{3+} state, which is not stable enough to be characterized experimentally. Besides, they have been able to predict how mutations of the active site residue Glu36 (Figure 3.1b) to either Asp or Ser affect the structure and flexibility of the active site and the overall protein.

Another example is the prediction of a third metal-binding site in RNase H (Ho et al. 2010). Kinetic experiments have shown that the catalytic activity of this enzyme depends on the magnesium concentration. RNase H contains two Mg^{2+} ions (MgA and MgB, see Figure 3.1c) in the catalytic site, which are essential to cleave the RNA strand of its RNA:DNA hybrid substrate. Hence, at low Mg^{2+} concentrations, the enzymatic activity increases with the metal ion concentration, until it reaches an optimal value in the mM range. Unexpectedly, further increase in the Mg^{2+} concentration results in enzyme inhibition, which seems to be at odds with magnesium being essential for RNase H activity. To explain this inhibition at high Mg^{2+} concentrations, an attenuation model has been proposed: an additional Mg^{2+} ion could bind near the catalytic site, tuning the enzymatic rate. However, it has not been possible to identify this third Mg^{2+} binding site by X-ray crystallography. To provide structural insight into the attenuation effect, our group has performed

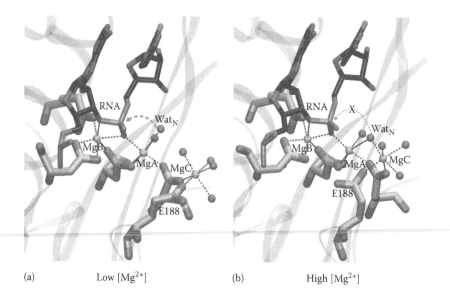

(a)   Low [Mg^{2+}]   (b)   High [Mg^{2+}]

FIGURE 3.3 Binding of a third magnesium ion (MgC) in ribonuclease H revealed by molecular dynamics. (a) Active conformation at low Mg^{2+} concentrations. (b) Inactive conformation at high Mg^{2+} concentrations. The position of the nucleophilic water (Wat$_N$) is affected by the magnesium concentration. (From Ho, M.-H. et al., *J. Am. Chem. Soc.*, 132 (39), 13702, 2010.)

classical MD simulations of RNase H at different Mg^{2+} concentrations (Ho et al. 2010). These simulations demonstrate that a third Mg^{2+} ion (MgC) can indeed bind near the catalytic site, in particular to the second-shell residue Glu188 (Figure 3.3). At the optimal, low Mg^{2+} concentration, MgC binding does not affect the geometry of the catalytic site (hereafter the active conformation; see Figure 3.3a) and thus does not impair the catalytic reaction. In contrast, at higher Mg^{2+} concentrations, MgC binding perturbs the position of the nucleophilic water (i.e., the inactive conformation; see Figure 3.3b), decreasing the catalytic efficiency. Additional free energy (adaptive biasing force; Hénin and Chipot 2004) simulations show that Mg^{2+} concentration (via MgC binding) modulates the population of these two catalytic site conformations. At low Mg^{2+} concentrations, the active conformation is more stable than the inactive, whereas at high concentration they are very similar in energy. Moreover, simulations have been able to rationalize why the activity of the E188A mutant is independent of the metal ion concentration. In the mutant, MgC binding is not possible and thus the active conformation of the catalytic site is maintained even at high Mg^{2+} concentrations.

3.3.3 ELECTRONIC CONFIGURATION AND SPIN STATES

Although UV–visible, EPR or Mössbauer spectroscopies can be used to study the electronic state of metalloproteins, some redox intermediates are too short lived to be investigated using these techniques. Therefore, the additional information provided by QM and QM/MM calculations can be very useful to unravel the electronic properties of redox metalloproteins. However, as mentioned in Section 3.2.2, this requires the theoretical method (usually DFT) to be accurate enough to describe the electronic configuration and energy ordering of the possible spin states.

For instance, superoxide reductase (SOR) is an iron-containing enzyme that reduces superoxide to hydrogen peroxide, preventing the formation of toxic oxygen radical species. Its reaction intermediate contains an iron–dioxygen complex (Figure 3.1f), whose protonation state and ground electronic configuration are not clear from experimental data. To elucidate these issues, our group has performed QM and QM/MM calculations (Sit et al. 2010). First, gas-phase calculations using different DFT approaches were carried out to assess the accuracy of the theoretical method used (see Section 3.2.2): a plane-wave basis set with a GGA exchange-correlation functional (PBE), the same level of theory with a Hubbard–U-correction, or an atomic basis set with different hybrid functionals (including different amounts of HF exchange). The geometry and spin-state energetics obtained with these disparate DFT schemes are quite similar. All methods predict superoxide to bind to the ferrous ion in an end-on (η^1) conformation. Moreover, the intermediate and high spin states are found to be nearly degenerated, explaining why different experimental studies have found the ground state to be either one or the other spin state. The spin densities obtained with the different methods are also similar, although the Hubbard–U-corrected and hybrid functional calculations seem to perform slightly better than the standard GGA functional. These unpaired spin density distributions, along with the calculated O–O distance, demonstrate that the superoxide anion is not reduced to peroxide upon binding to Fe^{2+}. Once established that the Hubbard–U-corrected DFT approach was accurate enough, the same active-site model was optimized by means of QM/MM. In the presence of the protein and solvent, superoxide binds to the ferrous ion in a side-on (η^2) conformation, different from the gas-phase model. This alternative conformation is stabilized by hydrogen bonds with the surrounding water molecules. Moreover, the side-on conformation is predicted to hinder fast proton transfer, which is critical to avoid undesirable side reactions in SOR. Therefore, in this case, inclusion of the environment is crucial to reproduce the correct structure of the Michaelis complex of SOR.

Heme catalase is another enzyme essential to avoid the toxicity of reactive oxygen species. As explained in Section 3.3.1, a combination of X-ray crystallography, UV–visible spectroscopy, and DFT calculations has demonstrated that the redox intermediate trapped for the heme d catalase PVC is an oxoferryl heme cation radical (Alfonso-Prieto et al. 2007). Therefore, EPR measurements are expected to show one signal, corresponding to the unpaired electron of the heme cation radical. However, the EPR spectrum turns out to be silent (A. Ivancich, personal communication). To understand this unexpected result, we performed QM calculations on a gas-phase model of the PVC redox intermediate. The obtained spin density distribution shows one unpaired electron delocalized on the heme macrocycle (Figure 3.4a), which contradicts the absence of signal in EPR. The accuracy of the DFT approach used (a plane-wave basis set with the Becke–Perdew GGA functional) does not seem to be the problem, since a similar calculation with a heme b model gives a spin density distribution and a ground spin-state consistent with the experimental data on heme b catalases (Alfonso-Prieto et al. 2007). To investigate if the environment can modulate the spin density distribution of the heme d redox intermediate, we carried

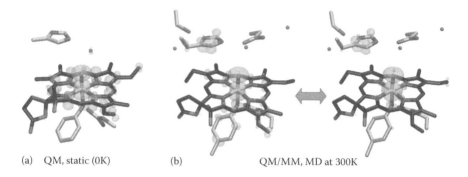

(a)   QM, static (0K)   (b)   QM/MM, MD at 300K

FIGURE 3.4 Electronic configuration of the reaction intermediate of *Penicillium vitale* catalase (PVC). (a) Quantum mechanical optimized geometry of the catalase active site in the gas phase. (b) Snapshots from a quantum mechanical/molecular mechanical molecular dynamics simulation. Unpaired spin density isosurfaces (at 0.004 e$^-$/Å^3) are shown. (From Alfonso-Prieto, M. et al., *J. Am. Chem. Soc.* 129 (14), 4193, 2007.)

out additional QM/MM calculations. These simulations show that the unpaired electron on the heme can delocalize over a nearby histidine residue (Figure 3.4b), due to ET (or orbital coupling) between the SOMO of the heme and the HOMO of the imidazole. This fluctuation of the radical between the two moieties rationalizes why the EPR of the PVC redox intermediate is silent (Alfonso-Prieto et al. 2007).

3.3.4   ELECTRON TRANSFER

As mentioned in Section 3.2.3, simulations are also useful to obtain ET parameters and to provide further insight into the structural and dynamical determinants of ET. For instance, Blumberger and coworkers have calculated the reorganization free energy of the ET between two heme cofactors (Blumberger and Klein 2006) or between a heme cofactor and a Ru complex (Blumberger 2008; Tipmanee et al. 2010) in different protein scaffolds. A modified version of the protocol presented in Section 3.2.3 was used, in which, after sampling with classical MD, energetics is calculated in an alternative way. Namely, reorganization energy (λ) is divided in three terms: the inner-sphere reorganization energy (λ^i), the outer-sphere reorganization energy (λ^o), and a bulk solvation correction term (λ^{fs}). λ^i is the contribution to the reorganization energy of the metal ion and the first-shell ligands, and is obtained from QM ionization energy calculations on a model of the metal-containing site. λ^o is the contribution of the protein and solvent environment, and is computed, using the FF point charges, as the difference in electrostatic potential energy between the two ET states (removing the self-contribution of the cofactors). Finally, λ^{fs} accounts for the error due to the finite number of explicit solvent molecules in the simulation, and is estimated from continuum electrostatic calculations or QM calculations on solution model complexes. The obtained reorganization energies (Blumberger and Klein 2006; Blumberger 2008; Tipmanee et al. 2010) are in good agreement with the experimental estimates, especially if a polarizable FF is used in the classical MD simulation. Moreover, the inner-sphere reorganization energy is found to be small, indicating that the protein and the solvent are the main contributors to the reorganization energy. Furthermore, the reorganization energy is larger for cofactors exposed to the solvent and the protein reorganization results from the collective contribution of many residues. Therefore, simulations predict that the reorganization energy, λ, of synthetic proteins can be reduced by designing less solvent-exposed cofactor binding sites, whereas introduction of protein mutations is expected to have a less dramatic effect.

As explained in Section 3.3.1, the heme catalases PVC and HPC form two different redox intermediates (Alfonso-Prieto et al. 2007). Experimental data suggest that HPC forms initially the same redox intermediate as PVC (i.e., an oxoferryl heme cation radical); however, this species is rapidly transformed into a hydroxoferryl neutral heme by transfer of one electron and one proton from the protein scaffold. To understand why the oxoferryl heme cation radical is stable for PVC, whereas it is easily reduced in HPC, we have calculated the heme redox potential for both catalases (Alfonso-Prieto et al. 2011), using the computational method outlined in Section 3.2.3. Surprisingly, the free energy of ET is found to be the same for both catalases, which seems to contradict the observation of two different redox intermediates. However, one has to consider that proton transfer to the oxo group coordinated to the ferryl ion accompanies ET to the heme. Additional free energy metadynamics (Laio and Parrinello 2002; Ensing et al. 2006) simulations show that protonation of the oxoferryl is more energetically favored in HPC than PVC (Figure 3.5). Therefore,

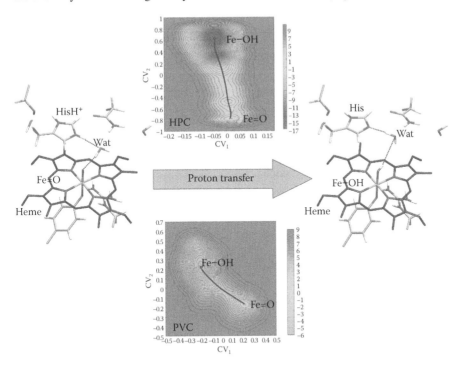

FIGURE 3.5 Proton transfer to the oxoferryl group in heme catalases. The free energy profiles (in kcal/mol) of the protonation reaction for *Helicobacter pylori* (HPC) and *Penicillium vitale* (PVC) catalases are shown. (From Alfonso-Prieto, M. et al., *J. Am. Chem. Soc.* 133 (12), 4285, 2011.)

the transformation into a different redox intermediate in HPC is driven by the proton transfer step and not by ET (Alfonso-Prieto et al. 2011). Indeed, maintaining similar redox potential for both enzymes ensures that the two catalases can catalyze the same reaction (i.e., dismutation of hydrogen peroxide) with similar efficiency.

3.3.5 CATALYTIC MECHANISM

Both QM and QM/MM calculations are powerful tools to study enzymatic reaction mechanisms, in particular for metalloproteins (Siegbahn and Borowski 2006; Dal Peraro et al. 2007c; Ramos and Fernandes 2008; Senn and Thiel 2009; van der Kamp and Mulholland 2013; Rovira 2013). They can complement experimental (structural and kinetic) studies to prove the proposed catalytic mechanism. On the one hand, theoretical characterization of the possible reaction intermediates and transition states can be compared with the experimental structures of the wild-type enzyme (e.g., in complex with a nonreactive substrate analog, a transition-like inhibitor or the product), as well as inactive mutants of the enzyme (e.g., in complex with the substrate). On the other hand, calculation of the reaction barriers for alternative reaction pathways can be compared with the measured reaction rates.

Two computational approaches are mainly used. The first one is the QM cluster method, in which a model of the (metal-containing) active site is studied in the gas phase (Siegbahn and Borowski 2006). The steric effects of the protein environment are modeled by fixing some atoms to their crystallographic position (typically the C_α atoms), and the electrostatic effects may be included with a continuum solvation model. This approach has the advantage of the smaller size and full control of the system. However, the lack (or the continuum description) of the environment and the restricted flexibility (due to the artificial constraints on the truncated atoms) may be problematic, especially in cases where the resolution of the initial crystal structure is not high enough or significant structural rearrangements occur in the active site and its surroundings during the reaction (Hu, Söderhjelm, and Ryde 2012; Liao and Thiel 2013). These limitations can be overcome by choosing a larger cluster model; however, this comes with a higher computational cost. The second method is QM/MM, in which the (metal-containing) active site is treated with quantum mechanics, and the protein and solvent environment are included explicitly and described with molecular mechanics (Dal Peraro et al. 2007c; Ramos and Fernandes 2008; Senn and Thiel 2009; van der Kamp and Mulholland 2013; Rovira 2013). Nevertheless, treating the QM–MM boundary atoms and the QM–MM energy coupling is technically difficult. Moreover, the larger size of the system (and thus of possible local minima) and the limited time scale computationally accessible (picoseconds) can complicate the conformational sampling. Performing classical MD simulations before the QM/MM study to select the most likely scenario(s) for the reaction can minimize this latter problem.

For instance, our group has applied the QM cluster approach to study the reaction mechanism of arginase (Ivanov and Klein 2004; Ivanov and Klein 2005), a dimanganese protein (Figure 3.1g) that catalyzes the hydrolysis of L-arginine into L-ornithine and urea, the last step of the urea cycle. First, QM geometry optimizations were performed on clusters of increasing size and using different constraints to assess the influence of the environment in the structure of the Michaelis complex (Ivanov and Klein 2004). Larger models, which include second-shell residues, were found to be necessary to improve the description of the dimanganese site. Then, the flexibility of the active site was studied (Ivanov and Klein 2005) by means of *ab initio* MD (Car and Parrinello 1985). During the simulations, the nucleophilic water molecule coordinated to both Mn ions (denoted MnA and MnB, see Figure 3.1g) oscillating between a bridging and a terminal position, and this switching is correlated with metastable proton transfer events to one of the ligands of MnA, Asp128. Indeed, Asp128 has been previously proposed to be the proton acceptor in the reaction mechanism, based on mutagenesis experiments. The catalytic role of Asp128 was further confirmed by QM constrained MD (Sprik and Ciccotti 1998), which shows a low barrier for proton transfer, consistent with the experimental results on a biomimetic solution complex. Moreover, the calculations provide structural and energetic insight into the lower activity of the metal-depleted form of arginase: a single Mn ion is not able to reduce the pK_a of the nucleophilic water as much as two ions.

Another application of the QM/MM approach in our group is in the study of the catalytic mechanism of the human soluble epoxide hydrolase (De Vivo, Ensing, and Klein 2005; De Vivo et al. 2007). This enzyme is an Mg^{2+}-dependent phosphatase (see Figure 3.1h) involved in fatty acid metabolism and, thus, is a possible drug target. Initial classical MD simulations (with the CD nonbonded model and explicit solvent) are used to sample the protein conformational space and the hydration of the active site to identify the most likely reactive configuration. Then, QM/MM calculations, in combination with a constrained MD free energy technique (Sprik and Ciccotti 1998), are used to investigate the feasibility of the reaction mechanism outlined in the literature. The calculated energy barriers are in agreement with the experimental values, confirming the proposed mechanism. Moreover, the simulations provide the molecular explanation of the high efficiency of the phosphoryl transfer. They show that a network of water molecules connecting the metal and the substrate act as a proton shuttle, helping activate the nucleophile and stabilize the leaving group.

We have also studied the phosphodiester hydrolysis reaction in RNase H (De Vivo, Dal Peraro, and Klein 2008) (Figure 3.1c). Similar to the previous example, classical MD simulations are initially performed to explore the hydration of the active site (see Section 3.3.1) and, in this case, also to revert the active-site mutation (D192N) present in the crystal structure. Subsequent QM/MM constrained MD (Sprik and Ciccotti 1998) simulations are used to calculate the free energy profile of the two possible reaction mechanisms, with either water or a hydroxide ion as a nucleophile. The energy barriers of the two pathways are similar and consistent with the experimentally measured rate. Therefore, both reaction mechanisms seem to be possible, in line with the RNase H activity being optimal at pH 8. Besides, the simulations give further insight into the cooperative role of the two Mg ions in the reaction. MgA stabilizes the attacking conformation of the nucleophile, whereas MgB destabilizes the Michaelis complex and stabilizes the leaving group.

Other hydrolytic enzymes studied in our group are the zinc-containing enzymes thermolysin (one of the best known metalloproteases) (Blumberger, Lamoureux, and Klein 2007), β-lactamase CcrA (involved in antibiotic resistance and thus a target for new inhibitors) (Dal Peraro et al. 2007b), and farnesyltransferase (a target for anticancer drug design) (Ho et al. 2009). A computational protocol similar to the previous examples has been used. A classical MD simulation is performed initially to relax the enzyme–substrate complex and explore the flexibility and water content of the active site, allowing one to identify the most likely reactive configuration. Then, QM/MM constrained MD (Sprik and Ciccotti 1998) or metadynamics (Laio and Parrinello 2002; Ensing et al. 2006) simulations are used to compute the free energy profiles for the different possible pathways. Comparison of the calculated energy barriers with the experimental values allows one to determine the actual reaction mechanism. Furthermore, the simulations provide the structure of the fleeting transition state (which can be used to design inhibitors) and a chemical explanation of why nature has selected that particular number and type of metal ions to catalyze a certain reaction more efficiently.

3.4 CONCLUDING REMARKS

In recent years, the number of applications of MD simulations in the field of metalloproteins has grown significantly, due to the progress in theoretical methods and the increase in computer power (Karplus and McCammon 2002; Banci 2003). On the one hand, the development of new bonded (Banci, Gori-Savellini, and Turano 1997; Autenrieth et al. 2004; Peters et al. 2010; Carvalho, Teixeira, and Ramos 2013) and nonbonded (Oelschlaeger et al. 2007; Dal Peraro et al. 2007a) FFs for metal-containing active sites has allowed one to complement the static picture provided by X-ray crystallography with a dynamical picture at near-physiological conditions. In addition, classical MD simulations have improved the refinement of structures derived from NMR (Calhoun et al. 2008), where metal–ligand restraints are silent. On the other hand, improvements in the quantum description of the metal within the framework of DFT (Cramer and Truhlar 2009; Sit et al. 2010) and the appearance of new and powerful free energy techniques (Sprik and Ciccotti 1998; Laio and Parrinello 2002; Hénin and Chipot 2004; Ensing et al. 2006) have allowed the study of the electronic properties and reactivity of metalloproteins with higher accuracy. Specifically, QM and QM/MM simulations have revealed the structure of reaction intermediates and redox states not accessible experimentally and have unraveled the catalytic mechanism of enzymatic reactions with unprecedented detail (Siegbahn and Borowski 2006; Dal Peraro et al. 2007c; Ramos and Fernandes 2008; Senn and Thiel 2009; McGeagh, Ranaghan, and Mulholland 2011; Rovira 2013). In other words, the interplay of experiments and theory has been shown to be essential to obtain a more complete understanding of metalloproteins (Figure 3.6). In particular, dynamics has been finally recognized as playing an essential role in protein function (Henzler-Wildman and Kern 2007), and it is here that MD simulations can provide a unique view on conformational transitions at atomic resolution (Karplus and McCammon 2002).

Notwithstanding the above-mentioned progress, computer simulation studies on metalloproteins still face several key challenges. Most importantly, the accuracy of the potential energy functions is still not completely satisfactory. FF parameterization of metal-containing sites could be improved by optimizing the van der Waals parameters of the metal ion (e.g., to better reproduce metal–ligand distances) and/or by

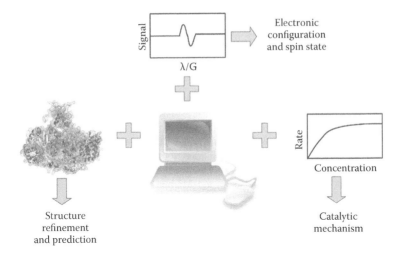

FIGURE 3.6   Interplay between theory and experiment used to study metalloproteins.

adding many body polarization effects. In this regard, an interesting advance is the force-matching algorithm (Maurer et al. 2007), which derives FF parameters based on QM/MM MD simulations; this strategy has been applied, e.g., to the KcsA potassium channel (Bucher et al. 2006; Bucher et al. 2009). Importantly, the quality of the DFT description needs to be enhanced by constructing better exchange-correlation functionals (Cramer and Truhlar 2009) and by including a correction for the missing dispersion interactions (Klimes and Michaelides 2012; Bankura, Carnevale, and Klein 2013). The further development of improved theoretical methods for handling metalloproteins will undoubtedly benefit from comparisons of any new computational results with the extensive body of experimental data.

REFERENCES

Alfonso-Prieto, Mercedes, Anton Borovik, Xavier Carpena, Garib Murshudov, William Melik-Adamyan, Ignacio Fita, Carme Rovira, and Peter C. Loewen. 2007. "The structures and electronic configuration of compound I intermediates of *Helicobacter pylori* and *Penicillium vitale* catalases determined by X-ray crystallography and QM/MM density functional theory calculations." *Journal of the American Chemical Society* 129 (14): 4193–4205.

Alfonso-Prieto, Mercedes, Harald Oberhofer, Michael L. Klein, Carme Rovira, and Jochen Blumberger. 2011. "Proton transfer drives protein radical formation in *Helicobacter pylori* catalase but not in *Penicillium vitale* catalase." *Journal of the American Chemical Society* 133 (12): 4285–4298.

Autenrieth, Felix, Emad Tajkhorshid, Klaus Schulten, and Zaida Luthey-Schulten. 2004. "Role of water in transient cytochrome c_2 docking." *Journal of Physical Chemistry B* 108 (52): 20376–20387.

Bader, Richard R. 1994. *Atoms in Molecules: A Quantum Theory*. Oxford: Oxford University Press.

Banci, Lucia. 2003. "Molecular dynamics simulations of metalloproteins." *Current Opinion in Chemical Biology* 7 (1): 143–149.

Banci, Lucia, Giovanni Gori-Savellini, and Paola Turano. 1997. "A molecular dynamics study in explicit water of the reduced and oxidized forms of yeast iso-1-cytochrome *c*." *European Journal of Biochemistry* 249 (3): 716–723.

Bankura, Arindam, Vincenzo Carnevale, and Michael L. Klein. 2013. "Hydration structure of salt solutions from *ab initio* molecular dynamics." *The Journal of Chemical Physics* 138: 014501.

Blumberger, Jochen. 2008. "Free energies for biological electron transfer from QM/MM calculation: Method, application and critical assessment." *Physical Chemistry Chemical Physics* 10 (37): 5651–5667.

Blumberger, Jochen, and Michael L. Klein. 2006. "Reorganization free energies for long-range electron transfer in a porphyrin-binding four-helix bundle protein." *Journal of the American Chemical Society* 128 (42): 13854–13867.

Blumberger, Jochen, Yoshitaka Tateyama, and Michiel Sprik. 2005. "*Ab initio* molecular dynamics simulation of redox reactions in solution." *Computer Physics Communications* 169 (1): 256–261.

Blumberger, Jochen, Guillaume Lamoureux, and Michael L. Klein. 2007. "Peptide hydrolysis in thermolysin: *Ab initio* QM/MM investigation of the Glu143-assisted water addition mechanism." *Journal of Chemical Theory and Computation* 3 (5): 1837–1850.

Bucher, Denis, Simone Raugei, Leonardo Guidoni, Matteo Dal Peraro, Ursula Rothlisberger, Paolo Carloni, and Michael L. Klein. 2006. "Polarization effects and charge transfer in the KcsA potassium channel." *Biophysical Chemistry* 124 (3): 292–301.

Bucher, Denis, Leonardo Guidoni, Patrick Maurer, and Ursula Rothlisberger. 2009. "Developing improved charge sets for the modeling of the KcsA K$^+$ channel using QM/MM electrostatic potentials." *Journal of Chemical Theory and Computation* 5 (8): 2173–2179.

Calhoun, Jennifer R., Weixia Liu, Katrin Spiegel, Matteo Dal Peraro, Michael L. Klein, Kathleen G. Valentine, A. Joshua Wand, and William F. DeGrado. 2008. "Solution NMR structure of a designed metalloprotein and complementary molecular dynamics refinement." *Structure* 16 (2): 210–215.

Car, Roberto, and Michele Parrinello. 1985. "Unified approach for molecular dynamics and density functional theory." *Physical Review Letters* 55 (22): 2471–2474.

Carvalho, Alexandra T. P., Ana F. S. Teixeira, and Maria J. Ramos. 2013. "Parameters for molecular dynamics simulations of iron–sulfur proteins." *Journal of Computational Chemistry* 34 (18): 1540–1548.

Cascella, Michele, Alessandra Magistrato, Ivano Tavernelli, Paolo Carloni, and Ursula Rothlisberger. 2006. "Role of protein frame and solvent for the redox properties of azurin from *Pseudomonas aeruginosa*." *Proceedings of the National Academy of Sciences United States of America* 103 (52): 19641–19646.

Ceccarelli, Matteo, and Massimo Marchi. 2003. "Simulation and modeling of the *Rhodobacter sphaeroides* bacterial reaction center II: Primary charge separation." *The Journal of Physical Chemistry B* 107 (23): 5630–5641.

Cramer, Christopher J., and Donald G. Truhlar. 2009. "Density functional theory for transition metals and transition metal chemistry." *Physical Chemistry Chemical Physics* 11 (46): 10757–10816.

Dal Peraro, Matteo, Katrin Spiegel, Guillaume Lamoureux, Marco De Vivo, William F. DeGrado, and Michael L. Klein. 2007a. "Modeling the charge distribution at metal sites in proteins for molecular dynamics simulations." *Journal of Structural Biology* 157 (3): 444–453.

Dal Peraro, Matteo, Alejandro J. Vila, Paolo Carloni, and Michael L. Klein. 2007b. "Role of zinc content on the catalytic efficiency of B1 metallo β-lactamases." *Journal of the American Chemical Society* 129 (10): 2808–2816.

Dal Peraro, Matteo, Paolo Ruggerone, Simone Raugei, Francesco Luigi Gervasio, and Paolo Carloni. 2007c. "Investigating biological systems using first principles Car–Parrinello molecular dynamics simulations." *Current Opinion in Structural Biology* 17 (2): 149–156.

De Vivo, Marco, Bernd Ensing, and Michael L. Klein. 2005. "Computational study of phosphatase activity in soluble epoxide hydrolase: High efficiency through a water bridge mediated proton shuttle." *Journal of the American Chemical Society* 127 (32): 11226–11227.

De Vivo, Marco, Bernd Ensing, Matteo Dal Peraro, German A. Gomez, David W. Christianson, and Michael L. Klein. 2007. "Proton shuttles and phosphatase activity in soluble epoxide hydrolase." *Journal of the American Chemical Society* 129 (2): 387–394.

De Vivo, Marco, Matteo Dal Peraro, and Michael L. Klein. 2008. "Phosphodiester cleavage in ribonuclease H occurs via an associative two-metal-aided catalytic mechanism." *Journal of the American Chemical Society* 130 (33): 10955–10962.

Ensing, Bernd, Marco De Vivo, Zhiwei Liu, Preston Moore, and Michael L. Klein. 2006. "Metadynamics as a tool for exploring free energy landscapes of chemical reactions." *Accounts of Chemical Research* 39 (2): 73–81.

Guidon, Manuel, Jürg Hutter, and Joost VandeVondele. 2010. "Auxiliary density matrix methods for Hartree–Fock exchange calculations." *Journal of Chemical Theory and Computation* 6 (8): 2348–2364.

Harding, Marjorie M., Matthew W. Nowicki, and Malcolm D. Walkinshaw. 2010. "Metals in protein structures: A review of their principal features." *Crystallography Reviews* 16 (4): 247–302.

Hénin, Jérôme, and Christophe Chipot. 2004. "Overcoming free energy barriers using unconstrained molecular dynamics simulations." *The Journal of Chemical Physics* 121: 2904.

Henzler-Wildman, Katherine, and Dorothee Kern. 2007. "Dynamic personalities of proteins." *Nature* 450 (7172): 964–972.

Ho, Ming-Hsun, Marco De Vivo, Matteo Dal Peraro, and Michael L. Klein. 2009. "Unraveling the catalytic pathway of metalloenzyme farnesyltransferase through QM/MM computation." *Journal of Chemical Theory and Computation* 5 (6): 1657–1666.

Ho, Ming-Hsun, Marco De Vivo, Matteo Dal Peraro, and Michael L. Klein. 2010. "Understanding the effect of magnesium ion concentration on the catalytic activity of ribonuclease H through computation: Does a third metal binding site modulate endonuclease catalysis?" *Journal of the American Chemical Society* 132 (39): 13702–13712.

Hu, LiHong, Pär Söderhjelm, and Ulf Ryde. 2012. "Accurate reaction energies in proteins obtained by combining QM/MM and large QM calculations." *Journal of Chemical Theory and Computation* 9 (1): 640–649.

Hubbard, John. 1963. "Electron correlations in narrow energy bands." *Proceedings of the Royal Society London A* 276 (1365): 238–257.

Ivanov, Ivaylo, and Michael L. Klein. 2004. "First principles computational study of the active site of arginase." *Proteins: Structure, Function, and Bioinformatics* 54 (1): 1–7.

Ivanov, Ivaylo, and Michael L. Klein. 2005. "Dynamical flexibility and proton transfer in the arginase active site probed by *ab initio* molecular dynamics." *Journal of the American Chemical Society* 127 (11): 4010–4020.

Karplus, Martin, and J. Andrew McCammon. 2002. "Molecular dynamics simulations of biomolecules." *Nature Structural Biology* 9 (9): 646–652.

Klimes, Jiri, and Angelos Michaelides. 2012. "Perspective: Advances and challenges in treating van der Waals dispersion forces in density functional theory." *Journal of Chemical Physics* 137: 120901.

Laio, Alessandro, and Michele Parrinello. 2002. "Escaping free-energy minima." *Proceedings of the National Academy of Sciences United States of America* 99 (20): 12562–12566.

Liao, Rong-Zhen, and Walter Thiel. 2013. "On the effect of varying constraints in the quantum mechanics only modeling of enzymatic reactions: The case of acetylene hydratase." *Journal of Physical Chemistry B* 117 (15): 3954–3961.

Lundberg, Marcus, and Per E. M. Siegbahn. 2005. "Quantifying the effects of the self-interaction error in DFT: When do the delocalized states appear?" *The Journal of Chemical Physics* 122: 224103.

Marcus, Rudolph A., and Norman Sutin. 1985. "Electron transfers in chemistry and biology." *Biochimica et Biophysica Acta—Reviews in Bioenergetics* 811 (3): 265–322.

Maurer, Patrick, Alessandro Laio, Haakan W. Hugosson, Maria Carola Colombo, and Ursula Rothlisberger. 2007. "Automated parametrization of biomolecular force fields from quantum mechanics/molecular mechanics (QM/MM) simulations through force matching." *Journal of Chemical Theory and Computation* 3 (2): 628–639.

McGeagh, John D., Kara E. Ranaghan, and Adrian J. Mulholland. 2011. "Protein dynamics and enzyme catalysis: Insights from simulations." *Biochimica et Biophysica Acta (BBA)—Proteins and Proteomics* 1814 (8): 1077–1092.

Oelschlaeger, Peter, Marco Klahn, William A. Beard, Samuel H. Wilson, and Arieh Warshel. 2007. "Magnesium-cationic dummy atom molecules enhance representation of DNA polymerase β in molecular dynamics simulations: Improved accuracy in studies of structural features and mutational effects." *Journal of Molecular Biology* 366 (2): 687–701.

Olsson, Mats H. M., Gongyi Hong, and Arieh Warshel. 2003. "Frozen density functional free energy simulations of redox proteins: Computational studies of the reduction potential of plastocyanin and rusticyanin." *Journal of the American Chemical Society* 125 (17): 5025–5039.

Peters, Martin B., Yue Yang, Bing Wang, Laszlo Fusti-Molnar, Michael N. Weaver, and Kenneth M. Merz Jr. 2010. "Structural survey of zinc-containing proteins and development of the zinc AMBER force field (ZAFF)." *Journal of Chemical Theory and Computation* 6 (9): 2935–2947.

Ramos, Maria J., and Pedro A. Fernandes. 2008. "Computational enzymatic catalysis." *Accounts of Chemical Research* 41 (6): 689–698.

Rokob, Tibor Andras, Martin Srnec, and Lubomir Rulivsek. 2012. "Theoretical calculations of physico–chemical and spectroscopic properties of bioinorganic systems: Current limits and perspectives." *Dalton Transactions* 41 (19): 5754–5768.

Rovira, Carme. 2013. "The description of electronic processes inside proteins from Car–Parrinello molecular dynamics: Chemical transformations." *Wiley Interdisciplinary Reviews: Computational Molecular Science* 3 (4): 393–407.

Ryde, Ulf. 2007. "Accurate metal-site structures in proteins obtained by combining experimental data and quantum chemistry." *Dalton Transactions* (6): 607–625.

Ryde, Ulf, and Kristina Nilsson. 2003. "Quantum refinement—A combination of quantum chemistry and protein crystallography." *Journal of Molecular Structure: Theochem* 632 (1–3): 259–275.

Senn, Hans Martin, and Walter Thiel. 2009. "QM/MM methods for biomolecular systems." *Angewandte Chemie International Edition* 48 (7): 1198–1229.

Siegbahn, Per E. M., and Tomasz Borowski. 2006. "Modeling enzymatic reactions involving transition metals." *Accounts of Chemical Research* 39 (10): 729–738.

Sit, Patrick H.-L., Agostino Migliore, Ming-Hsun Ho, and Michael L. Klein. 2010. "Quantum mechanical and quantum mechanical/molecular mechanical studies of the iron–dioxygen intermediates and proton transfer in superoxide reductase." *Journal of Chemical Theory and Computation* 6 (9): 2896–2909.

Spiegel, Katrin, William F. DeGrado, and Michael L. Klein. 2006. "Structural and dynamical properties of manganese catalase and the synthetic protein DF1 and their implication for reactivity from classical molecular dynamics calculations." *Proteins: Structure, Function, and Bioinformatics* 65 (2): 317–330.

Sprik, Michiel, and Giovanni Ciccotti. 1998. "Free energy from constrained molecular dynamics." *The Journal of Chemical Physics* 109: 7737.

Sulpizi, Marialore, Simone Raugei, Joost VandeVondele, Paolo Carloni, and Michiel Sprik. 2007. "Calculation of redox properties: Understanding short-and long-range effects in rubredoxin." *The Journal of Physical Chemistry B* 111 (15): 3969–3976.

Thomson, Andrew J., and Harry B. Gray. 1998. "Bio-inorganic chemistry." *Current Opinion in Chemical Biology* 2 (2): 155–158.

Tipmanee, Varomyalin, Harald Oberhofer, Mina Park, Kwang S. Kim, and Jochen Blumberger. 2010. "Prediction of reorganization free energies for biological electron transfer: A comparative study of Ru-modified cytochromes and a 4-helix bundle protein." *Journal of the American Chemical Society* 132 (47): 17032–17040.

van der Kamp, Marc W., and Adrian J. Mulholland. 2013. "Combined quantum mechanics/molecular mechanics (QM/MM) methods in computational enzymology." *Biochemistry* 52 (16): 2708–2728.

Warshel, Arieh. 1982. "Dynamics of reactions in polar solvents. Semiclassical trajectory studies of electron-transfer and proton-transfer reactions." *The Journal of Physical Chemistry* 86 (12): 2218–2224.

4 Catalysis of Methane Oxidation by the Tricopper Cluster in the Particulate Methane Monooxygenase and Biomimetic Tricopper Complexes

Suman Maji, Steve S.-F. Yu, and Sunney I. Chan

CONTENTS

4.1   INTRODUCTION

In this chapter, we summarize our current understanding of the structure and function of particulate methane monooxygenase (pMMO), a tantalizing enzyme found
in methanotrophic bacteria that nature has created to mediate the facile conversion
of methane to methanol at room temperature and under ambient pressures of air.
We focus on the catalytic site of the enzyme, and show that it consists of a unique
$Cu^ICu^ICu^I$ tricopper cluster, which upon activation by dioxygen harnesses a highly
reactive "singlet oxene" that can be directly inserted across a C–H bond of the hydrocarbon substrate in the transition state of the complex formed during the reaction of
the substrate with the tricopper cluster. The oxidation of methane to methanol is difficult chemistry, as the C–H bond in methane is extremely inert. To accomplish this
task efficiently, nature has apparently invented new chemistry and exploited it for
this difficult chemical transformation. Armed with these new chemical principles,
we have designed and synthesized mimics of the active site of pMMO. We show that
these models can mediate efficient methane oxidation in the laboratory as well.

We begin this chapter by introducing the general issue of methane oxidation. Following a review of our present understanding on the structure and function of pMMO, we will discuss biomimetic models of the tricopper cluster site in the enzyme, including the design of ligand scaffolds to support a triad of Cu^I ions for oxidation of various compounds, the dioxygen chemistry of these tricopper complexes, the oxidation of simple organic substrates mediated by these complexes under noncatalytic and catalytic conditions, and the mechanism of hydrocarbon oxidation. We conclude with a comparison of the catalytic oxidation chemistry mediated by the model tricopper complexes with the catalytic turnover in the enzyme.

4.1.1 METHANE OXIDATION

The major source of methane is natural gas from geological deposits.[1] Methane is produced in enormous quantity from anaerobic decomposition of organic material by methanogenic bacteria.[2,3] It is also produced from human-induced sectors like livestock, landfill, and rice cultivation.[1,3,4]

Methane activation is a very difficult process. The inertness of the C–H bond necessitates a significant amount of energy for both the homolytic and heterolytic C–H bond cleavage. Its extremely high pK_a value makes methane an unsuitable candidate for conventional acid-base type reactions. Methane is also nonpolar with a negligible polarizability so that it is relatively unreactive for substitution or addition-elimination reactions. Given its high ionization potential, negligible electron affinity, and high HOMO–LUMO gap, methane is unfavorable for a redox-type reaction.

Till now, there are no laboratory reactions that can transform methane directly in an ecologically or economically friendly process under ambient conditions. Oxygenation of the C–H bonds in alkanes, especially methane oxidation to produce methanol, still remains one of the "holy grails in chemistry."[5]

In nature, methanotrophic bacteria use methane as the terminal electron donor during their metabolic process. They use methane as their only source of carbon and energy.[2,6] While part of the carbon is directly inducted into biomass through assimilation, the rest is converted ultimately to CO_2 and used by organisms capable of photosynthesis.

Methanotrophs occur mostly in soils, and are especially common near environments where methane is produced, including muds, marshes, swamps, landfill, rice paddies, and oceans.[2,3] They form a subclass of a large variety of microorganisms called methylotrophs, which are able to grow at the expense of reduced carbon compounds containing one or more carbon atoms but containing no carbon–carbon bonds.[2,7] Obligate methylotrophs grow only on such compounds, whereas facultative methylotrophs are also able to grow on a variety of other organic multicarbon compounds.[2]

There are essentially two different ways to oxidize methane in the natural environment, anaerobic and aerobic. Methanotrophs uses molecular oxygen to oxidize methane to methanol using methane monooxygenases (MMOs). Anaerobic oxidation of methane (AOM) occurs mostly in anoxic marine sediments by syntrophic methane-oxidizing archaea and sulfate-reducing bacteria,[8] or through coupling of denitrification to methane oxidation to form CO_2.[9]

4.1.2 Impetus for the Development of a Catalyst for Efficient Conversion of Methane to Methanol and Controlled Oxidation of Hydrocarbons at Room Temperature

Oxidation of unactivated C–H, C–C, and C=C bonds in hydrocarbons are important for the synthesis of fine chemicals in synthetic organic chemistry, as well as for the industrial conversion of basic building-block materials into useful end products or for synthesis of intermediates that are used for the production of high-value end products.[10] The planet has a huge reserve of naturally occurring hydrocarbons, and more are constantly being produced by natural processes, which can be transformed into useful oxygen derivatives through hydroxylation, epoxidation, or through degradation of oxygenated intermediates. Conventional methods use high-valent metal oxidants like permanganate or dichromate to accomplish the oxidation.[11–13] A catalytic process can offer a quick, economical, and environment-friendly way to achieve the same goal with higher yield, better substrate conversion, and greater selectivity in lesser time, manageable reaction conditions, and at lower cost.

In nature, similar processes of chemical transformations are taken care of by specialized proteins called enzymes, especially metalloenzymes, which have been optimized with time through evolution. These enzymatic systems perform efficiently at ambient temperatures and pressures with high turnover frequency (TOF) and minimal or without generation of undesirable by-products harmful to the biological system. Model catalytic systems could be developed with knowledge of the active-site structure and function of these enzymes, which can be applicable for useful chemical transformations with high TOF even at ambient conditions and with chemospecificity or regiospecificity and stereoselectivity. These designed "bioinspired" catalysts could be tuned to the need of a specific substrate or scale of production, and will find applications in the development of environment-friendly catalytic processes, which specifically need to avoid undesirable by-products; use of expensive and potentially toxic metal reagents, oxidants, and solvents; and reduction of high-energy-consuming multiple steps.

There are huge deposits of methane on Earth. Natural gas reserves yield a lot of methane, and through improved coal-bed methane extraction methods or extraction of methane hydrates or clathrates, it is becoming an attractive source of fuel.[14] However, since methane is a very potent greenhouse gas, there is growing interest in using this vast reserve of stored energy in a controlled and useful way through chemical transformation.[15] The first oxidation product, methanol, can be used as a starting material for other high-value industrial products or as a fuel for producing clean electricity through fuel cell technology.[10] Other hydrocarbons can also be used in an identical process to derive selective oxidation products. Catalytic oxidations of halogenated hydrocarbons are also of special interest for the detoxification of potential environmental pollutants.[16,17]

An industrial process of catalysis requires the system to be at ambient conditions for the maximum benefit in the utilization of resources. Thus, room-temperature catalytic oxidation of hydrocarbons has become the most sought-after reaction by developers of chemical processes.

4.1.3   Biological Methane Oxidation

Methane is the simplest hydrocarbon but with the highest C–H bond energy of 105 kcal/mol.[18] The activation of the C–H bond in CH_4 is a very difficult chemical process. However, biological systems accomplish this feat readily under ambient conditions using metalloenzymes called methane monooxygenases (MMOs). The MMOs catalyze oxidation of methane to methanol utilizing O_2 and two reducing equivalents from NAD(P)H[19] or quinols[20] to split the O–O bond. One oxygen atom of the O_2 is incorporated into the hydrocarbon substrate to yield methanol, and the other oxygen atom is reduced to water with the uptake of two protons.

MMOs are found only in methanotrophic bacteria, which are gram-negative bacteria belonging to the α- and γ-Proteobacteria phyla. Methanotrophs have been isolated from both mesophilic and extreme environmental conditions, including frozen tundra, thermal hot springs, marine and freshwater systems, and soil sources. Two types of MMOs have been isolated from methanotrophs: the soluble methane monooxygenase (sMMO)[16] and the pMMO,[21,22] which are mutually exclusive. Many methanotrophs, such as *Methylomicrobium album* BG8 and *Methylomonas methanica*, produce only the membrane-associated pMMO.[23] Others, such as *Methylococcus capsulatus* (Bath) and *Methylosinus trichosporium* OB3b, can generate either form, depending on the copper-to-biomass ratio of the culture.[24] Cells containing pMMO have demonstrated higher growth capabilities and higher affinity for methane than sMMO-containing cells.[3] It is suspected that copper ions may play a key role in both pMMO regulation and the enzyme catalysis, thus limiting pMMO cells to more copper-rich environments than sMMO-producing cells.[25]

4.2   PARTICULATE METHANE MONOOXYGENASE (pMMO) OF *METHYLOCOCCUS CAPSULATUS* (BATH)

Among many monooxygenases, the MMOs are unique in that they are the only ones that are able to catalyze efficient oxidation of methane to methanol, although they can also use a range of other hydrocarbons as potential substrates. The sMMO in particular can also oxidize a remarkably large number of hydrocarbons, including saturated, unsaturated, linear, branched, and cyclic hydrocarbons up to approximately C8 in size.[6,26–30] This relatively nonspecific but powerful monooxygenase enzyme can even oxidize halogenated alkenes and alkanes, one- and two-ring aromatics, and even heterocycles. pMMO, on the other hand, can oxidize only normal alkanes and alkenes up to C5.[31]

The hydrocarbon oxidation chemistry mediated by pMMO appears to be more straightforward because of its more limited substrate specificity. More interestingly, the oxidization proceeds with unusual regiospecificity and stereoselectivity.[31–34] The pMMO in *M. capsulatus* (Bath) can hydroxylate straight-chain alkanes from C1 to C5 and epoxidize related alkenes, and the dominant products are propan-2-ol, butan-2-ol, and pentan-2-ol for propane, butane, and pentane, respectively.[31] The related alkene 1-butene is hydroxylated to give 3-buten-2-ol and epoxidized to 1-epoxybutane, with a product distribution of ~1:1.[31] Studies on cryptically chiral ethanes (Scheme 4.1) have shown that the hydroxylation of C–H and C–D bonds proceeds with total retention of configuration at the carbon center oxidized with a k_H/k_D kinetic isotope effect (KIE) of

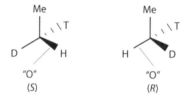

SCHEME 4.1   Cryptically chiral ethanes.

5.2–5.5.[33] Similarly, experiments on *n*-butane and *d,l*-[2-²H₁,3-²H₁]butanes have shown that the hydroxylation of the secondary carbon in *n*-butane also proceeds with a total retention of configuration, with a similar k_H/k_D isotope effect.[34] These observations rule out a radical mechanism for the hydroxylation chemistry. Instead, it has been proposed that the O-atom transfer occurs by concerted oxenoid insertion[33] (Scheme 4.2). A transition state involving singlet oxene insertion into the C–H bond should result in facile bond closure of the C–O bond following formation of the O–H bond, and the process should proceed with full retention of configuration at the carbon center oxidized.[34]

When propene and 1-butene are used as epoxidation substrates, the enantiomeric excess (ee) of the enzymatic products is only 18% and 37%, respectively.[31] However, epoxidation of *trans*-2-butene by the enzyme gives only the *d,l*-2,3-dimethyloxirane products and oxidation of *cis*-2-butene yields only the *meso* isomer.[32] These latter observations indicate that the enzymatic epoxidation also proceeds via the concerted electrophilic *syn* addition, and the relatively poor stereoselectivity in the enzymatic epoxidation of propene and 1-butene merely reflects the low stereochemical differentiation between the *re* and *si* faces in the hydrophobic pocket of the active site (Scheme 4.3).

"Singlet oxene"

SCHEME 4.2   Insertion of a "singlet oxene" across a C–H bond.

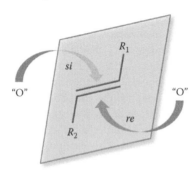

SCHEME 4.3   Different presentations of the *re* and *si* faces of the alkene to the O-atom.

4.2.1 SUBUNIT STRUCTURE

pMMO is a membrane protein residing in the intracytoplasmic membrane of the cell. It comprises three distinct subunits encoded by the genes *pmoA*, *pmoB*, and *pmoC* with an αβγ arrangement.[22,35] The protein has been highly produced in cells of *M. capsulatus* (Bath), solubilized in the detergent dodecyl-β-maltoside from pMMO-enriched (80%–90%) membranes, and the protein–detergent complex has been purified by size-exclusion chromatography.[36] The purified micelle contains one copy each of the three subunits encapsulated in ca. 240 detergent molecules. Matrix-assisted laser desorption/ionization time-of-flight mass spectrometry (MALDI–TOF–MS) has yielded three peptides with molecular masses of 42,785.52, 29,733.31, and 28,328.25 Da in excellent agreement with the subunit masses predicted by the gene sequences *pmoB* (α), *pmoC* (γ), and *pmoA* (β), respectively. The assignment has been confirmed by peptide-mass fingerprinting of in-gel digests of the subunits on SDS gels.[37,38] The purified protein–detergent complex exhibits high specific activity toward propene epoxidation using either NADH or duroquinol to provide the reducing equivalents for the catalytic turnover.[36]

4.2.2 METAL COFACTORS

The pMMO of *M. capsulatus* (Bath) is a multicopper protein containing up to ca.15 copper ions. Analysis of the purified protein–detergent micelles for metal content using atomic absorption, inductively coupled plasma mass spectrometry (ICP–MS), and x-ray absorption edge spectroscopy reveals 13.6 copper atoms per protein complex (MW 100 kDa). Only traces of iron are found (Cu/Fe atom ratio 80:1), and there is no evidence of any Zn.[36,37]

The 12–15 copper ions are sequestered into two major groupings, one containing six copper ions arranged in the form of a tricopper cluster, one dicopper center, and a type 2 copper site.[38] This set of coppers is responsible for dioxygen chemistry and alkane hydroxylation, and are dubbed the catalytic or C-clusters. The second group containing around six to nine copper ions mediate the electron input from NADH or other external reductants to the catalytic clusters, providing a buffer of reducing equivalents to re-reduce the C-clusters following turnover, and are called the electron transfer or E-clusters. The tricopper cluster is associated primarily with the PmoA and PmoC subunits.[39] The remaining copper ions are embedded within the PmoB subunit, with the dinuclear site at the membrane–aqueous interface[35] and the E-clusters residing in the water-exposed hydrophilic domain in the C-terminus of this subunit.[40] The functional form of the enzyme is the fully or partially reduced copper hydroxylase.[38]

With almost 15 copper ions occupying various copper sites in the protein, there are far too many coppers to discriminate among them by spectroscopy. To begin with, the copper ions appear to be fully reduced in the functional form of the enzyme. Fortunately, it is possible to access different redox states of the enzyme under varying conditions. For example, the "as-isolated" pMMO is obtained when the enzyme is purified under ambient aerobic conditions without methane, where only several of the copper centers become oxidized.[36] The remaining copper ions could also be fully oxidized by incubation of the enzyme with ferricyanide, and with hydrogen peroxide or dioxygen when the enzyme is modified by suicide substrates.[41]

While the electron paramagnetic resonance (EPR) spectrum of the "as-isolated" pMMO show a type 2 Cu(II) center, it was not possible to account for this spectrum on the basis of a type 2 Cu(II) site alone. EPR spectral simulation suggests that the spectrum is a superposition of components from two distinct set of copper ions, one corresponding to the type 2 Cu(II) site and the other a trinuclear Cu(II) cluster[41,42] (Figure 4.1). This conclusion has been confirmed by redox potentiometry/EPR titration experiments, in which the protein is titrated at different cell potentials in an electrochemical cell and the EPR spectroscopic features for each site could be distinguished and individually assigned.[39,41]

These EPR results have been also corroborated by Cu K-edge x-ray absorption spectroscopy measurements. Starting with the fully reduced enzyme, Nguyen et al.[43] has incubated the enzyme in the presence of dioxygen at various times and recorded the Cu K-edge spectrum. In this study, they have used the intensity of the unique near-edge (XANES) feature at 8984 eV as an indication of the level of Cu(I) in the sample.[44] According to these measurements, no more than 30%–40% of the copper ions in the protein, or about six copper ions, are reoxidized by this procedure even after prolonged incubation of the sample. Essentially the same level of oxidation of the copper ions is observed in the EPR of the enzyme "as-isolated," when we take into consideration that the two Cu(II) ions of the dinuclear site are EPR-silent due to antiferromagnetic coupling even if this site of the enzyme is oxidized. Thus, there is a thermodynamic/kinetic barrier to further oxidation of the remaining copper ions in the protein when the enzyme is turned over by dioxygen in the absence of methane. The six to nine E-cluster copper ions are inert to oxidation by dioxygen.[40] Moreover, they have very high redox potentials.[39]

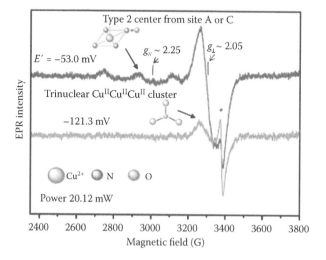

FIGURE 4.1 Redox potentiometry and EPR titration of C-cluster copper ions. Asterisks in the spectra at $g \approx 2.002$ denote signals originating from free radicals associated with dithionite and redox mediators. (From Chan, S.I. et al., *Angew. Chem. Int. Ed.*, 46, 1992, 2007.)

Taken together, the previously mentioned observations formed the basis for dividing the ~15 copper ions in the pMMO of *M. capsulatus* (Bath) into catalytic and electron transfer functions: six copper ions for dioxygen chemistry and methane hydroxylation, and the remaining six to nine copper ions as a buffer of reducing equivalents within the enzyme to replenish the electrons at the catalytic copper sites following the dioxygen chemistry and the alkane hydroxylation in the turnover.

4.2.3 Crystal Structure of pMMO: Disagreement over the Number and Type of Copper Sites

In the purification of pMMO, the strategy to start with membranes highly enriched in the enzyme, then solubilize the pMMO-enriched membranes in detergent, and fractionate the detergent–protein micelle particles by size-exclusion chromatography has been critical to our success in obtaining high-quality protein for in-depth biophysical and biochemical study.[36] Compared with conventional methods of membrane–protein isolation and purification by detergent solubilization, ammonium sulfate fractionation, and/or affinity chromatography, our approach is simple and essentially nondisruptive of the protein fold, and hence less prone to the loss of metal cofactors. It is now apparent that ammonium sulfate precipitation can leech out many of the copper ions from pMMO.[45] Although it is apparent that the addition of methanobactin can replenish these copper cofactors,[46] this copper-binding compound is not an integral component of the pMMO system. The pMMO with the full complement of copper ions does not require methanobactin for activity.[40]

The x-ray crystal structure of pMMO isolated and purified from *M. capsulatus* (Bath) was reported by the laboratory of Amy Rosenzweig in 2005.[35] According to this work, the protein crystallizes as a trimer of αβγ monomers with three copper ions and one zinc ion per monomer. A ribbon diagram of the structure is shown in Figure 4.2, with different colors used to depict the three subunits. PmoA and PmoC are mostly transmembrane, each with approximately six transmembrane segments. The N- and C-terminal subdomains of PmoB are exposed to the cytosol and are anchored at the water–membrane interface by two α-helical transmembrane segments inserted into the membrane. This protein architecture is consistent with the membrane topology predicted earlier by protease digestion of membrane fragments followed by mass fingerprinting of the peptides released by MALDI–TOF–MS.[37] The three copper ions and the zinc ion reported in the crystal structure are also shown in Figure 4.2, with the three copper ions at site A and site B in dark gray and the zinc ion at site C in light gray. Highlighted in the transmembrane domain of the crystal structure at site D is a cavity consisting of a hydrophilic cluster of potential metal-ligating residues, including His38, Met42, Met45, Asp47, Trp48, Asp49, and Glu100 from PmoA and Glu154 from PmoC. This "cluster of hydrophilic residues" has been discounted as a metal-binding site. However, the electrostatic energy of sustaining this cavity would be extremely high without metal counterions to balance the negative charges of the hydrophilic residues. The metal ions must have been stripped away from site D during the purification of the protein for crystallographic analysis.[45] Repeat of the purification procedures adopted by the Northwestern group indicates that as many as 12 of the ca. 15 copper ions are removed from the protein during the ammonium

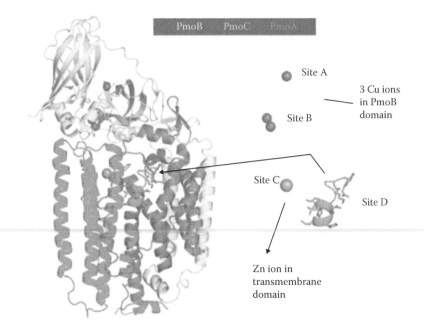

FIGURE 4.2 Architecture of pMMO isolated and purified from *Methylococcus capsulatus* (Bath) according to the x-ray crystal structure of Lieberman and Rosenzweig. (Adapted from Lieberman, R.L. and Rosenzweig, A.C., *Nature*, 434, 177, 2005.)

sulfate fractionation with concomitant loss of enzymatic activity[37] (Table 4.1). The protein preparation on which the x-ray crystal structure has been obtained exhibits essentially no pMMO activity.

4.2.4 CASE FOR THE TRICOPPER SITE

On the basis of the x-ray structure, Rosenzweig and coworkers have ruled out the possibility of a tricopper cluster in the enzyme.[35,47] There are now additional spectroscopic data to reinforce the idea that this tricopper cluster is indeed missing from the x-ray structure. First, the isotropic EPR signal attributed to the putative $Cu^{II}Cu^{II}Cu^{II}$ cluster has now been resolved from the type 2 Cu^{II} signal by redox potentiometry/ EPR.[39] As noted earlier, these copper sites have distinct redox potentials, and thus by titration of the protein at different cell potentials in an electrochemical cell, the spectroscopic features for each site could be distinguished and individually assigned. For example, the intensity of the signal at +53 mV (vs. SHE) corresponds to approximately two copper ions per protein monomer, as expected for contributions from one type 2 Cu^{II} site and one $Cu^{II}Cu^{II}Cu^{II}$ cluster (Figure 4.1). With increasingly more negative cell potentials, the type 2 Cu^{II} EPR decreases gradually in intensity, beginning at +18.3 mV. At −121.3 mV, the spectrum eventually gives way to an isotropic signal centered at $g = 2.05$ corresponding to the $Cu^{II}Cu^{II}Cu^{II}$ cluster. These results

TABLE 4.1
Repeat of Purification of pMMO

pMMO Sample, Status of Purification	Specific Activity (nmol/ [min · mg Protein])		Copper Content (Atoms/Protein)		Iron Content (Atoms/Protein)	
	This Work	Refs. 35, 47	This Work	Refs. 35, 47	This Work	Refs. 35, 47
Membranes	88.9	19.0	12–15	a	0.9	a
Membrane fragments solubilized in detergent	93.5	3.9	13.3	10.4	b	1.1
Purified detergent–protein complex (before $(NH_4)_2SO_4$ precipitation)	21.5	a	10.0	a	b	a
Purified detergent–protein complex (following $(NH_4)_2SO_4$ precipitation, detergent resolubilization, and anion exchange chromatography)	c	c	2.5	2.4	b	0.8

Note: According to procedures of Lieberman et al.[35,47] and comparison of specific activity and metal contents obtained with those previously reported for protein preparation used in crystallographic x-ray analysis.

a No data provided or available.

b Not detected.

c Inactive.

provide unequivocal evidence for a tricopper cluster in a preparation of pMMO containing the full complement of copper cofactors.

Second, the essentially identical tricopper cluster EPR signal has also been recorded for several model ferromagnetically coupled trinuclear $Cu^{II}Cu^{II}Cu^{II}$ clusters constructed with different trinucleating ligands.[48,49] These ligands can not only trap three Cu(I) ions, but the $Cu^{I}Cu^{I}Cu^{I}$ tricopper complexes can also mediate efficient O-atom transfer to hydrocarbon substrates upon dioxygen activation.[48–51]

Assuming that the overall fold of the protein structure has not been dramatically compromised by the loss of the copper ions, we have made an attempt to rebuild the coppers back into the protein scaffold.[39] This effort has led to a tricopper cluster at site D (Figure 4.3). The ligands to the copper atoms in the model are as follows: PmoC Glu154 and PmoA His38 for Cu1, PmoA Met42 and Asp47 for Cu2, and PmoA Asp49 and Glu100 for Cu3. The $Cu^{II}Cu^{II}Cu^{II}$ cluster is also capped with a μ-oxo (O^{2-}) to ensure electrical neutrality at the site. The coordinated ligands and the geometry of the cluster, including the Cu–Cu and Cu–O distances, are all reasonable, demonstrating the feasibility of pMMO to accommodate a tricopper cluster. The trinuclear $Cu^{II}Cu^{II}Cu^{II}$ structure modeled here would correspond to that of

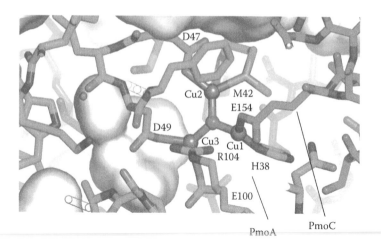

FIGURE 4.3 Tricopper cluster modeled as $Cu^{II}Cu^{II}Cu^{II}$ with capping "oxo" at site D of the crystal structure. The amino acid side chains coordinating the three copper ions are as follows: Cu1, PmoA His38 and PmoC Glu154; Cu2, PmoA Asp47 and Met42; Cu3, PmoA Asp49 and Glu100. (From Chan, S.I. et al., *Angew. Chem. Int. Ed.*, 46, 1992, 2007.)

the fully oxidized cluster after turnover by dioxygen in the absence of hydrocarbon substrate, the so-called dead-end species.[38]

4.2.5 Hydrocarbon Substrate–Binding Site

To account for the substrate specificity of pMMO, not to mention the unprecedented regiospecificity and stereoselectivity observed for the alkane oxidation and alkene epoxidation, there must be a unique hydrocarbon substrate-binding site in the enzyme in close proximity to the site of oxidation of the substrate. In a concerted mechanism of hydroxylation of a C–H bond, for example, the substrate must form a complex to allow the O-atom "harnessed" by the activated tricopper cluster to become inserted across the C–H bond in the transition state. This criterion has been exploited to locate the hydroxylation site in the enzyme.[32]

The possible binding site(s) for the hydrocarbon substrate within the three-subunit αβγ monomer of pMMO has been searched, and the information was used to home in on the oxidation site.[32] The search has been carried out using the binding sites prediction program Dockligand (LigandFit) on Discovery Studio 1.7 (Accelrys Software Inc.). On the basis of the alkane substrate specificity of the enzyme, we have selected pentane as the ligand to search for possible binding site(s) in the receptor protein. Apparently, all substrates of pMMO use the same binding pocket, including the suicide substrate acetylene. Pentane is chosen because it represents the substrate with the largest surface area and volume that could be oxidized by the enzyme.

To accomplish this search, the published crystal structure is subjected to the Global Protein Surface Survey (GPSS) analysis on the GPSS website (http://gpss.mcsg.anl.gov).

The GPSS PyMOL plugin is applied to the pMMO protein monomer (PDB ID WS_1YEW1), which is constructed from the PDB model of 1YEW. The calculations yield 122 CASTp surfaces; however, on the basis of the surface area and volume of the cavity required to accommodate pMMO substrates, the most probable site was determined to be the hydrophobic pocket previously identified near site D of the protein adjacent to the tricopper cluster that we have modeled into the crystal structure.[35,39]

The same pentane binding site is predicted by Dockligand (LigandFit) in support of the GPSS analysis. This putative hydrophobic pocket is sufficiently long to bind only C1–C5 hydrocarbons, and it is wide enough to accommodate only straight-chain alkanes. The hydrophobic "channel" is lined by the aromatic residues Trp48, Phe50, Trp51, and Trp54 of PmoA, and is "closed" at one end (Figure 4.4). Located at the open substrate entrance to the pocket is Gly46. With the "probing" hydrocarbon substrate fully inserted into the pocket, the putative tricopper cluster is directed at the secondary carbon of the substrate near the depth of the pocket, perfectly poised for O-atom transfer to this secondary carbon when the tricopper cluster is activated by dioxygen. Thus, this structural arrangement would account for the regiospecificity observed for the hydroxylation of this substrate. Experimentally, only (2R)-pentan-2-ol and (2S)-pentan-2-ol are observed (ee = 80%).[31]

With the hydrophobic substrate-binding site located within 6 Å of site D in the protein structure, it is evident that the tricopper site is the location of the catalytic chemistry in pMMO. Sites A, B, and C are, respectively, ~40, 26, and 13 Å away from the hydrophobic pocket identified here, too far for direct concerted O-atom insertion into a C–H bond.

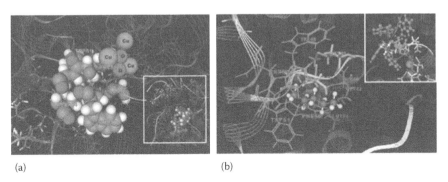

(a)   (b)

FIGURE 4.4 (a) Hydrophobic pocket adjacent to the site of the putative tricopper cluster (site D), with the amino acid residues forming the cavity highlighted by CPK model (inset, the most probable CASTp surface calculated from the GPSS program). (b) Discovery Studio modeling of the pentane molecule within the hydrophobic pocket (lined by Gly46, Trp48, Phe50, Trp51, and Trp54), showing an orientation of the hydrocarbon (side view) with the H_R atom of the C2 carbon directed at site D of pMMO (inset: an end view of the site showing an orientation of the activated oxygen of the tricopper cluster directed at C2–H_R of the pentane). The amino acid residues associated with the hydrophobic cavity are denoted in dark gray, and those associated with the tricopper cluster in site D are denoted in white. (From Ng, K.-Y. et al., *Chem. Biol. Chem.*, 9, 1116, 2008.)

4.2.6   ENGINEERING THE CATALYTIC SITE: TRICOPPER–PEPTIDE COMPLEX

To underscore the significance of the presence of the putative tricopper cluster in pMMO, a tricopper–peptide complex based on the HIHAMLTMGDWD fragment of PmoA that lines the empty hydrophilic cavity at site D site in the crystal structure of pMMO purified from *M. capsulatus* (Bath) has been prepared and shown to be capable of mediating facile methane oxidation at room temperature as well.[49]

4.2.6.1   Formation of the Tricopper–Peptide Complex

The HIHAMLTMGDWD peptide binds three copper ions to form $Cu^I Cu^I Cu^I$– and $Cu^{II} Cu^{II} Cu^{II}$–peptide complexes in the presence of excess acetate or chloride.[49] Both complexes are insoluble in aqueous buffer. The $Cu^{II} Cu^{II} Cu^{II}$–peptide complex forms a blue precipitate, and the precipitate for the $Cu^I Cu^I Cu^I$–peptide complex is white. The copper contents are 3.00 ± 0.05 and 3.08 ± 0.09 Cu/peptide for the $Cu^{II} Cu^{II} Cu^{II}$– and $Cu^I Cu^I Cu^I$–peptide complexes, respectively, as determined by inductively coupled plasma optical emission spectrometry. Formation of the $Cu^I Cu^I Cu^I$– and $Cu^{II} Cu^{II} Cu^{II}$–peptide complexes is confirmed by MS. The $Cu^{II} Cu^{II} Cu^{II}$–peptide complex exhibits the featureless isotropic 77K EPR signal near $g \sim 2.1$ (Figure 4.5), as previously reported for pMMO-enriched membranes and pMMO isolated from *M. capsulatus* (Bath),[41] providing a direct link between the tricopper–peptide complex with the putative tricopper cluster implicated at site D in the enzyme.

4.2.6.2   Ligand Structure of the Tricopper–Peptide Complex

Structural elucidation of the $Cu^{II} Cu^{II} Cu^{II}$–peptide complex is provided by the ligand structure deduced from K-edge extended x-ray absorption fine structure (EXAFS).[49] To obtain high-quality data, the peptide is covalently linked to the C-terminus of the TEL-SAM protein. EXAFS of the $Cu^{II} Cu^{II} Cu^{II}$–protein complex formed by treating

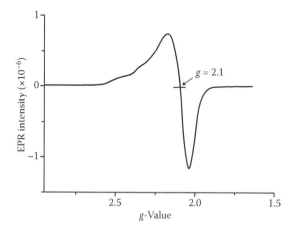

FIGURE 4.5   77K EPR spectrum of the $Cu^{II} Cu^{II} Cu^{II}$–peptide complex. The blue precipitate is dispersed in KCl (1% by weight) for EPR measurements. (From Chan, S.I. et al., *Angew. Chem. Intl. Ed.*, 52, 3731, 2013.)

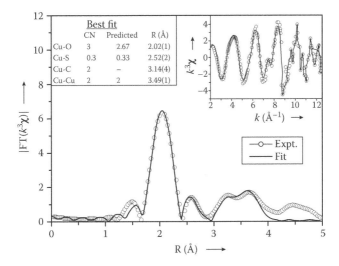

FIGURE 4.6 Cu K-edge EXAFS spectrum for the TEL-SAM-PmoA(38–49) fusion protein with the peptide sequence HIHAMLTMGDWDPD. Left inset: best fit of the ligand structure for the tricopper cluster. Right inset: fitted $k^3\chi(k)$ data. (From Chan, S.I. et al., *Angew. Chem. Intl. Ed.*, 52, 3731, 2013.)

the TEL-SAM-PmoA(38–49) fusion protein with $Cu(OAc)_2$ is summarized in Figure 4.6. The average first-shell (O/N), S, and Cu coordination numbers of the three copper ions are 3, 0.3, and 2, respectively, in excellent agreement with the ligand structure of the tricopper cluster modeled earlier into site D of the enzyme (Table 4.2). The fitting required a capping "oxo," which would account for the ferromagnetic coupling among the three Cu^{II} ions inferred for the cluster from EPR. The assignment of the observed EPR to the $S_T = 3/2$ ground state of a triad of ferromagnetically coupled Cu^{II} spins is also confirmed by magnetic susceptibility measurements on the $Cu^{II}Cu^{II}Cu^{II}$–peptide complex.[49]

The EXAFS of the $Cu^{I}Cu^{I}Cu^{I}$–peptide complex (without the chimeric TEL-SAM protein) indicates that the structure of the $Cu^{I}Cu^{I}Cu^{I}$ triad in the complex is similar to that of the oxidized cluster, but without the capping oxo. The Cu–Cu distance increases to ca. 4 Å from 3.5 Å in the $Cu^{II}Cu^{II}Cu^{II}$–peptide complex, as expected.

TABLE 4.2

Average Coordination Numbers (CN) Predicted for Ligand Structure of Site D as Modeled in Figure 4.3

		O, N	S	Cu
Cu K-edge EXAFS	Cu1	3	0	2
predictions	Cu2	2	1	2
	Cu3	3	0	2
	Average	2.67	0.33	2

The peptide **HIHAMLTMGDWD** provides the bulk of the potential metal-ligating residues of the PmoA fragment that have been used to assemble the tricopper cluster into site D of the pMMO crystal structure. The residues highlighted in bold in the peptide sequence correspond (from left to right) to His38, Met42, Asp47, and Asp49 of PmoA.[39] These residues are found to be absolutely required for the formation of the tricopper–cluster complex according to studies on peptide mutants with single amino acid replacements. In the tricopper cluster modeled into the protein, the remaining ligands are provided by the carboxylate side chains of Glu100 from PmoA and of Glu154 from PmoC. In the biomimetic tricopper–peptide complex, these carboxylates are replaced by exogenous anionic acetate or chloride in the peptide/copper solution during the formation of the tricopper–peptide complex.

4.2.6.3   Reactivity of the $Cu^I Cu^I Cu^I$ Tricopper–Peptide Complex

X-ray absorption spectroscopy of the solid $Cu^I Cu^I Cu^I$–peptide complex confirms that the copper ions in the complex are fully reduced.[49] Cu XANES and EPR reveal that the copper ions in the $Cu^I Cu^I Cu^I$–peptide complex are rapidly reoxidized in the presence of air or dioxygen, as observed for the putative tricopper cluster in pMMO. Moreover, upon activation of the $Cu^I Cu^I Cu^I$–peptide complex by dioxygen in the presence of propylene or methane, rapid oxidation of these substrates is observed. Methane is converted to methanol, and propylene is epoxidated to propylene oxide. Since the $Cu^I Cu^I Cu^I$–peptide complex is insoluble in aqueous buffer, the substrate oxidation is carried out with the $Cu^I Cu^I Cu^I$–peptide complex encapsulated in mesoporous carbon. Thus, only a single-turnover experiment is performed. Nevertheless, the oxidation chemistry is facile and stoichiometric. These experiments clearly demonstrate that the $Cu^I Cu^I Cu^I$–peptide complex is capable of mediating facile O-atom transfer to organic substrates upon activation by dioxygen.[49]

4.3   BIOMIMETIC MODELS OF THE TRICOPPER CLUSTER SITE

For the simplest form of methane oxidation, the reaction can be written formally as

$$CH_4 + O_2 \rightarrow CH_3OH + <O>$$

When the oxidation is mediated by a monooxygenase, as in pMMO and sMMO, the $<O>$ that is not incorporated into the substrate picks up two reducing equivalents and two protons to form a molecule of H_2O. As noted earlier, the insertion of an O-atom across the C–H bond is a difficult reaction to begin with because of the high bond energy of the C–H bond. However, there are additional challenges associated with the design of a catalyst for the conversion of methane to methanol. First, in practice, the product CH_3OH is prone to further oxidation to other products. In fact, in the bacterium, CH_3OH is readily converted to formaldehyde (CH_2O) by another enzyme, methanol dehydrogenase.[2,52] This oxidation reaction is more favorable than for the original substrate CH_4, which has thermodynamically stronger and kinetically more inert C–H bonds. In light of this, the controlled oxidation of methane to methanol is challenging to say the least. Second, both CH_4 and CH_3OH exist as singlet electronic ground states; thus, when O_2, a triplet in its ground state, is employed as the terminal oxidant, the reaction is spin forbidden. Thus, the conversion is sluggish unless a metal center or cluster of metal ions is used to activate the O_2 in the first place.

In recent years, there has been considerable interest in developing copper complexes to study their reactivity toward dioxygen and direct subsequent controlled oxidation of organic substrates.[53] Indeed, new complexes with different ligands can give new insights into the mechanism of oxygenation and can even help in identifying and/or trapping intermediates. Both dicopper and tricopper complexes have been synthesized and characterized, their dioxygen chemistry studied, and the dioxygen adducts have been evaluated for their propensity toward oxygenation of organic substrates. By and large, these complexes have shown low reactivity toward hydrocarbon oxidation.[54,55] Although a fragment of the PmoB subunit of pMMO containing the dinuclear site has been shown to mediate methane oxidation, the reactivity is very low.[56] Thus, the unusual reactivity of the tricopper site toward methane oxidation, as manifested by the tricopper–peptide complex derived from site D of pMMO discussed earlier, must reside in the special chemical and/or electronic structure of the dioxygen-activated tricopper cluster.[49]

4.3.1 DESIGNING A LIGAND SCAFFOLD TO SUPPORT A TRIAD OF CU[I] IONS

To mimic the tricopper site in pMMO, we have deployed the following criteria in the design of the ligand scaffold to support the triad of copper ions[48]:

- The ligand scaffold must be able to trap three Cu(I) ions.
- The Cu[I] ions must be coordinately unsaturated, except for solvent molecules that might be weakly coordinated.
- At least two of the three Cu[I] ions must be in a geometric juxtaposition so that they can participate in the coordination and activation of a dioxygen molecule.
- Adjacent Cu[II] ions must be strongly antiferromagnetically coupled after the cluster is activated by dioxygen to maintain the cluster in an overall singlet state.

The first generation of $[Cu^ICu^ICu^I(L)]^{1+}$ tricopper complexes are based on the multidentate ligands (L) 3,3′-(1,4-diazepane-1,4-diyl)bis(1-{[2-(dimethylamino) ethyl] (methyl)amino}propan-2-ol) (**7-Me**) and 3,3′-(1,4-diazepane-1,4-diyl)bis(1-{[2-(diethylamino)ethyl](ethyl)amino}propan-2-ol) (**7-Et**)[48] (Scheme 4.4). Upon activation by dioxygen, these $[Cu^ICu^ICu^I(L)]^{1+}$ tricopper complexes are demonstrated to facilitate facile O-atom insertion into the C–H bond of acetonitrile (MeCN) and the central C–C bond in benzil and 2,3-butanedione under mild conditions (Scheme 4.5). Experiments with isotopically labeled oxygen ($^{18}O_2$) together with electrospray ionization–MS analysis of the product distributions confirm that the reaction proceeds according to a mechanism that involves concerted insertion of a "singlet oxene." Thus, these model tricopper complexes perform the same kind of O-atom transfer chemistry as the tricopper cluster site of pMMO, except that hydrocarbons are not substrates.

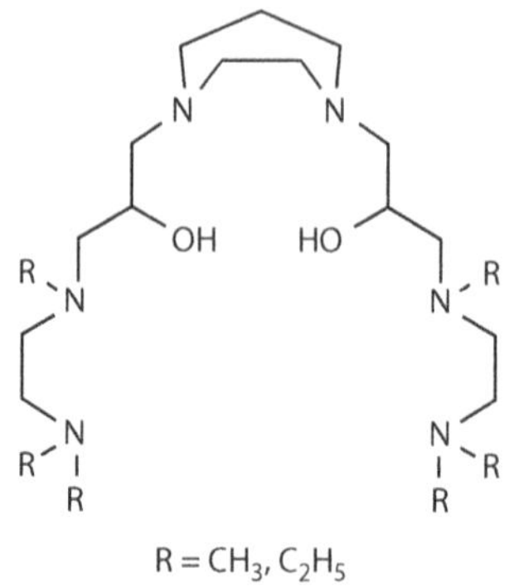

R = CH$_3$, C$_2$H$_5$

SCHEME 4.4 Ligands **7-Me** and **7-Et**.

The ligand environment of a given tricopper complex affects the formation and/or stability of the activated tricopper complex, as well as the approach of a potential substrate to the active site of the catalyst. For productive O-atom transfer, the substrate must form at least a weak transient complex with the catalyst so that the appropriate transition state can be reached. For this reason, the tricopper complexes

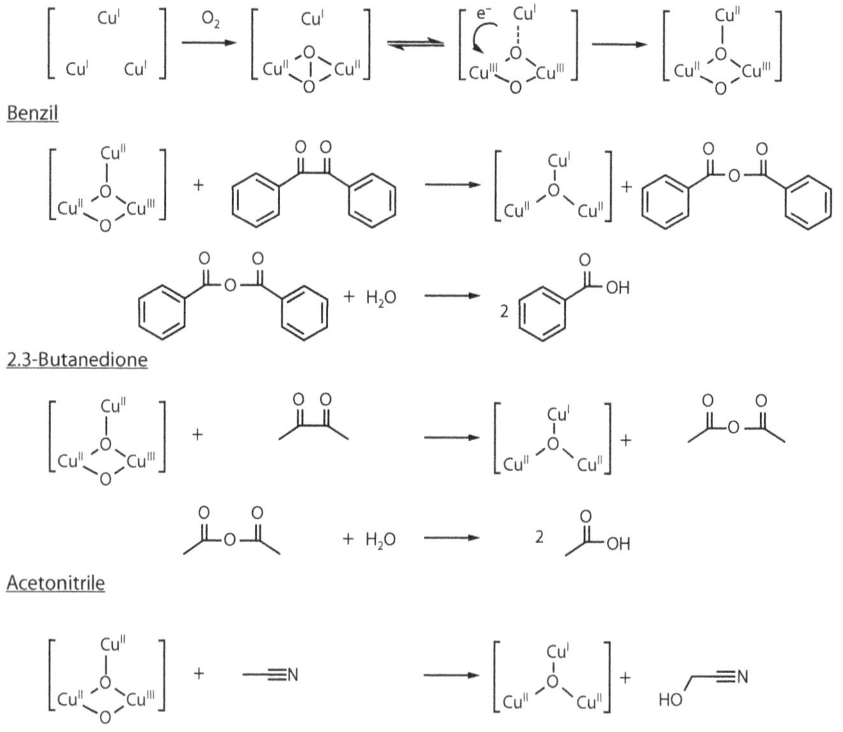

SCHEME 4.5 O-atom transfers mediated by the activated tricopper complex to various organic substrates mediated by **7-Me** and **7-Et**.

assembled by **7-Met** and **7-Et** are designed to allow the β-dicarbonyls of benzil or 2,3-butanedione to interact with the two copper ions at the base of the isosceles triad to facilitate O-atom insertion across the central C–C bond in these β-diketones. Similarly, CH_3CN coordinates weakly with these copper ions, which allows the methyl group of the solvent molecule to approach the "harnessed" oxene provided that there is no steric hindrance from the alkyl substituents attached to the base of the tricopper complex.

To accommodate the oxidation of hydrocarbon substrates, a second generation of ligands have been designed and synthesized with different donor atoms (N, O, and S) and steric environments around the active site of the tricopper complexes (Scheme 4.6).[49–51] The apical copper in the tricopper complexes with the new ligands **7-Dipy**, **7-Ethppz**, **7-Mehppz**, **7-Bn**, **7-Thio**, and **7-Morph** shares the same environment as **7-Met** and **7-Et**, with two neutral amines and two hydroxyl groups. However, the steric environment and molecular surface surrounding the two basal coppers are designed to suppress the coordination of CH_3CN (solvent) and to promote van der Waals interactions with hydrocarbon substrates. The ligands **7-Dipy**, **7-Ethppz**, **7-Mehppz**, and **7-Bn** provide amine nitrogens along with the bridging hydroxy oxygens for coordination of the basal copper ions. In the case of **7-Thio** and **7-Morph**, there are coordinating sulfur or oxygen atoms from thioethers or ethers, in addition to the amine nitrogens and bridging hydroxy oxygens. In any case, all these various ligand scaffolds can trap three Cu^I ions to form an isosceles triad without

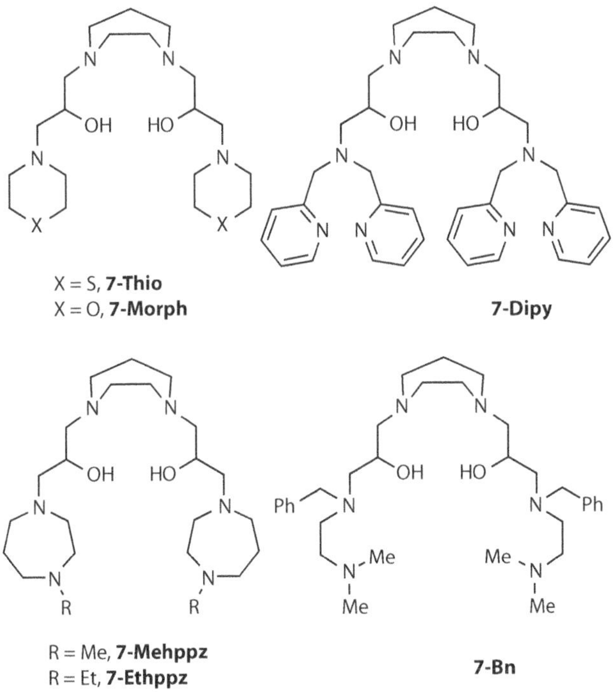

X = S, **7-Thio**
X = O, **7-Morph**

7-Dipy

R = Me, **7-Mehppz**
R = Et, **7-Ethppz**

7-Bn

SCHEME 4.6 Ligands **7-Thio**, **7-Dipy**, **7-Morph**, **7-Mehppz**, **7-Ethppz**, and **7-Bn**.

the metal sites becoming coordinately saturated, except for solvent molecules that might be readily displaced during activation of the complex by dioxygen. Indeed, the $[Cu^ICu^ICu^I(L)]^{1+}$ tricopper complexes formed with these ligands are capable of mediating facile oxidation of hydrocarbons at room temperature.[49-51]

4.3.2   DIOXYGEN CHEMISTRY OF THE TRICOPPER COMPLEXES

The $[Cu^ICu^ICu^I(L)]^{1+}$ systems are expected to react with dioxygen to form a dioxygen adduct at the base of the copper triad in the structure. Rapid intracluster electron transfer from the remaining Cu^I within the cluster (at the apex of the triad) to the high-potential $[Cu^{III}(\mu-O)_2Cu^{III}]$ or the $[Cu^{II}(\mu-\eta^2:\eta^2\text{-peroxyl})Cu^{II}]$ center at the base of the structure would yield the putative $[Cu^{II}Cu^{II}(\mu-O)_2Cu^{III}(L)]^{1+}$ species (Scheme 4.7). To elucidate the factors that might influence this outcome of the activation of the tricopper complexes by dioxygen, we have examined the dioxygen chemistry of these systems at low temperatures ($-80°C$, $-55°C$, and $-35°C$), under conditions in which the O-atom transfer chemistry is quenched.[57]

The experimental observations are indeed consistent with the above scenario. However, the outcome of the experiments indicates that the putative $[Cu^{II}Cu^{II}(\mu-O)_2Cu^{III}(L)]^{1+}$ species that is formed is readily aborted by a reducing equivalent in the solution to give the $[Cu^{II}Cu^{II}(\mu-O)Cu^{II}(L)]^{2+}$ species. At low temperatures, the dioxygen chemistry is sufficiently slow that before the bulk of the starting $[Cu^ICu^ICu^I(L)]^{1+}$ complexes have been consumed, there is an excess of the starting complex, which can quickly reduce a $[Cu^{II}Cu^{II}(\mu-O)_2Cu^{III}(L)]^{1+}$ species already formed. This intercomplex electron transfer culminates in the production of two final Cu(II) species: (i) the blue $[Cu^{II}Cu^{II}(\mu-O)Cu^{II}(L)]^{2+}$ complex and (ii) and the green $[Cu^{II}Cu^{II}(\mu-\eta^2:\eta^2\text{-peroxo})Cu^{II}(L)]^{2+}$ species, possibly in equilibrium with the colorless $[Cu^{II}Cu^{III}(\mu-O)_2Cu^{III}(L)]^{2+}$. The species (ii) are formed when the partially oxidized $[Cu^{II}Cu^ICu^I(L)]^{2+}$ formed following the donation of the reducing equivalent to the original $[Cu^{II}Cu^{II}(\mu-O)_2Cu^{III}(L)]^{1+}$ species subsequently forms an adduct with dioxygen in the solution (Scheme 4.8). These conclusions are corroborated by spectroscopic and other physical characterization of the oxygenated reaction medium as well as isolated products. Two lines of evidence have allowed us to confirm that the green species is $[Cu^{II}Cu^{II}(\mu-\eta^2:\eta^2\text{-peroxo})Cu^{II}(L)]^{2+}$: (i) the reactivity of the green species with H_2O in MeCN to give H_2O_2, which transforms the green dioxygen adduct to a blue solution, and (ii) the propensity of the green species to photoreduction by extended exposure to the x-ray beam in the x-ray absorption experiment under ambient conditions.[57]

We have followed the time course of the dioxygen chemistry of the $[Cu^ICu^ICu^I(L)]^{1+}$ complexes under ambient conditions by quick manual mixing experiments.[57]

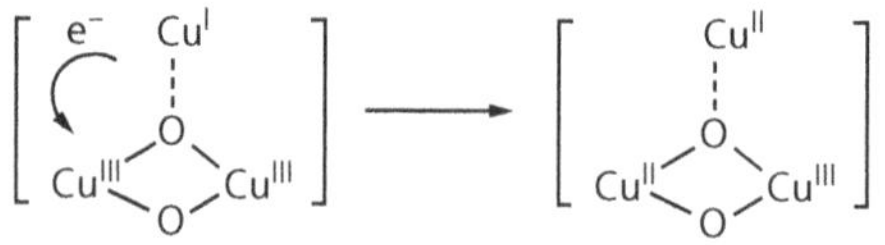

SCHEME 4.7 Dioxygen activation of the $[Cu^ICu^ICu^I(L)]^{1+}$ complex to form the $[Cu^{II}Cu^{II}(\mu-O)_2Cu^{III}(L)]^{1+}$ species.

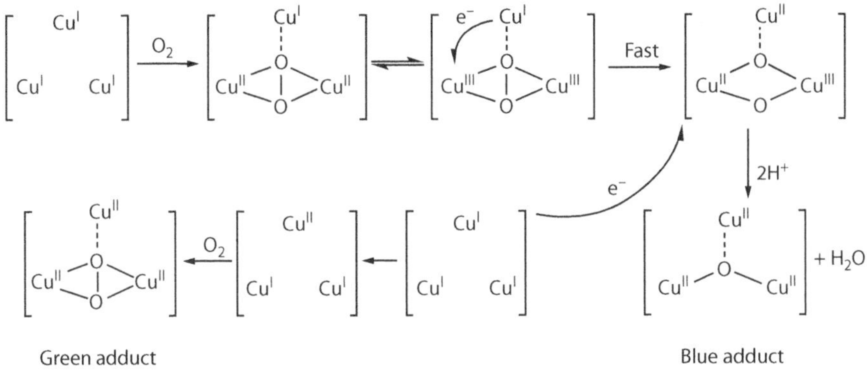

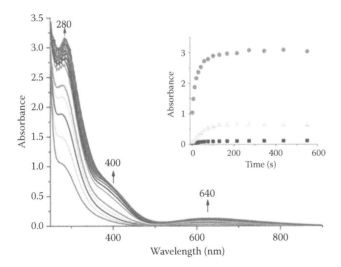

SCHEME 4.8 Outcome of the dioxygen chemistry of the $[Cu^ICu^ICu^I(L)]^{1+}$ complex at low temperatures.

Changes in the absorption spectra are followed by diode array spectrophotometry. The results for the $[Cu^ICu^ICu^I(\textbf{7-N-Meppz})]^{1+}$ complex are summarized in Figure 4.7. Comparison of the time course of the absorbance at 280, 400, and 640 nm indicates that the blue species is formed earlier than the green species, with the onset of the green adduct (400 nm) delayed by some 30 s relative to the blue species. A study of the concentration dependence reveals that the kinetics is first order with respect to $[Cu^ICu^ICu^I(\textbf{7-N-Meppz})]^{1+}$. The pseudo-first-order rate constants (k_1) are 4.68 (0.31) × 10^{-2}, 1.52 (0.12) × 10^{-2}, and 3.87 (0.38) × 10^{-2} s^{-1} for the 280, 400, and 640 nm absorptions, respectively. Since $k_1 = k_2[O_2]$, the corresponding second-order rate

FIGURE 4.7 Rapid mixing of an anaerobic solution of the $[(Cu^ICu^ICu^I)(\textbf{7-N-Meppz})]^{1+}$ complex with MeCN saturated with dioxygen followed by diode array spectrophotometry. (From Maji, S. et al., *Chem. Eur. J.*, 18, 3955, 2012.)

constants (k_2) are 5.77, 1.87, and 4.77 M^{-1} s^{-1}, respectively. The solubility of oxygen in MeCN at room temperature is 8.1 mM.[58] From these rate constants, it is evident that the formation of the green adduct occurs after the formation of the blue species. A similar study with the $[Cu^ICu^ICu^I(\textbf{7-Me})]^{1+}$ complex reveals that the dioxygen chemistry is significantly faster than that of the $[(Cu^ICu^ICu^I)(\textbf{7-NMeppz})]^{1+}$ system under ambient conditions and, in fact, only $[Cu^{II}Cu^{II}(\mu\text{-}O)Cu^{II}(\textbf{7-Me})]^{2+}$ is identified in the reaction mixture. Consistent with this observation, the $[Cu^ICu^ICu^I(\textbf{7-Me})]^{1+}$ system is capable of mediating facile O-atom transfer to acetonitrile to produce glycolonitrile upon dioxygen activation at room temperature.[48] In contrast, the $[(Cu^ICu^ICu^I)(\textbf{7-NMeppz})]^{1+}$ complex does not seem to mediate O-atom transfer to oxidize the solvent under the same conditions. These results underscore the importance of designing a tricopper system with a sufficiently rapid dioxygen chemistry to mitigate abortive processes so that the active $[Cu^{II}Cu^{II}(\mu\text{-}O)_2Cu^{III}(\textbf{L})]^{1+}$ species can be maintained at sufficiently high levels to sustain efficient O-atom transfer chemistry.

4.3.3 OXIDATION OF METHANE MEDIATED BY TRICOPPER COMPLEXES: SINGLE TURNOVER VERSUS CATALYTIC TURNOVERS AND PRODUCTIVE VERSUS ABORTIVE CYCLING

The oxidation of CH_4 mediated by the tricopper complex $[Cu^ICu^ICu^I(\textbf{7-N-Etppz})]^{1+}$ in acetonitrile (MeCN), where **7-N-Etppz** corresponds to the ligand 3,3′-(1,4-diazepane-1,4-diyl)bis[1-(4-ethylpiperazine-1-yl)propan-2-ol], is summarized in Figure 4.8.[49] A single turnover (turnover number [TON] = 0.92) is obtained when this $[Cu^ICu^ICu^I(\textbf{L})]^{1+}$ complex is activated by excess dioxygen in the presence of excess CH_4. The chemistry is completed within 10 min, clearly indicating that the oxidation is very rapid. In accordance with the single turnover, the kinetics of the overall process is pseudo

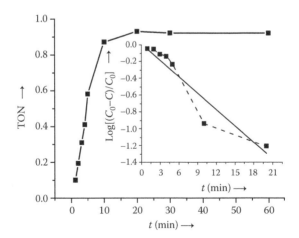

FIGURE 4.8 Time course of methane oxidation; inset: pseudo-first-order kinetic plot with rate constant $k_1 = 0.065$ min^{-1} (C_0 = initial concentration of the fully reduced tricopper complex; C = concentration of methanol produced at any given time), with the best straight-line fit to the data. (From Nagababu, P. et al., *Adv. Synth. Catal.*, 354, 3275, 2012.)

first order with respect to the concentration of the fully reduced tricopper complex with a rate constant $k_1 = 0.065$ min^{-1}. If we assume that the kinetics is limited by the dioxygen activation of the [Cu^{I}Cu^{I}Cu^I(**L**)]$^{1+}$ complex with the subsequent O-atom transfer to the substrate molecule being rapid, then $k_1 = k_2$ [O$_2$]$_0$, and from the solubility of dioxygen in MeCN at 25°C (8.1 mM),[58] we obtain the bimolecular rate constant k_2 of 1.33×10^{-1} M^{-1} s^{-1} for the dioxygen activation of the [Cu^{I}Cu^{I}Cu^I(**7-N-Etppz**)]$^{1+}$ complex. This second-order rate constant is similar to values that we have previously determined for the dioxygen activation of other model tricopper clusters at room temperature.[48,57]

The process can be rendered catalytic by adding the appropriate amounts of H$_2$O$_2$ to regenerate the "spent" catalyst after O-atom transfer from the activated tricopper complex to CH$_4$.[49] In these multiple turnover experiments, the [Cu^{I}Cu^{I}Cu^I(**7-N-Etppz**)]$^{1+}$ catalyst is activated by O$_2$ as in the single-turnover experiment described previously; however, the spent catalyst is regenerated by two-electron reduction by a molecule of H$_2$O$_2$ (Figure 4.9). Since the TON peaks at ~6 when the turnover is initiated with 20 equiv. H$_2$O$_2$, it is evident that abortive cycling begins to kick in when the steady-state concentration of the H$_2$O$_2$ concentration exceeds ~10 equiv. When the steady-state H$_2$O$_2$ concentration is above this level, reductive abortion of the activated catalyst becomes competitive with the O-atom transfer to methane to produce methanol.

The catalytic reaction is perceived to proceed by way of the putative [Cu^{II}Cu^{II}(μ-O)$_2$Cu^{III}(**7-N-Etppz**)]$^{1+}$, which is believed to be the active oxidizing species. Space-filling and ball-and-stick models of the structure of [Cu^{II}Cu^{II}(μ-O)$_2$Cu^{III}(**7-N-Etppz**)]$^{1+}$ intermediate optimized by the semiempirical PM6 method are shown in Figure 4.10, together with a depiction of the singlet oxene transfer across one of the C–H bonds in methane. This transient species transfers the singlet oxene to the substrate in the transition state when it forms a bimolecular complex with the substrate.

However, because of the high redox potential of the [Cu^{II}Cu^{II}(μ-O)$_2$Cu^{III}(**7-N-Etppz**)]$^{1+}$, it can also be readily deactivated by reduction with a molecule of H$_2$O$_2$ present in the reaction mixture. This is an abortive process that competes with

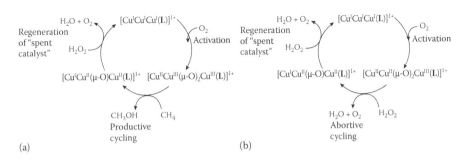

FIGURE 4.9 (a) Productive cycling and (b) abortive cycling in the oxidation of methane by O$_2$, mediated by the Cu^{I}Cu^{I}Cu^I(**7-N-Etppz**)$^{1+}$ complex in the presence of H$_2$O$_2$ as the sacrificial reductant. BDE, bond dissociation energy. (From Chan, S.I. et al., *Angew. Chem. Intl. Ed.*, 52, 3731, 2013.)

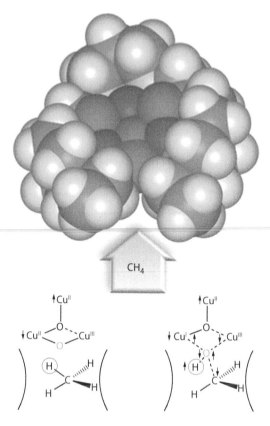

FIGURE 4.10   Space-filling model of the optimized structure of [Cu^{II}Cu^{II}(μ-O)$_2$Cu^{III}(**7-N-Etppz**)]$^{1+}$ showing the funnel-like opening or cleft at the bottom for a hydrocarbon substrate to access the "hot" oxene group. Formation of the transition-state complex during facile singlet oxene transfer to methane from a dioxygen-activated tricopper complex is shown at the bottom. (From Chan, S.I. et al., *Angew. Chem. Intl. Ed.*, 52, 3731, 2013.)

productive transfer of the harnessed singlet oxene to a substrate molecule to form product. These competing processes can be described by the following two chemical equations:

"O-atom" transfer from the activated catalyst to substrate:

$$[Cu^{II}Cu^{II}(\mu\text{-}O)_2Cu^{III}(L)]^{1+} + S \xrightarrow[\text{fast}]{k_2 \text{ or } k_{OT}} [Cu^{I}Cu^{II}(\mu\text{-}O)Cu^{II}(L)]^{1+} + SO$$

Abortion of the activated catalyst:

$$[Cu^{II}Cu^{II}(\mu\text{-}O)_2Cu^{III}(L)]^{1+} + H_2O_2 \xrightarrow{k_{abortive}} [Cu^{I}Cu^{II}(\mu\text{-}O)Cu^{II}(L)]^{1+} + H_2O + O_2$$

Thus, unless k_2 [substrate] = k_{OT} [substrate] >> $k_{abortive}$ [H_2O_2], abortive cycling will compete effectively with productive cycling. With the limited solubility of CH_4 in MeCN and the lower k_2 expected for CH_4 because of its high bond-dissociation energy relative to the other substrates that we have studied, e.g., cyclohexane, benzene, and styrene (Section 4.3.4), this condition is difficult to fulfill in our methane oxidation experiments.

With our catalyst, the regeneration of the spent catalyst is the rate-limiting step, and the TOFs of the catalysts are found to depend linearly on the H_2O_2 concentration with a second-order rate constant of ca. 2×10^{-2} M^{-1} s^{-1} regardless of the catalyst or the substrate.

Regeneration of the spent catalyst:

$$[Cu^I Cu^{II} (\mu\text{-O})_2 Cu^{II} (L)]^{1+} + H_2O_2 \xrightarrow[\text{rate-limiting}]{k_3} [Cu^I Cu^I Cu^I (L)]^{1+} + H_2O + O_2$$

The productive and abortive pathways consume H_2O_2 differently. In the oxidation of hydrocarbons, 3 equiv. H_2O_2 will be consumed during a productive cycle, two molecules of H_2O_2 to activate the tricopper complex and another H_2O_2 molecule to regenerate the spent catalyst. An abortive cycle will consume an additional H_2O_2 molecule. In effect, the tricopper complex is acting merely as a catalyst to disproportionate the four molecules of H_2O_2 to produce two O_2 and four H_2O. In practice, substrate oxidation prevails at high substrate concentrations relative to the H_2O_2 concentration present in the reaction mixture, and abortive cycling taking over at high H_2O_2 and low substrate concentrations. When productive cycling is obtained, the TON is effectively controlled by the amount of H_2O_2 used to drive the catalytic turnover.[49–51]

4.3.4   OTHER HYDROCARBON SUBSTRATES

For the oxidation of CH_4 by the homogeneous catalytic process described in the previous Section 4.3.3, the rate of O-atom transfer is limited by the relatively low solubility of CH_4 in MeCN under ambient conditions of temperature and pressure. Because of the low concentration of the substrate, the abortive process kicks in even under the fairly low concentration of the H_2O_2 used to drive the catalysis. This scenario is not obtained for liquid substrates like cyclohexane, benzene, and styrene (C_6H_{12}, C_6H_6 and C_8H_8), which are sufficiently soluble in MeCN. With these substrates, the substrate concentration can be as high as 500 equiv. (relative to the concentration of the catalyst), if not higher. Under these concentrations, the bimolecular O-atom transfer reaction competes favorably against abortive cycling. In fact, there is no deactivation of the active $[Cu^{II} Cu^{II} (\mu\text{-O})_2 Cu^{III} (L)]^{1+}$ catalytic species in these experiments, and the TONs are limited by the amount of H_2O_2 used to drive the catalysis.

Figure 4.11 compares the catalytic oxidation of cyclohexane, benzene, and styrene mediated by the tricopper cluster complexes supported by the series of multidentate ligands (**L = 7-Dipy, 7-Ethppz, 7-Mehppz, 7-Bn, 7-Thio,** and **7-Morph**) that have been developed for hydrocarbons.[50] These complexes are clearly capable of mediating facile oxygen atom transfer to these aliphatic and aromatic hydrocarbons

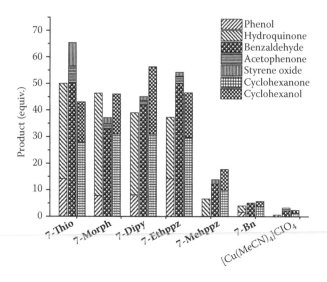

FIGURE 4.11 Product analysis after a 1-h catalytic oxidation of C_6H_6, C_8H_8, and C_6H_{12} by H_2O_2 in CH_3CN mediated by various tricopper complexes at room temperature. The products are given in equivalents produced. Reaction mixture: 500 equiv. substrate, 200 equiv. H_2O_2, and 1 equiv. (24 mmol) $[Cu^ICu^ICu^I(L)]^{1+}$. As a control, with the complex $[Cu(CH_3CN)_4]$ ClO_4 as the catalyst, only very low levels of substrate oxidation are obtained, if any. (From Nagababu, P. et al., *Adv. Synth. Catal.*, 354, 3275, 2012.)

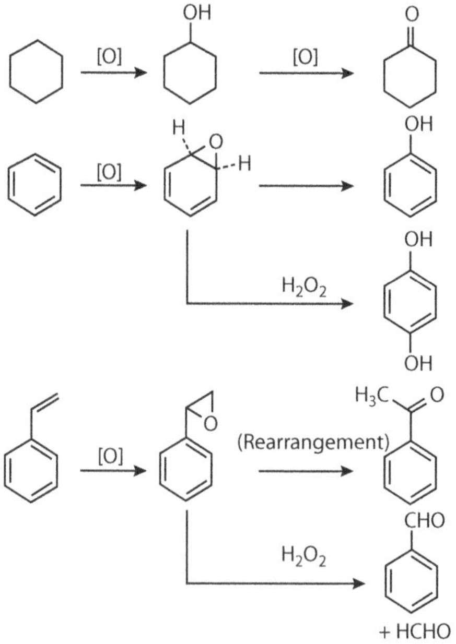

SCHEME 4.9 Oxidation of C_6H_{12}, C_6H_6, and C_8H_8 by H_2O_2 mediated by the tricopper complexes constructed using the ligands **L = 7-Dipy, 7-Ethppz, 7-Mehppz, 7-Bn, 7-Thio,** and **7-Morph.** (From Nagababu, P. et al., *Adv. Synth. Catal.*, 354, 3275, 2012.)

upon activation by hydrogen peroxide at room temperature. The processes are catalytic with high TOFs, efficiently oxidizing the substrates to their corresponding alcohols and ketones in moderate to high yields (Scheme 4.9). The catalysts are robust with TON limited only by the availability of hydrogen peroxide used to drive the catalytic turnover. The TON is independent of the substrate concentration, and the TOF depends linearly on the hydrogen peroxide concentration when the oxidation of the substrate mediated by the activated tricopper complex is rapid. At low substrate concentrations, however, the catalytic system also exhibits abortive cycling resulting from competing reduction of the activated catalyst by hydrogen peroxide, as noted for CH_4.

4.4 CHEMICAL MECHANISMS OF METHANE OXIDATION

Oxidation reactions involving O–O bond cleavage with concomitant formation of C–O and O–H bonds are energetically very favorable. However, even these highly exothermic reactions are kinetically very slow at ambient temperatures. The unique electronic structure of dioxygen keeps at bay the strong thermodynamic driving force. Otherwise, life would not exist under aerobic conditions, as organic compounds would be oxidized spontaneously to CO_2 and H_2O. The two oxygen atoms in dioxygen share six electrons in the $\sigma 2p_z$, $\pi 2p_x$, and $\pi 2p_y$ molecular orbitals and two unpaired electrons reside in the two degenerate *anti*-bonding $\pi 2p_x^*$ and $\pi 2p_y^*$ molecular orbitals, making the formal bond order of two in O_2.[59,60] Thus, it is a biradical with two unpaired electrons and the ground electronic state is a triplet ($S = 1$). Reactions of most organic molecules in their spin-paired singlet state ($S = 0$) with the triplet ground state of O_2 are thus formally spin-forbidden processes.[61]

However, in nature, dioxygen is required to drive metabolic processes for all systems living in aerobic conditions. Simply reversing the spin of one of the unpaired electrons occupying the *anti*-bonding molecular orbitals to yield a singlet state requires significant amounts of energy. Therefore, O_2 has to be reductively activated from its abundant triplet ground state to reactive singlet or doublet species $\left(O_2^-, O_2^{2-}\right)$. Enzymes in biological systems like monooxygenases and dioxygenases perform this task easily, and in many cases can generate species of even greater reactivity by cleaving the O–O bond.[60,62,63] In most cases, the help of a transition metal (mostly Cu and Fe) is sought, which, in the appropriate oxidation state, reacts directly with triplet O_2 to form the reactive oxygen species.[64] All of these reactions are highly specific and tightly regulated.

4.4.1 Radical Mechanisms: Proton-Coupled Electron Transfer followed by Radical Rebound

The oxidation mechanism of methane mediated by sMMO is thought to be similar to that for the oxidation of organic substrates by cytochrome P450. The reactive intermediate of the di-iron active site in sMMO is compound **Q**,[65] the $(Fe^{IV})_2(\mu\text{-}O)_2$ species that is considered to be the symmetrical analog of the heme $Fe^{IV} = O$ π cation radical species of cytochrome P450.[66] In the case of cytochrome P450, the initial reduction of the substrate-bound ferric heme to the ferrous form, followed by dioxygen activation and rearrangement, produces the active ferryl species.[67] This active intermediate then abstracts hydrogen from the substrate to form the substrate radical and a hydroxyl radical bound to the metal center. Rebound of the radicals will ultimately produce the oxygenated product.

Similar to the cytochrome P450 mechanism, the di-iron center in sMMO in the diferrous state will bind oxygen to form the peroxo compound P^{65} (Scheme 4.10). The second iron atom serves the same purpose of the porphyrin ring in supplying the second electron for heterolytic cleavage of O–O bond during the transformation of the peroxo species (compound P) to the ferryl species (compound Q). The compound Q is the active intermediate and abstracts hydrogen from the substrate to produce the substrate radical and metal hydroxyl species. Radical rebound will form compound T, which eventually breaks down to release the product. A number of observations seem to be consistent with this mechanism:

1. The unique methyl radical formed during the oxidation of methane by sMMO has been inferred by radical traps.
2. The enzyme from *M. trichosporium* OB3b catalyzes the oxidation of a range of radical clock reagents with varying extents of rearranged products.
3. In some cases, complete or partial change of configuration, i.e., racemization, occurs through the formation of an intermediate alkyl radical.
4. For the MMO enzyme from *M. trichosporium* OB3b, a large deuterium isotope effect has been observed between normal and deuterated methanes, consistent with a radical mechanism.

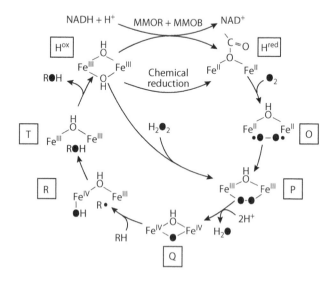

SCHEME 4.10 Mechanism of oxidation of methane and other hydrocarbons mediated by sMMO. (From Wallar, B.J. and Lipscomb, J.D., *Chem. Rev.*, 96, 2625, 1996.)

A parallel mechanism has been proposed for the oxidation of methane by pMMO. The dicopper center observed in the crystal structure of the protein has been proposed to be the active site of this enzyme.[56,68] However, according to DFT calculations and reactivity studies of model dioxo–dicopper complexes, the bis-μ-oxo-Cu^{III}_2 is less favored than its one electron-reduced counterpart $Cu^{II}-(O)_2-Cu^{III}$ species.[69–72] A similar species, $Cu^{II}-(O)(OH)-Cu^{III}$, has also been implicated as a necessary intermediate of oxidation/abstraction reactions; however, it has never been observed.[53,73] Exposed terminal cupryl species, such as $\{Cu^{II}-OH...Cu^{II}-O\cdot\}$, has also been suggested by DFT calculations to be reactive toward methane.[74] All these species are supposed to abstract hydrogen from the hydrocarbon substrate to form an alkyl radical and a metal-bound hydroxyl group. The alkyl radical standing nearby the active site can then rearrange or reorient and rebound with the metal hydroxyl group. The resulting intermediate ultimately releases the hydroxylated product and the metal center reverts back to its resting state.

4.4.2 Direct O-Atom Insertion

In the case of pMMO from *M. capsulatus* (Bath), experiments on cryptically chiral ethanes have shown that the insertion of the active oxygen atom species into the C–H bond during the hydroxylation occurs with 100% retention of configuration. This result, together with the relatively normal KIE observed for hydroxylation of the C–H and C–D bonds, has provided strong argument in support of a concerted reaction pathway in the case of the membrane-bound form of the MMO from methanotrophic bacteria.

The isolated O-atom or the oxene with the four p-electrons could exist in the ^3P, ^1D, and ^1S electronic states.[67] The ^1S state has all four p-electrons paired in two of the p atomic orbitals and has the highest energy. The ^1S and ^3P states are biradicals with two electrons sharing one p-orbital, but with the remaining two electrons occupying the other two p-orbitals with their spins in the opposite or same direction, respectively. The ^3P is the ground electronic state. However, when a copper complex mediate an O-atom to the organic substrate, it is not the "naked" oxene that is being transferred. The "oxene" is harnessed by the copper complex and is being transferred from the Cu–O bonds of the copper complex to the C–H bond of the organic substrates in the transition state. The nature of the harnessed "oxene" that is transferred depends on the number of copper ions associated with the O-atom as well as the nature of the chemical bonding in the copper complex.

4.4.3 Electronic Structure of the Activated Tricopper Cluster

4.4.3.1 Ground State

When the tricopper cluster site in pMMO or one of our $Cu^ICu^ICu^I$ complexes is activated by molecular oxygen, the highly reactive $Cu^{II}Cu^{II}(\mu-O)_2Cu^{III}$ species is formed. In this bis-μ-oxo complex, the oxo (O^{2-}) is bridging between two copper ions with different formal charges, one a d^9 Cu^{II} and the other a d^8 Cu^{III}, while the apical copper remains in the d^9 Cu^{II} state. The unpaired electrons of the two Cu^{II}

ions are antiferromagnetically coupled through bridging ligands, while the Cu^{III} ion is diamagnetic with all its electron spins paired up. Thus, the overall spin of the $Cu^{II}Cu^{II}(\mu\text{-}O)_2Cu^{III}$ species is a singlet. Note that Cu^{III} is only formally a d^8 ion; as with hard ligands, Cu^{III} possesses a similar electronegativity to the bridging O-atom so that $Cu^{III}O^{2-}$ is really a covalent bond between a Cu^{II} ion and an oxyl. In pMMO, the ligands of the putative Cu^{III} are carboxylates of aspartate and glutamate residues,[39,49] and in the design of our biomimetic tricopper complexes, we have also chosen hard ligands.[48–51]

4.4.3.2 Transition State during Methane Oxidation

When substrates like methane, which itself is a singlet species without any unpaired electrons, approaches the active $Cu^{II}Cu^{II}(\mu\text{-}O)_2Cu^{III}$ species sideways, a transient complex can be formed and facile O-atom transfer can take place in the transition state.[70] The transition state can be four centered or five centered depending on whether the carbon center to be oxidized interacts with the "empty" d-orbital of the diamagnetic Cu^{III}. In any case, before O-atom transfer, the bridging oxo contains a full octet of electrons with its own six valence electrons and the two electrons donated by two copper ions for the formation of the two Cu–O bonds in the active species. Upon formation of the transition state, the basal $Cu^{II}\text{–}O\text{–}Cu^{III}$ will interact with the C–H bond, and this interaction can lead to insertion of an oxene into the C–H bond upon homolytic cleavage of the two Cu–O bonds to give the $Cu^{II}Cu^{I}(\mu\text{-}O)Cu^{II}$ species.[57] With the $Cu^{II}Cu^{II}(\mu\text{-}O)_2Cu^{III}$ species in an overall electronic singlet state, the initial Cu^{II} ions as well as the Cu^{II} ions in the product tricopper species are antiferromagnetically coupled, and this ensures that the oxene will be released as a singlet, and not a triplet. Insertion of a singlet oxene across the C–H bond ensures facile bond closure to form the hydroxylated product methanol. Otherwise, the formation of the C–H bond will take place in two successive distinct kinetic steps, or via spin cross from the triplet to the singlet potential surface during the time course of the reaction, processes that are necessarily slower and might not lead to total retention of the configuration at the carbon center oxidized.

4.4.4 DFT CALCULATIONS OF THE TRANSITION STATE DURING O-ATOM TRANSFER TO CH$_4$

DFT studies have been performed on three possible models of the pMMO active site: (i) the trinuclear copper cluster, $Cu^{II}Cu^{II}(\mu\text{-}O)_2Cu^{III}$ complex **1** proposed by the Chan laboratory[37]; (ii) the most frequently used model $Cu^{III}(\mu\text{-}O)_2Cu^{III}$ complex **2**; and (iii) the mixed-valent $Cu^{II}(\mu\text{-}O)_2Cu^{III}$ complex **3** (Scheme 4.11).[70]

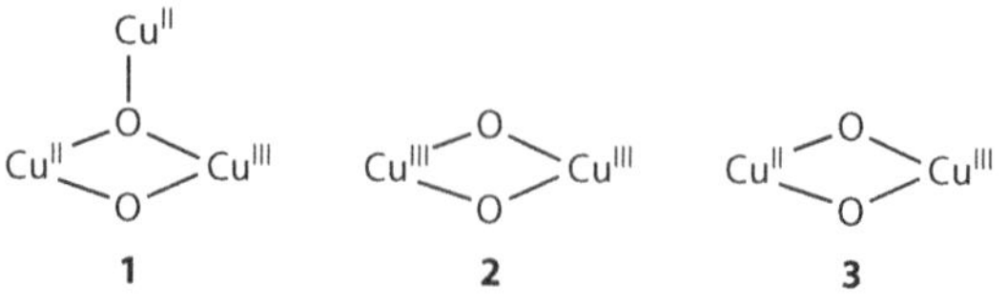

SCHEME 4.11 Models of the catalytic site in pMMO.

Two neutral ammonia molecules are used as supporting ligands around each of the copper ions. On the basis of the spectroscopic data on related complexes, both the $Cu^{II}Cu^{II}(\mu\text{-}O)_2Cu^{III}$ tricopper and $Cu^{III}(\mu\text{-}O)_2Cu^{III}$ dicopper complexes are taken to be in their singlet ground states ($S_T = 0$), assuming strong antiferromagnetic coupling of adjacent Cu(II) ions. The mixed-valent $Cu^{II}(\mu\text{-}O)_2Cu^{III}$ dicopper complex is treated as a doublet state ($S_T = 1/2$).

The results obtained using the $Cu^{II}Cu^{II}(\mu\text{-}O)_2Cu^{III}$ complex as the active site are highlighted in Figure 4.12. The tricopper complex transition state is connected to a methanol and a $Cu^{II}Cu^{II}(\mu\text{-}O)_2Cu^{III}$ complex by intrinsic reaction coordinate (IRC) calculations following the transition mode. The conversion of methane to methanol mediated by the trinuclear copper cluster proceeds by a concerted side-on O-atom insertion step, exactly as predicted by the Chan et al.[37] Although the oxygen in $Cu^{II}Cu^{II}(\mu\text{-}O)_2Cu^{III}$ complex is more "μ-oxo" in its ground state, in the transition state it is more like a singlet oxene. The optimized transition state (TS1) structure has an imaginary mode of 593i cm^{-1}, which is far different for those obtained for direct hydrogen abstraction by iron-containing or copper-containing model complexes. The activation barrier for the "oxo-transfer" process is calculated to be 15 kcal/mol.

The hydroxylation of methane by oxo-transfer from a $Cu^{III}(\mu\text{-}O)_2Cu^{III}$ dicopper complex is found to be similar to the tricopper $Cu^{II}Cu^{II}(\mu\text{-}O)_2Cu^{III}$ complex as described previously. Again, hydroxylation via concerted side-on "oxene"

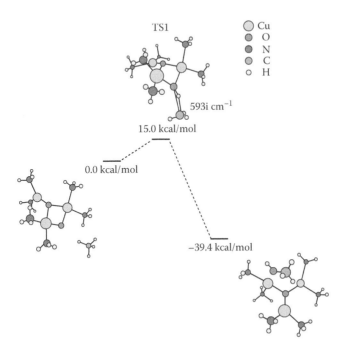

FIGURE 4.12 Energetics and computed structures for the conversion of methane to methanol mediated by the $Cu^{II}Cu^{II}(\mu\text{-}O)_2Cu^{III}$ trinuclear copper cluster **1**. (From Chen, P.P.Y. and Chan, S.I., *J. Inorg. Biochem.*, 100, 801, 2006.)

insertion is predicted (Figure 4.13), and the reaction occurs exclusively within the singlet manifold. However, the transition state (TS2) occurs at a much higher barrier (20 kcal/mol for TS2 compared with 15 kcal/mol compared with TS1). The imaginary mode of the optimized TS2 structure of hydroxylation is calculated to be 799i cm^{-1}.

The mixed-valent Cu^{II}(μ-O)$_2$Cu^{III} dicopper cluster has often been adopted as a pMMO model. As a doublet state, the hydroxylation of alkane mediated by the mixed-valent model should proceed via a radical mechanism. According to conservation of spin multiplicity, the process should proceed by hydrogen abstraction, yielding both a doublet product (methyl radical) and a singlet product (the hydroxylated Cu^{II}(μ-O)(μ-OH)Cu^{II} dicopper cluster) derived from the hydrogen abstraction. This step should be followed by dissociation of a hydroxyl radical and radical recombination to give a diamagnetic methanol (two radicals could combine to give a singlet product). The activation barrier associated with the hydrogen abstraction is calculated to be 19.1 kcal/mol. The imaginary mode of the optimized transition state TS3 (Figure 4.14) structure of hydrogen abstraction is calculated to be 1298i cm^{-1}.

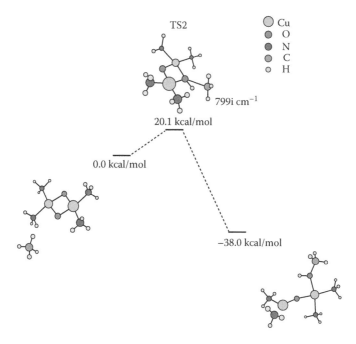

FIGURE 4.13 Energetics and computed structures for the conversion of methane to methanol mediated by the symmetrical Cu^{III}(μ-O)$_2$Cu^{III} cluster **2** through the concerted mechanism. (From Chen, P.P.Y. and Chan, S.I., *J. Inorg. Biochem.*, 100, 801, 2006.)

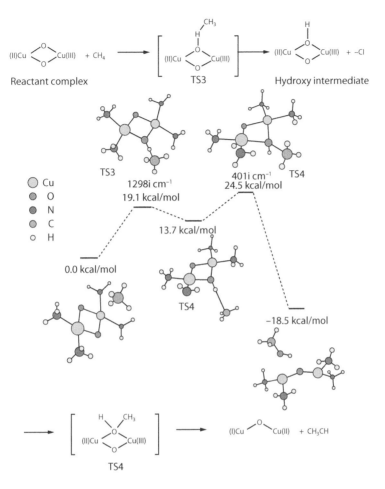

FIGURE 4.14 Energetics and computed structures for the conversion of methane to methanol by the mixed-valent $Cu^{II}(\mu\text{-}O)_2Cu^{III}$ through hydrogen atom abstraction and radical rebound mechanism. (From Chen, P.P.Y. and Chan, S.I., *J. Inorg. Biochem.*, 100, 801, 2006.)

4.4.4.1 Calculated Rate Constants and Kinetic Isotope Effects

Table 4.3 summarizes the calculated rate constants and the KIEs of the O-atom transfer[70] according to models **1** and **2** in the DFT calculations, as well for the hydrogen abstraction step according to model **3** at 300 K. It is clear from these results that the hydroxylation mediated by the $Cu^{II}Cu^{II}(\mu\text{-}O)_2Cu^{III}$ complex **1** is the most facile. Methane hydroxylation mediated by either $Cu^{III}(\mu\text{-}O)_2Cu^{III}$ dicopper complex **2** or the mixed-valent $Cu^{II}(\mu\text{-}O)_2Cu^{III}$ dicopper cluster **3** (we may assume that the radical rebound step is significantly faster than the hydrogen abstraction step) are slower by several orders of magnitude. In addition, only a small KIE of ~5

TABLE 4.3

First-Order Rate Constants and KIE (k_H/k_D) Values for Hydroxylation of Methane by a $Cu^{II}Cu^{II}(\mu\text{-}O)_2Cu^{III}$ Species, a $Cu^{III}(\mu\text{-}O)_2Cu^{III}$ Dicopper Species, and a $Cu^{II}(\mu\text{-}O)_2Cu^{III}$ Dicopper Species in Transient Binary Complexes Formed between Methane and Corresponding Copper Species

	T (K)	k (CH_4)[a] (s^{-1})	k (CD_4)[b] (s^{-1})	k_H/k_D
		$[Cu^{II}Cu^{II}(\mu\text{-}O)_2Cu^{III}]^{3+}$		
Facile	300	2.91×10^4	6.83×10^3	4.3 (5.2)[c]
		$bis(\mu\text{-oxo})Cu^{III}Cu^{III}$		
Inert	300	5.67×10^{-1}	1.49×10^{-1}	3.8 (4.9)[c]
		$bis(\mu\text{-oxo})Cu^{II}Cu^{III}$		
Slow	300	1.19×10^2	7.65	16 (49)[c]

Source: Chen, P.P.Y. and Chan, S.I., *J. Inorg. Biochem.*, 100, 801, 2006.

[a] Rate constant for CH_4 as substrate.

[b] Rate constant for CD_4 as substrate.

[c] Tunneling effects are included.

is predicted for the O-atom transfer chemistry mediated by $Cu^{II}Cu^{II}(\mu\text{-}O)_2Cu^{III}$ complex **1** (also for $Cu^{III}(\mu\text{-}O)_2Cu^{III}$ dicopper complex **2**) under ambient temperatures, in good agreement with the experimental value. In contrast, the radical mechanism deployed by the mixed-valent $Cu^{II}(\mu\text{-}O)_2Cu^{III}$ dicopper cluster **3** predicts a significantly larger KIE of 15 for the hydrogen-abstraction step at 300 K, even without taking hydrogen tunneling into consideration in the transition state theory calculations. When hydrogen tunneling is included in the calculation of the rate constant, the KIE becomes ~49, as expected. In contrast, hydrogen tunneling makes only a small contribution to the KIE for the methane hydroxylation mediated by either $Cu^{II}Cu^{II}(\mu\text{-}O)_2Cu^{III}$ complex **1** or the $Cu^{III}(\mu\text{-}O)_2Cu^{III}$ dicopper complex **2**. Interestingly, the calculated KIE including hydrogen tunneling for the tricopper model **1** is in essential agreement with the experimental value.

In summary, the DFT analysis of the catalytic pathways for methane hydroxylation mediated by three different models indicate that the trinuclear copper cluster, $Cu^{II}Cu^{II}(\mu\text{-}O)_2Cu^{III}$ complex **1** proposed by Chan et al.,[37] offers the most facile pathway, and the mechanism yields k_{cat} and $^1H/^2H$ KIE values that are in close agreement with experimental values.

4.4.5 CHEMISTRY ON A SINGLET POTENTIAL SURFACE

The observed KIEs for the hydroxylation chemistry of alkanes do not seem to be consistent with the classical radical mechanism in case of pMMO. If the hydroxylation proceeds via a direct concerted O-atom insertion mechanism, the overall rate should be determined by the single oxene transfer step, as the rates of substrate reorientations to present the stereoheterotopic faces of the C–H or C–D bonds to the hot O-atom or the singlet oxene species should be rapid. The relatively consistent k_H/k_D observed for the C–H and C–D bonds of chiral deuterated alkanes (~5), the intrinsic k_H/k_D of 5.5 determined for [2,2-^{2}H$_2$]butane, and the k_H/k_D observed for the primary carbon (5.2–5.5) in cryptically chiral ethanes, is not consistent with the end-on hydrogen-abstraction radical-rebound mechanism. Owing to extensive hydrogen tunneling in a linear transition state to form the hydroxyl radical, a much larger k_H/k_D would have been expected. Rather, these data are more in line with product formation from a nonlinear early transition state, such as one that is formed from side-on O-atom attack across a C–H or C–D bond (Scheme 4.12).

In our proposed concerted mechanism for the O-atom transfer step in the hydroxylation chemistry, the active oxene is delivered side-on from the $Cu^{II}Cu^{II}(\mu-O)_2Cu^{III}$ active intermediate to the C–H or C–D bond following an overall singlet potential surface. This transition state could be a four- or five-centered one, depending on whether there is interaction between the metal center(s) delivering the oxene and the carbon center or the C–H bond accepting the hot O-atom. In any case, since the process starts from an overall diamagnetic tricopper cluster, conservation of spin multiplicity dictates that the "O" atom be delivered as a singlet oxene. The transient OH species that is moving away from the carbon would then be generated with the spin of the odd electron antiparallel to the odd spin of the electron localized on the carbon center. These two spins would thus be favorably aligned for rapid bond closure to form the C–O bond upon product formation. For such a scenario, there should only be a modest potential barrier to hinder the motion of the (H or D) atom during elongation of the C–H or C–D bond to form the final product in the transition state, and we would expect only a modest H/D KIE. Also, there should only be minor structural changes about the carbon center attacked during the formation of the transition state, in accordance with the observed lack of a ^{12}C/^{13}C KIE on the rate of the hydroxylation.[75] This chemistry is analogous to the mechanism of singlet carbine insertion across C–H bonds, for which there has been a celebrated history.

In light of this discussion, the transition state involving the oxene of the copper cluster and the C–H bond of the alkane should be primarily singlet in character. This feature of the transition state promotes rapid bond closure, and the resultant process that ensues should be a concerted one.

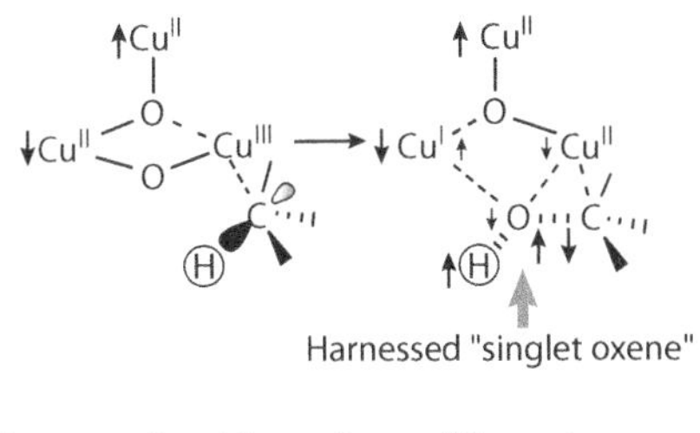

Harnessed "singlet oxene"

SCHEME 4.12　Transition state for side-on "oxene" insertion across a C–H bond.

4.5 HARNESSING THE SINGLET OXENE: CHALLENGES TO THEORY AND EXPERIMENT

4.5.1 NEW COPPER DIOXYGEN CHEMISTRY

Copper has been identified as an important constituent of many proteins and enzymes that mediate crucial biological processes like electron transfer, dioxygen transport and activation, reduction of nitrogen oxides, disproportionation of superoxide ion, and so on.[61,76,77] Subsequent x-ray crystal structure analyses and detailed spectroscopic methods have revealed structural information showing the predominant presence of histidine imidazoles ligating d^{10} copper at the active site. Many biochemical and synthetic modeling studies have been performed to elucidate how dioxygen interacts with the reduced active sites, the fate of the resulting intermediates, and to shed light on the structure/function relationships in these copper proteins.[78,79]

Dioxygen can bind to copper in different stoichiometric ratios from reaction of mononuclear copper complexes to bi-, tri-, and even tetra-nuclear copper complexes.[55,80] A range of different binding modes are also available, as dioxygen can bind in one- or two-electron reduced superoxide $\left(O_2^-\right)$ or peroxide $\left(O_2^{2-}\right)$ formed in either the end-on or side-on fashion with a single metal ion.[80,81–86] Protonation can occur to yield hydroperoxides bound to copper (II).[87] These monocopper systems can eventually bind to another copper ion to create dicopper complexes with bridging oxygen $\mu\text{-}\eta^1$, $\mu\text{-}\eta^1$: η^1 (*cis* or *trans*) and $\mu\text{-}\eta^2$: η^2 bridging mode.[80,88–93] μ-Oxo bridged complexes may be derived from O–O bond cleavage in these bridged dicopper systems.[94,95] Well-established structural and spectroscopic evidences have confirmed a vast number of these complexes, most of which are only detected at low temperature[80] (Scheme 4.13).

With the suggestion of the presence of trinuclear copper species at the active site of pMMO, a few tricopper complexes have been reported with trinucleating or macrocyclic ligands; however, they are unable to react with dioxygen.[96–99] There is report of a mononuclear copper complex that reacts with oxygen and can form a mixed-valent Cu_3O_2 core (T) but disintegrates upon reduction.[100] A tripodal ligand has also been reported to bind three copper ions and can reduce dioxygen to water; however, the oxygen adduct has not been characterized.[101]

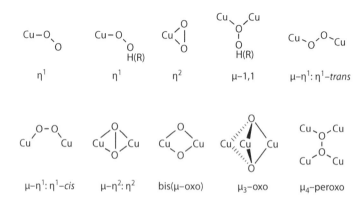

SCHEME 4.13 Binding modes of dioxygen with copper ions.

The Chan laboratory has introduced a new type of trinucleating ligand for the preparation of tricopper complexes, which upon dioxygen activation are able to oxidize aliphatic and aromatic systems at room temperature by mediating the transfer of singlet oxene.[48–51] Although the mixed-valent active species could not be isolated for crystal structural analysis because of its high reactivity, spectroscopic and kinetic studies suggest that the dioxygen first bind to the dicopper at the base of the triad to form a peroxo species with concomitant intramolecular electron transfer from the third copper to produce the highly reactive $Cu^{II}Cu^{II}(\mu\text{-}O)_2Cu^{III}$ species (Scheme 4.7). The electronic structure of this active species ensures that the oxene is transferred to the organic substrate in the singlet potential surface, and the chemistry is facile, regiospecific, and stereoselective. Thus, this research has opened a new direction in copper dioxygen chemistry.

4.5.2 SINGLET OXENE IS THE MOST POWERFUL OXIDIZING AGENT FOR DIRECT O-ATOM INSERTION CHEMISTRY

Atomic oxygen is termed "oxene" because of its similarity in reactivity to carbene, the corresponding bivalent carbon equivalent. It can react with alkenes in the triplet state as a free radical and can abstract hydrogen.[73,102,103] It can also be inserted into the C–H bond in the singlet state to produce alcohols.[104,105] From data collected during gasphase reactions of alkanes with singlet oxene, the reaction pathways can be interpreted in two different ways. One is the direct insertion of the O-atom into the C–H bond, and the other is the abstraction of the hydrogen from the bond. When the substrate is a small alkane with strong C–H bonds, the reaction occurs predominantly through the insertion mechanism, and for higher hydrocarbons with steric hindrance or low C–H bond activation energy, the second pathway, i.e., hydrogen abstraction, takes place. The oxygen atom can be bound to a metal ion and still maintain its activity toward insertion into a C–H bond or abstraction of hydrogen from a hydrocarbon.[106] The insertion reaction occurs through a single transition state with no intermediates; thus, the reaction is faster, while the removal of hydrogen produces a radical intermediate, which can undergo configurational changes followed by rebound or recombination.

The chemistry mediated by the tricopper cluster in pMMO is similar, except that the singlet oxene is not naked but harnessed. The reactivity of the naked singlet oxene is well known. However, the use of the tricopper system to harness the singlet oxene is novel chemistry. When the singlet oxene is harnessed, the O-atom transfer can only take place in the transition state so its reactivity can be tightly controlled by the details of the interactions of the substrate with the tricopper catalyst.

4.5.3 NEED TO UNDERSTANDING HOW HARD AND SOFT LIGANDS TO THE COPPER IONS CONTROL HOMOLYTIC O-ATOM TRANSFER VERSUS HETEROLYSIS TO FORM WATER UPON PROTONATION

The proposed site D that harbors the tricopper center in pMMO is lined by hydrophilic amino acid residues. The amino acid side chains coordinating the three copper ions are as follows: Cu1, PmoA His38 and PmoC Glu154; Cu2, PmoA Asp47 and Met42; Cu3, PmoA Asp49 and Glu100.[39,49] Copper enzymes in nature generally prefer histidine imidazole-type moderate ligands, which help them maintain low Cu^{II}/Cu^I redox potentials.[76] Oxidation states of +2 and +1 are mostly accessed by nature to perform electron transfer or oxygen transfer chemistry by copper proteins. In multicopper oxidases, the ligands to the copper ions are imidazoles of histidines, and dioxygen binds to these multicopper sites.[107] As a moderate ligand, the imidazole is capable of donating a significant amount of electron density to the copper ion to facilitate heterolytic Cu–O bond cleavage to produce H_2O in the presence of protons upon reduction of the Cu^{II} ions. Thus, these enzymes can readily perform the four-electron reduction of molecular oxygen to water. However, in the case of the active site of pMMO, the ligands to the copper ions are hard carboxylate ligands from Asp or Glu, in addition to His, which render the higher oxidation states of copper (+3) accessible to the catalytic process instead. This facilitates the formation of the active intermediate $Cu^{II}Cu^{II}(\mu\text{-}O)_2Cu^{III}$, which can mediate the facile O-atom transfer chemistry as we have described here.

4.5.4 Differences in the Catalytic Turnover Mediated by the Biomimetic Tricopper Complexes and the Tricopper Active Site in the pMMO Enzyme

In pMMO, the active site is buried within the transmembrane domain of the protein scaffold.[35,39,49] Given that the oxidation of the hydrocarbon mediated by pMMO involves a direct concerted O-atom insertion mechanism, it follows that there must be a binding pocket for the hydrocarbon substrate in close proximity to the catalytic site. Thus, the hydrophobic pocket, or the "aromatic box" created by the aromatic amino acids Trp48, Phe50, Trp51, and Trp54 of PmoA, provides a pathway for the substrate to gain access to the active site.[32] That this pocket is capable of accommodating only the limited number of substrates known to be oxidized by the enzyme supports this picture. With this arrangement of the catalytic site and the hydrocarbon-binding pocket, it follows that during turnover, there must be conformational change in the enzyme that brings the substrate and activated tricopper cluster together to form the transient complex to form the transition state for the O-atom transfer. We have previously proposed that Gly46 of PmoA is involved in this process.[32]

In the case of the biomimetic tricopper complexes, there is no such control by a protein scaffold, and the results obtained when these complexes are employed as catalyst for hydrocarbon oxidation are somewhat different from the enzyme itself. To accomplish substrate specificity, a substrate must exhibit specific intrinsic binding affinity for a particular tricopper catalyst.[49–51] Moreover, the binding must occur in such a way that the complex forms the appropriate transition state to facilitate singlet oxene insertion to the C–H that is being oxidized in the substrate. In other words, the substrate must recognize the tricopper catalyst, and it must bind near the basal copper ions of the complex so that the appropriate transition state could be formed to facilitate catalysis. Moreover, depending on the binding affinity, the level of the substrate in the solution, and the kinetics of O-atom transfer, abortion of the active catalyst by reducing equivalents in the solution can compete with product formation to limit the TON.[108]

While we have had some success in designing tricopper clusters for the oxidation of different kinds of chemical substrates, we have had no or little control over the substrate specificity of hydrocarbons at the present state of ligand design. Our model complexes are able to oxidize C1–C5 hydrocarbons similar to the enzyme itself, but are also able to oxidize a variety of other substrates like aromatic and aliphatic hydrocarbons. Moreover, although the desired stereoselectivity and regiospecificity have been achieved with our model catalysts in certain instances, it is difficult to rule out overoxidation, although the biomimetic tricopper catalysts that we have designed for hydrocarbons oxidize methane, ethane, and propane to their corresponding alcohols exclusively.[108] In general, the first oxidation product, i.e., the alcohol, is kinetically more prone to be oxidized to their corresponding aldehydes or ketones from their parent hydrocarbons. In the enzyme, there are specific product channels to extrude the product from the active site and the protein as soon as it is produced; thus, overoxidation is not an issue.

Our model tricopper catalysts can support only single turnovers when they become activated by dioxygen. The system can be rendered catalytic by use of H_2O_2, which acts both as the oxidant to produce the active $Cu^{II}Cu^{II}(\mu\text{-}O)_2Cu^{III}$ species from $Cu^ICu^ICu^I$ as well as the reductant to reduce the end product $Cu^{II}Cu^{II}(\mu\text{-}O)Cu^{II}$ to its initial $Cu^ICu^ICu^I$ state so that the cycling of the catalyst can continue. Accordingly, multiple catalytic turnovers can be sustained using hydrogen peroxide to drive the process. However, at sufficiently high concentrations of hydrogen peroxide, abortive cycling will compete effectively against productive cycling of the catalytic system. Otherwise, the TOF of the catalyst is determined by the rate of regeneration of the catalyst and the TONs dictated by the amount of hydrogen peroxide used to drive the catalysis.

The catalytic cycle in the pMMO enzyme is totally different.[38] Following O-atom transfer from the activated tricopper cluster to the hydrocarbon substrate, one reducing equivalent is left behind at the catalytic site. The fate of this electron is not clear at this time. Either it remains at the site waiting for two additional reducing equivalents to regenerate the $Cu^ICu^ICu^I$ species, as in the case of biomimetic model catalysts that are turning over as described earlier,[51] or the electron is transferred out of the site to participate in the dioxygen reduction at the dinuclear site. It is now established that the dinuclear center at the B site binds dioxygen and can reduce it to water.[109] With a total of six reducing equivalents from the A, B, and D sites of the enzyme at the beginning of the turnover, together with the two additional reducing equivalents from the hydrocarbon substrate and six protons, the enzyme produces one methanol and three water molecules at the end of the oxidative phase of each turnover.[37,38] It is then poised for re-reduction to complete the turnover cycle. It is thought that the reducing equivalents for this comes from the 9–10 Cu(I) ions of the E-clusters occupying the water-exposed C-terminal domain of PmoB.[40,41] These E-clusters take electrons from the NAD(P)H or quinols and maintain a pool or reservoir of reducing equivalents to re-reduce the copper ions at the active site after the oxidative phase of the turnover.

The flow of reducing equivalents within pMMO is tightly regulated. The enzyme has to modulate the availability of the electrons at the active site so that the putative $Cu^{II}Cu^{II}(\mu\text{-}O)_2Cu^{III}$ species formed by activation of the $Cu^ICu^ICu^I$ by molecular oxygen is not aborted by reducing equivalents even before the oxene transfer chemistry takes place. As a case in point, in the absence of hydrocarbon substrate, the activated tricopper center ends up in the dead-end $Cu^{II}Cu^{II}(\mu\text{-}O)Cu^{II}$ species, the aborting reducing equivalent coming presumably from the Cu(I) occupying site A or site C.[37,38] As noted earlier, there is a reservoir of reducing equivalents occupying the E-cluster site.[40,43] These copper ions have usually high redox potentials,[39] and they are inert to reoxidation by O_2.[40] Presumably, the dioxygen chemistry at the dinuclear copper center occupying site B plays a role in controlling the flow of these electrons to the active site of the enzyme (C-clusters). However, the details remain to be worked out.

The availability of dioxygen for the hydrocarbon oxidation is also different for the enzyme and the biomimetic tricopper catalysts. Recently, a native bacteriohemerythrin (McHr) has been identified in *M. capsulatus* (Bath).[110] Both the pMMO and McHr are overexpressed in cells of this bacterium when this strain of methanotroph is cultured and grown under high copper to biomass conditions.[111] It has been suggested that the role of the McHr is to provide a shuttle to transport dioxygen from the cytoplasm of the cell to the intracytoplasmic membranes for consumption by the pMMO.[111] Activity assays of the purified pMMO solubilized in dodecyl-β-maltoside micelles have been shown to be dependent on the McHr/pMMO concentration ratio.[112] The activity of pMMO is usually determined by the propene epoxidation assay. For pMMO-enriched membranes, membrane fragments solubilized in dodecyl-β-maltoside, and the purified protein–detergent complex in micelles, the specific activity is typically 10–100 nmol/(min·mg protein).[36,45]

4.6 SUMMARY

In this chapter, we have summarized our current understanding of the structure and function of pMMO. This is a fascinating enzyme with the unique capability to mediate the facile conversion of methane to methanol. With a TOF approaching 1 s^{-1}, pMMO is the most efficient methane oxidizer discovered to date. After years of intense experimentation and debate, there is now little question that pMMO is a multicopper protein. In addition, the protein contains a unique tricopper cluster that uses new chemistry to transfer a singlet oxene to the C–H bond of the hydrocarbon substrate. Mimics of the tricopper cluster in the enzyme have now been developed, and we show here that these model tricopper complexes are capable of mediating the same oxidation chemistry toward hydrocarbon substrates as the enzyme as well. However, challenges remain toward attaining a more in-depth understanding of how the catalysis is controlled in the enzyme, and in obtaining greater physical/chemical insights on how to design an even more efficient biomimetic tricopper cluster for industrial applications.

ACKNOWLEDGMENTS

This work is supported by funds from Academia Sinica and grants from the National Science Council of the Republic of China (NSC 95-2113M-001-046, 97-2113M-001-027, and 98-2113M-001-026 to S.I.C.; 96-2627M-001-006 and 97-2113M-001-006-MY3 to S.S.F.Y.). S.M. is grateful to Academia Sinica for postdoctoral fellowship support that has enabled him to participate in this research project. We are also grateful to many Chan/Yu group members, past and present, who have made contributions to our understanding of pMMO as well as their inputs into the development of the model mimics of the active site of the enzyme, specifically, Michael K. Chan, Hoa H.-L. Nguyen, Sean Elliott, Kelvin H.-C. Chen, C.-L. Chen, Peter P.-Y. Chen, Yu-Jhang Lu, Penumaka Nagababu, Mu-Cheng Hung, Marianne M. Lee, I-Jui Hsu, Pham Dinh Minh, Jeff C.-H. Lai, and Kok Yoah Ng.

REFERENCES

1. USEPA. April 2011. *Inventory of U.S. Greenhouse Gas Emissions and Sinks 1990–2009*. Available at http://www.epa.gov/climatechange/Downloads/ghgemissions/US-GHG -Inventory-2011-Complete_Report.pdf (accessed December 12, 2014).
2. Anthony, C. 1982. *The Biochemistry of Methylotrophs*. Academic Press, London.
3. Hanson, R. S. and Hanson, T. E. 1996. Methanotrophic bacteria. *Microbiol. Rev.* 60:439–471.
4. EPA. April 2010. *Methane and Nitrous Oxide Emissions from Natural Sources*. Available at http://www.epa.gov/outreach/pdfs/Methane-and-Nitrous-Oxide-Emissions-From-Natural -Sources.pdf (accessed December 12, 2014).
5. Arndtsen, B. A.; Bergman, R. G.; Mobley, T. A. and Peterson, T. H. 1995. Selective intermolecular carbon–hydrogen bond activation by synthetic metal complexes in homogeneous solution. *Acc. Chem. Res.* 28:154–162.
6. Dalton, H. 1980. Oxidation of hydrocarbons by methane monooxygenases from a variety of microbes. *Adv. Appl. Microbiol.* 26:71–87.
7. Colby, J. and Zatman, L. J. 1972. Hexose phosphate synthese and tricarboxylic acid-cycle enzymes in bacterium 4B6, an obligate methylotroph. *Biochem. J.* 128:1373–1376.
8. Valentine, D. L. and Reeburgh, W. S. 2000. New perspectives on anaerobic methane oxidation. *Environ. Microbiol.* 2:477–484.
9. Raghoebarsing, A. A.; Pol, A.; van de Pas-Schoonen, K. T. et al. 2006. Microbial consortium couples anaerobic methane oxidation to denitrification. *Nature* 440:918–921.
10. Olah, G. A.; Goeppert, A. and Prakash, G. K. S. 2009. *Beyond Oil and Gas: The Methanol Economy*. Wiley-VCH, Weinheim.
11. Benson, D. 1975. *Mechanisms of Oxidation by Metal Ions*. Elsevier, New York.
12. Arndt, D. 1981. *Manganese Compounds as Oxidizing Agents in Organic Chemistry*. Open Court Publishing Company, La Salle, IL.
13. Cainelli, G. and Cardillo, G. 1984. *Chromium Oxidations in Organic Chemistry*. Springer-Verlag, Berlin.
14. O. R. N. Laboratory. 2003. *Basic Research Needs to Assure a Secure Energy Future*. Oak Ridge. Available at http://science.energy.gov/~/media/bes/besac/pdf/Basic_research _needs_to_assure_a_secure_energy_future_feb_2003.pdf (accessed on December 12, 2014).
15. Park, S.; Brown, K. W. and Thomas, J. C. 2002. The effect of various environmental and design parameters on methane oxidation in a model biofilter. *Waste Manag. Res.* 20:434–444.
16. Colby, J.; Stirling, D. I. and Dalton, H. 1977. The soluble methane mono-oxygenase of *Methylococcus capsulatus* (Bath). Its ability to oxygenate *n*-alkanes, *n*-alkenes, ethers, and alicyclic, aromatic and heterocyclic compounds. *Biochem. J.* 165:395–402.
17. Sullivan, J. P.; Dickinson, D. and Chase, C. A. 1998. Methanotrophs, *Methylosinus trichosporium* OB3b, sMMO, and their application to bioremediation. *Crit. Rev. Microbiol.* 24:335–373.

18. David, R. L. ed. 2003. *CRC Handbook of Chemistry and Physics*, 84th ed. CRC Press, Boca Raton, FL.

19. Bodrossy, L. L.; Kovacs, K. L.; McDonald, I. R. and Murrell, J. C. 1999. A novel thermophilic methane oxidising γ proteobacterium. *FEMS Microbiol. Lett.* 170:335–341.

20. Cook, S. A. and Shiemke, A. K. 2002. Evidence that a type-2NADH:quinone oxidoreductase mediates electron transfer to pMMO in *Methylococcus capsulatus. Arch. Biochem. Biophys.* 398:32–40.

21. Nguyen, H.-H. T.; Shiemke, A. K.; Jacobs, S. J.; Hales, B. J.; Lidstrom, M. E. and Chan, S. I. 1994. The nature of the copper ions in the membranes containing the particulate methane monooxygenase from *Methylococcus capsulatus* (Bath). *J. Biol. Chem.* 269:14995–15005.

22. Nguyen, H.-H. T.; Elliott, S. J.; Yip, J. H.-K. and Chan, S. I. 1998. The particulate methane monooxygenase from *Methylococcus capsulatus* (Bath) is a novel copper-containing three-subunit enzyme. Isolation and characterization. *J. Biol. Chem.* 273:7957–7966.

23. Tavormina, P. L.; Orphan, V. J.; Kalyuzhnaya, M. G.; Jetten, M. and Klotz, M. G. 2011. A novel family of functional operons encoding methane/ammonia monooxygenase-related proteins in gammaproteobacterial methanotrophs. *Environ. Microbiol. Rep.* 3:91–100.

24. Lipscomb, J. D. 1994. Biochemistry of the soluble methane monooxygenase. *Annu. Rev. Microbiol.* 48:371–399.

25. Lieberman, R. L. and Rosenzweig, A. C. 2004. Biological methane oxidation: Regulation, biochemistry, and active site structure of particulate methane monooxygenase. *Crit. Rev. Biochem. Mol. Biol.* 39:147–164.

26. Andersson, K. K.; Froland, W. A.; Lee, S.-K. and Lipscomb, J. D. 1991. Dioxygen independent oxygenation of hydrocarbons by methane monooxygenase hydroxylase component. *New J. Chem.* 15:411–415.

27. Fox, B. G.; Borneman, J. G.; Wackett, L. P. and Lipscomb, J. D. 1990. Haloalkene oxidation by the soluble methane monooxygenase from *Methylosinus trichosporium OB3b*: Mechanistic and environmental implications. *Biochemistry* 29:6419–6427.

28. Green, J. and Dalton, H. 1989. Substrate specificity of soluble methane monooxygenase. Mechanistic implications. *J. Biol. Chem.* 264:17698–17703.

29. Higgins, I. J.; Best, D. J. and Hammond, R. C. 1980. New findings in methane-utilizing bacteria highlight their importance in the biosphere and their commercial potential. *Nature* 286:561–564.

30. Rataj, M. J.; Kauth, J. E. and Donnelly, M. I. 1991. Oxidation of deuterated compounds by high specific activity methane monooxygenase from *Methylosinus trichosporium*. Mechanistic implications. *J. Biol. Chem.* 266:18684–18690.

31. Elliott, S. J.; Zhu, M.; Nguyen, H. H. T.; Yip, J. H. K. and Chan, S. I. 1997. The regio- and stereoselectivity of particulate methane monooxygenase from *Methylococcus capsulatus* (Bath). *J. Am. Chem. Soc.* 119:9949–9955.

32. Ng, K.-Y.; Tu, L.-C.; Wang, Y.-S.; Chan, S. I. and Yu, S. S.-F. 2008. Probing the hydrophobic pocket of the active site in the particulate methane monooxygenase (pMMO) from *Methylococcus capsulatus* (Bath) by variable stereo-selective alkane hydroxylation and olefin epoxidation. *Chem. Biol. Chem.* 9:1116–1123.

33. Wilkinson, B.; Zhu, M.; Priestley, N. D. et al. 1996. A concerted mechanism for ethane hydroxylation by the particulate methane monooxygenase from *Methylococcus capsulatus* (Bath). *J. Am. Chem. Soc.* 118:921–922.

34. Yu, S. S.-F.; Wu, L.-Y.; Chen, K. H.-C.; Luo, W.-I.; Huang, D.-S. and Chan, S. I. 2003. The stereospecific hydroxylation of [2,2-^{2}H$_2$]butane and chiral dideuteriobutanes by the particulate methane monooxygenase from *Methylococcus capsulatus* (Bath). *J. Biol. Chem.* 278:40658–40669.

35. Lieberman, R. L. and Rosenzweig, A. C. 2005. Crystal structure of a membrane-bound metalloenzyme that catalyses the biological oxidation of methane. *Nature* 434:177–182.

36. Yu, S. S.-F.; Chen, K. H.-C.; Tseng, M. Y.-H. et al. 2003. Production of high-quality particulate methane monooxygenase in high yields from *Methylococcus capsulatus* (Bath) with a hollow-fiber membrane bioreactor. *J. Bacteriol.* 185:5915–5924.

37. Chan, S. I.; Chen, K. H.-C.; Yu, S. S.-F.; Chen, C.-L. and Kuo, S. S.-J. 2004. Toward delineating the structure and function of the particulate methane monooxygenase from methanotrophic bacteria. *Biochemistry* 43:4421–4430.

38. Chan, S. I. and Yu, S. S.-F. 2008. Controlled oxidation of hydrocarbons by the membrane-bound methane monooxygenase: The case for a tricopper cluster. *Acc. Chem. Res.* 41:969–979.

39. Chan, S. I.; Wang, V. C.-C.; Lai, J. C.-H. et al. 2007. Redox potentiometry studies of particulate methane monooxygenase: Support for a trinuclear copper cluster active site. *Angew. Chem. Int. Ed.* 46:1992–1994.

40. Yu, S. S.-F.; Ji, C. Z.; Wu, Y. P. et al. 2007. The C-terminal aqueous-exposed domain of the 45 kDa subunit of the particulate methane monooxygenase in *Methylococcus capsulatus* (Bath) is a Cu(I) sponge. *Biochemistry* 46:13762–13774.

41. Chen, K. H.-C.; Chen, C.-L.; Tseng, C.-F. et al. 2004. The copper clusters in the particulate methane monooxygenase (pMMO) from *Methylococcus capsulatus* (Bath). *J. Chin. Chem. Soc.* 51:1081–1098.

42. Hung, S.-C.; Chen, C.-L.; Chen, K. H.-C.; Yu, S. S.-F. and Chan, S. I. 2004. The catalytic copper clusters of the particulate methane monooxygenase from methanotrophic bacteria: Electron paramagnetic resonance spectral simulations. *J. Chin. Chem. Soc.* 51:1229–1244.

43. Nguyen, H.-H. T.; Nakagawa, K. H.; Hedman, B. et al. 1996. X-ray absorption and EPR studies on the copper ions associated with the particulate methane monooxygenase from *Methylococcus capsulatus* (Bath). Cu(I) ions and their implications. *J. Am. Chem. Soc.* 118:12766–12776.

44. Kau, L.-S.; Spira-Solomon, D. J.; Penner-Hahn, J. E.; Hodgson, K. O. and Solomon, E. I. 1987. X-ray absorption edge determination of the oxidation state and coordination number of copper. Application to the type 3 site in *Rhus vernicifera* laccase and its reaction with oxygen. *J. Am. Chem. Soc.* 109:6433–6442.

45. Chan, S. I.; Nguyen, H.-H. T.; Chen, K. H.-C. and Yu, S. S.-F. 2011. Overexpression and purification of the particulate methane monooxygenase from *Methylococcus capsulatus* (Bath). In *Methods in Enzymology*, Vol. 495, eds. Rosenzweig, A. C. and Ragsdale, S. W. Academic Press, Burlington, MA, 177–193.

46. Choi, D. W.; Antholine, W. E.; Do, Y. S. et al. 2005. Effect of methanobactin on the activity and electron paramagnetic resonance spectra of the membrane-associated methane monooxygenase in *Methylococcus capsulatus* Bath. *Microbiology* 151:3417–3426.

47. Lieberman, R. L.; Shrestha, D. B.; Doan, P. E.; Hoffman, B. M.; Stemmler, T. L. and Rosenzweig, A. C. 2003. Purified particulate methane monooxygenase from *Methylococcus capsulatus* (Bath) is a dimer with both mononuclear copper and a copper-containing cluster. *Proc. Natl. Acad. Sci. U.S.A.* 100:3820–3825.

48. Chen, P. P.-Y.; Yang, R. B.-G.; Lee, J. C.-M. and Chan, S. I. 2007. Preparation of a bis(μ_3-oxo)trinuclear copper(II, II, III) complex to harness a "singlet oxene" for facile insertion across a C–C and a C–H bond of an exogenous substrate. *Proc. Natl. Acad. Sci. U.S.A.* 104:14570–14575.

49. Chan, S. I.; Lu, Y.-J.; Nagababu, P. et al. 2013. Efficient oxidation of methane to methanol by dioxygen mediated by tricopper clusters. *Angew. Chem. Intl. Ed.* 52:3731–3735.

50. Nagababu, P.; Maji, S.; Kumar, M. P.; Chen, P. P.-Y.; Yu. S. S.-F. and Chan, S. I. 2012. Efficient room-temperature oxidation of hydrocarbons mediated by tricopper cluster complexes with different ligands. *Adv. Synth. Catal.* 354:3275–3282.

51. Chan, S. I.; Chien, C. Y.-C.; Yu, C. S.-C.; Nagababu, P.; Maji, S. and Chen, P. 2012. Efficient catalytic oxidation of hydrocarbons mediated by tricopper clusters under mild conditions. *J. Catal.* 293:186–194.

52. Myronova, N.; Kitmitto, A.; Collins, R. F.; Miyaji, A. and Dalton, H. 2006. Three-dimensional structure determination of a protein supercomplex that oxidizes methane to formaldehyde in *Methylococcus capsulatus* (Bath). *Biochemistry* 45:11905–11914.

53. Himes, R. A. and Karlin, K. D. 2009. Copper–dioxygen complex mediated C–H bond oxygenation: Relevance for particulate methane monooxygenase (pMMO). *Curr. Opin. Chem. Biol.* 13:119–131.

54. Lewis, E. A. and Tolman, W. B. 2004. Reactivity of dioxygen–copper systems. *Chem. Rev.* 104:1047–1076.

55. Que, L. and Tolman, W. B. 2008. Biologically inspired oxidation catalysis. *Nature* 455:333–340.

56. Balasubramanian, R.; Smith, S. M.; Rawat, S.; Yatsunyk, L. A.; Stemmler, T. L. and Rosenzweig, A. C. 2010. Oxidation of methane by a biological dicopper centre. *Nature* 465:115–131.

57. Maji, S.; Lee, J. C.-M.; Lu, Y.-J. et al. 2012. Dioxygen activation of a trinuclear $Cu^ICu^ICu^I$ cluster capable of mediating facile oxidation of organic substrates: Competition between O-atom transfer and abortive inter-complex reduction. *Chem. Eur. J.* 18:3955–3968.

58. Sawyer, D. T.; Sobkowiak, A. and Roberts, J. L. Jr. 1995. *Electrochemistry for Chemists*, 2nd ed. John Wiley & Sons, New York, 358–371.

59. Ingraham, L. L. and Meyer, D. L. 1985. *Biochemistry of Dioxygen*. Plenum, New York.

60. Valentine, J. S.; Foote, C. S.; Liebman, J. and Greenberg, A. eds. 1995. *Active Oxygen in Biochemistry*. Blackie, Academic & Professional, Bishopbriggs, Glasgow, UK.

61. Bertini, I., Gray, H. B. and Valentine, J. S. 2007. *Biological Inorganic Chemistry: Structure and Reactivity*. University Science Books, Sausalito, CA.

62. Malmstrom, B. G. 1982. Enzymology of oxygen. *Annu. Rev. Biochem.* 51:21–59.

63. Hamilton, G. A. 1974. Chemical Models and Mechanisms for Oxygenases. In *Molecular Mechanisms of Oxygen Activation*. Hayaishi, O. (ed.), Academic Press, New York, pp. 405–451.

64. Spiro, T. G. ed. 1980. *Metal Ion Activation of Dioxygen: Metal Ions in Biology*, Vol. 2. Wiley-Interscience, New York.

65. Wallar, B. J. and Lipscomb, J. D. 1996. Dioxygen activation by enzymes containing binuclear non-heme iron clusters. *Chem. Rev.* 96:2625–2657.

66. Ortiz de Montellano, P. R. and De Voss, J. J. 2005. Substrate Oxidation by Cytochrome P450 Enzymes. In *Cytochrome P450: Structure, Mechanism, and Biochemistry*. Ortiz de Montellano, P. R. (ed.), Plenum Press, New York, pp. 183–245.

67. Sono, M.; Roach, M. P.; Coulter, E. D. and Dawson, J. H. 1996. Heme-containing oxygenases. *Chem. Rev.* 96:2841–2887.

68. Woertinka, J. S.; Smeetsa, P. J.; Groothaert, M. H. et al. 2009. A $[Cu_2O]^{2+}$ core in Cu-ZSM-5, the active site in the oxidation of methane to methanol. *Proc. Natl. Acad. Sci. U.S.A.* 106:18908–18913.

69. Yoshizawa, K.; Suzuki, A.; Shiota, Y. and Yamabe, T. 2000. Conversion of methane to methanol on diiron and dicopper enzyme models of methane monooxygenase: A theoretical study on a concerted reaction pathway. *Bull. Chem. Soc. Jpn.* 73:815–827.

70. Chen, P. P. Y. and Chan, S. I. 2006. Theoretical modeling of the hydroxylation of methane as mediated by the particulate methane monooxygenase. *J. Inorg. Biochem.* 100:801–809.

71. Yoshizawa, K. and Shiota, Y. 2006. Conversion of methane to methanol at the mononuclear and dinuclear copper sites of particulate methane monooxygenase (pMMO): A DFT and QM/MM study. *J. Am. Chem. Soc.* 128:9873–9881.

72. Shiota, Y. and Yoshizawa, K. 2009. Comparison of the reactivity of bis(μ-oxo)Cu^{II}Cu^{III} and Cu^{III}Cu^{III} species to methane. *Inorg. Chem.* 48:838–845.

73. Himes, R. A. and Karlin, K. D. 2009. A new copper-oxo player in methane oxidation. *Proc. Natl. Acad. Sci. U.S.A.* 106:18877–18878.

74. Blain, I.; Giorgi, M.; De Riggi, I. and Reglier, M. 2000. Substrate-binding ligand approach in chemical modeling of copper-containing monooxygenases. 1: Intramolecular stereoselective oxygen atom insertion into a non-activated C–H bond. *Eur J Inorg Chem.* 2: 393–398.

75. Huang, D.-S.; Wu, S.-H.; Wang, Y.-S.; Yu, S. S.-F. and Chan, S. I. 2002. Determination of the carbon kinetic isotope effects on propane hydroxylation mediated by the methane monooxygenases from *Methylococcus capsulatus* (Bath) by using stable carbon isotopic analysis. *Chem. Biol. Chem.* 3:760–765.

76. Lippard, S. J. and Berg, J. M. 1994. *Principles of Bioinorganic Chemistry.* University Science Books, Mill Valley, CA.

77. Solomon, E. I.; Chen, P.; Metz, M.; Lee, S.-K. and Palmer, A. E. 2001. Oxygen binding, activation, and reduction to water by copper proteins. *Angew. Chem. Int. Ed.* 40: 4570–4590.

78. Zhang, C. X.; Liang, H.-C.; Humphreys, K. J. and Karlin, K. D. 2003. Copper–dioxygen complexes and their roles in biomimetic oxidation reactions. In *Advances in Catalytic Activation of Dioxygen by Metal Complexes*, ed. Simándi, L. I. Kluwer Academic Publishers, Netherlands, 79–121.

79. Karlin, K. D. and Zuberbuhler, A. D. 1999. Formation, Structure, and Reactivity of Copper Dioxygen Complexes. In *Bioinorganic Catalysis*, Reedijk, J. and Iouwman, E. (eds.), Marcel Dekker, New York, pp. 469–534.

80. Mirica, L. M.; Ottenwaelder, X. and Stack, T. D. P. 2004. Structure and spectroscopy of copper–dioxygen complexes. *Chem. Rev.* 104:1013–1045.

81. Weitzer, M.; Schindler, S.; Brehm, G. et al. 2003. Reversible binding of dioxygen by the copper(I) complex with tris(2-dimethylaminoethyl)amine (Me6tren) ligand. *Inorg. Chem.* 42:1800–1806.

82. Börzel, H.; Comba, P.; Hagen, K. S. et al. 2002. Copper–bispidine coordination chemistry: Syntheses, structures, solution properties, and oxygenation reactivity. *Inorg. Chem.* 41:5440–5452.

83. Fujisawa, K.; Tanaka, M.; Moro-Oka, Y. and Kitajima, N. A. 1994. A monomeric side-on superoxocopper(II) complex: Cu(O$_2$)(HB(3-tBu-5-iPrpz)$_3$). *J. Am. Chem. Soc.* 116:12079–12080.

84. Chen, P.; Root, D. E.; Campochiaro, C.; Fujisawa, K. and Solomon, E. I. 2003. Spectroscopic and electronic structure studies of the diamagnetic side-on CuII-superoxo complex Cu(O$_2$)[HB(3-R-5-iPrpz)$_3$]: Antiferromagnetic coupling versus covalent delocalization. *J. Am. Chem. Soc.* 125:466–474.

85. Wurtele, C.; Gaoutchenova, E.; Harms, K.; Holthausen, M. C.; Sundermeyer, J. and Schindler, S. 2006. Crystallographic characterization of a synthetic 1:1 end-on copper dioxygen adduct complex. *Angew. Chem. Int. Ed.* 45:3867–3869.

86. Spencer, D. J. E.; Aboelella, N. W.; Reynolds, A. M.; Holland, P. L. and Tolman, W. B. 2002. β-Diketiminate ligand backbone structural effects on Cu(I)/O$_2$ reactivity: Unique copper–superoxo and bis(μ-oxo) complexes. *J. Am. Chem. Soc.* 124:2108–2109.

87. Wada, A.; Harata, M.; Hasegawa, K. et al. 1998. Structural and spectroscopic characterization of a mononuclear hydroperoxo–copper(II) complex with tripodal pyridylamine ligands. *Angew. Chem. Int. Ed.* 37:798–799.

88. Jacobson, R. R.; Tyeklar, Z.; Farooq, A.; Karlin, K. D.; Liu, S. and Zubieta, J. 1988. A copper–oxygen (Cu$_2$–O$_2$) complex. Crystal structure and characterization of a reversible dioxygen binding system. *J. Am. Chem. Soc.* 110:3690–3692.

89. Tyeklár, Z.; Jacobson, R. R.; Wei, N.; Murthy, N. N.; Zubieta, J. and Karlin, K. D. 1993. Reversible reaction of dioxygen (and carbon monoxide) with a copper(I) complex. X-ray structures of relevant mononuclear Cu(I) precursor adducts and the *trans*-(μ-1,2-peroxo)dicopper(II) product. *J. Am. Chem. Soc.* 115:2677–2689.

90. Becker, M.; Heinemann, F. W. and Schindler, S. 1999. Reversible binding of dioxygen by a copper(I) complex with tris(2-dimethylaminoethyl)amine (Me6tren) as a ligand. *Chem. Eur. J.* 5:3124–3129.

91. Weitzer, M.; Schatz, M.; Hampel, F.; Heinemann, F. W. and Schindler, S. 2002. Low temperature stopped-flow studies in inorganic chemistry. *J. Chem. Soc. Dalton Trans.* 5:686–694.

92. Kitajima, N.; Fujisawa, K.; Moro-oka, Y. and Toriumi, K. 1989. .mu.-.eta.2:.eta.2-Peroxo binuclear copper complex, [Cu(HB(3,5-(Me$_2$CH)$_2$pz)$_3$)]$_2$(O$_2$). *J. Am. Chem. Soc.* 111:8975–8976.

93. Kitajima, N.; Fujisawa, K.; Fujimoto, C. et al. 1992. A new model for dioxygen binding in hemocyanin. Synthesis, characterization, and molecular structure of the μ-η$_2$:η$_2$ peroxo dinuclear copper(II) complexes, [Cu(HB(3,5-R$_2$pz)$_3$)]$_2$(O$_2$) (R = isopropyl and Ph). *J. Am. Chem. Soc.* 114:1277–1291.

94. Mahapatra, S.; Halfen, J. A.; Wilkinson, E. C. et al. 1996. Structural, spectroscopic, and theoretical characterization of bis(μ-oxo)dicopper complexes, novel intermediates in copper-mediated dioxygen activation. *J. Am. Chem. Soc.* 118:11555–11574.

95. Halfen, J. A.; Mahapatra, S.; Wilkinson, E. C. et al. 1996. Reversible cleavage and formation of the dioxygen O–O bond within a dicopper complex. *Science* 271:1397–1400.

96. Adams, H.; Bailey, N. A.; Dwyer, M. J. S. et al. 1993. Synthesis and crystal structure of a first-generation model for the trinuclear copper site in ascorbate oxidase and of a dinuclear silver precursor. *J. Chem. Soc. Dalton Trans.* 8:1207–1216.

97. Gonzalez-Alvarez, A.; Alfonso, I.; Cano, J. et al. 2009. A ferromagnetic [Cu$_3$(OH)$_2$]$^{4+}$ cluster formed inside a tritopic nonaazapyridinophane: Crystal structure and solution studies. *Angew. Chem. Int. Ed.* 48:6055–6058.

98. Suh, M. P.; Han, M. Y.; Lee, J. H.; Min, K. S. and Hyeon, C. 1998. One-pot template synthesis and properties of a molecular bowl: Dodecaaza macrotetracycle with μ$_3$-oxo and μ$_3$-hydroxo tricopper(II) cores. *J. Am. Chem. Soc.* 120:3819–3820.

99. Inoue, M.; Ikeda, C.; Kawata, Y.; Venkatraman, S.; Furukawa, K. and Osuka, A. 2007. Synthesis of calix[3]dipyrrins by a modified Lindsey protocol. *Angew. Chem. Int. Ed.* 46:2306–2309.

100. Cole, A. P.; Root, D. E.; Mukherjee, P.; Solomon, E. I. and Stack, T. D. P. 1996. A trinuclear intermediate in the copper-mediated reduction of O$_2$: Four electrons from three coppers. *Science* 273:1848–1850.

101. Tsui, E. Y.; Day, M. W. and Agapie, T. 2011. Trinucleating copper: Synthesis and magnetostructural characterization of complexes supported by a hexapyridyl 1,3,5-triarylbenzene ligand. *Angew. Chem. Int. Ed.* 50:1668–1672.

102. Groves, J. T. 1985. Key elements of the chemistry of cytochrome P-450: The oxygen rebound mechanism. *J. Chem. Educ.* 62:928.

103. Groves, J. T. and Viski, P. 1989. Asymmetric hydroxylation by a chiral iron porphyrin. *J. Am. Chem. Soc.* 111:8537–8538.

104. Yamazaki, H. and Cvetanovich, R. J. 1964. Collisional deactivation of the excited singlet oxygen atoms and their insertion into the CH bonds of propane. *J. Chem. Phys.* 41:3703–3710.

105. Paraskovopoules, G. and Cvetanovich, R. J. 1969. Reaction of the excited oxygen atoms (O^{1}D$_2$) with isobutane. *J. Chem. Phys.* 50:590–600.

106. Karasevich, E. I.; Kulikova, V. S.; Shilov, A. E. and Shteinman, A. A. 1998. Biomimetic alkane oxidation involving metal complexes. *Russ. Chem. Rev.* 67:335–355.

107. Bento, I.; Martins, L. O.; Lopes, G. G.; Carrondo, M. A. and Lindley, P. F. 2005. Dioxygen reduction by multi-copper oxidases; a structural perspective. *Dalton Trans.* 21:3507–3513.

108. Nagababu, P.; Yu, S. S.-F.; Maji, S.; Ramu, R. and Chan, S. I. 2014. An efficient catalyst for controlled oxidation of hydrocarbons under ambient conditions. *Catal. Sci. Technol.* 4:930–935.

109. Culpepper, M. A.; Cutsail, G. E. III; Hoffman, B. M. and Rosenzweig, A. C. 2012. Evidence for oxygen binding at the active site of particulate methane monooxygenase. *J. Am. Chem. Soc.* 134:7640–7643.

110. Kao, W. C.; Wang, V. C. C.; Huang, Y. C.; Yu, S. S.-F.; Chang, T. C. and Chan, S. I. 2008. Isolation, purification and characterization of hemerythrin from *Methylococcus capsulatus* (Bath). *J. Inorg. Biochem.* 102:1607–1614.

111. Kao, W. C.; Chen, Y. R.; Yi, E. C. et al. 2004. Quantitative proteomic analysis of metabolic regulation by copper ions in *Methylococcus capsulatus* (Bath). *J. Biol. Chem.* 279:51554–51560.

112. Chen, K. H.-C.; Wu, H.-H.; Ke, S.-F. et al. 2012. Bacteriohemerythrin bolsters the activity of the particulate methane monooxygenase (pMMO) in *Methylococcus capsulatus* (Bath). *J. Inorg. Biochem.* 111:10–17.

5 Oxygen-Evolving Complex of Photosystem II
Insights from Computation and Synthetic Models

*Jacob S. Kanady, Jose Mendoza-Cortes,
William A. Goddard III, and Theodor Agapie*

CONTENTS

5.1 PHOTOSYNTHESIS AND PHOTOSYSTEM II

One of the most fascinating and important transformations in nature is the biological generation of O_2 by the oxygen-evolving complex (OEC) of photosystem II (PSII) in cyanobacteria and plants.[1-8] This transformation was responsible for the formation of the oxygenic atmosphere that has shaped the evolution of life on Earth as we know it. In this process, solar energy is converted to the reducing equivalents and proton gradient necessary to power carbon dioxide fixation and other processes of life, while forming dioxygen as a by-product. The biological catalyst PSII has been studied in detail for >50 years. Progress in understanding the site of catalysis, the OEC, has depended on advances in several fields, including biochemistry, biophysics, spectroscopy, inorganic chemistry, and computational chemistry. While many properties of the OEC are well documented and generally agreed on, many aspects of the catalytic site remain controversial, with computational and experimental chemists still pushing the boundaries of our understanding of the OEC. Herein, we provide an overview of the structural and mechanistic proposals of the OEC as they were reported chronologically, and the technologies and methods that supported them, with a focus on the insight gained from recent synthetic inorganic and computational work in the field.

Photosystem II is a 350-kDa homodimer in the thylakoid membrane with ca. 20 protein subunits.[6,8] PSII absorbs photons that drive the separation of charge, which is transferred through several redox cofactors. The ultimate electron donor is water, being oxidized to O_2 and releasing four electrons and four protons. The chemiosmotic gradient generated by the released protons powers ATP synthesis. The electrons are transferred from the site of catalysis, the OEC, through tyrosine D1-Tyr161 (Y_Z) to chlorophyll a P680, pheophytin a, quinone A, and quinone B (Figure 5.1). Structurally, the D1 and D2 subunits make up the main membrane-bound core of PSII, with D1 containing much of the electron transfer pathway.[9,10] The other membrane-bound subunits mainly function as a light absorption antenna via a multitude of cofactors to transfer the exciton to P680. There are also a number of extrinsic, water-soluble subunits that bind to the lumenal side of PSII that are proposed to stabilize the binding of the Ca^{2+} and Cl^- cofactors necessary for efficient oxygen evolution.[11]

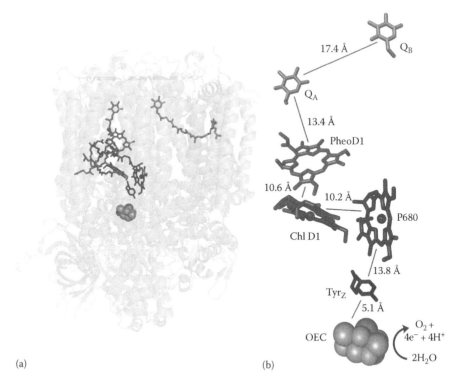

(a)   (b)

FIGURE 5.1 Electron transfer pathway shown in the overall PSII structure given by the 1.9 Å diffraction data[10] (a) and with distances (b). Aliphatic tails of the quinones, PheoD1, and chlorophylls are not shown for clarity.

5.2 OXYGEN-EVOLVING COMPLEX: COMPOSITION AND KOK CYCLE

The OEC is located on the lumenal face of PSII with the majority of ligating side chains from the D1 subunit, positioning it approximately 5 Å away from Y_Z.[10] Manganese, calcium, and chloride are all necessary for OEC function. The OEC has been known to contain Mn since the 1950s,[12,13] although oxygenic photosynthesis has been known to be Mn dependent for much longer.[14] As PSII isolation and purification methods improved, the stoichiometry of four Mn ions was verified by a number of methods, including quantitative electron paramagnetic resonance (EPR) spectroscopy of released Mn^{2+},[15] and atomic absorption spectroscopy.[16,17] The specific importance of Ca^{2+} over other dications was proposed in the 1970s on the basis of O_2 evolution activities at variable Ca^{2+} concentrations and the catalytic ineffectiveness or inhibitory effects of other dications.[11,18–21] Given the redox nature of the catalytic reaction, the role of the redox-inactive Ca^{2+} has been debated. Notably, the only metal to substitute for Ca^{2+} and generate a catalytic system, albeit with lower activity, is Sr^{2+}.[22] A single Ca^{2+} center is required for the restoration of the catalytic activity.[23] The close association of the redox-inactive metal with the OEC was supported by early EPR data on Sr^{2+}-substituted samples.[24] Removal of Ca^{2+} was shown by spectroscopy to arrest the catalytic cycle at intermediate states and affect electron transfer, further supporting the role of Ca^{2+} in catalysis.[25–28] More recently, EPR and XAS studies indicate that Ca^{2+} is part of the OEC.[29–31]

Until recently, Cl^- was also thought to be part of the OEC, as it is a native cofactor for O_2 production[32,33] and found to have a 1:1 stoichiometry with the OEC, based on $^{36}Cl^-$ labeling analysis.[34] However, more recent structural work suggests a role as H-bond acceptor near the OEC,[10,35] but not directly coordinated to it.

The OEC must be reassembled frequently under full solar flux due to photooxidative damage.[36] The assembly of the OEC, called photoactivation,[13,37] requires Mn^{2+}, Ca^{2+}, Cl^-, bicarbonate, water, and photogenerated oxidizing equivalents from P680.[38–42] A mechanism has been proposed based on kinetic and spectroscopic data: Mn^{2+} first binds to a "high-affinity" site proposed to contain D1-Asp170,[43] and is photooxidized in low quantum yield to Mn^{3+}, giving intermediate 1 (IM_1). The quantum efficiency of this initial oxidation is dependent on the presence of Ca^{2+}, which can bind either before or after the initial Mn^{2+}.[44] Ca^{2+} is proposed to bridge to the Mn^{3+} center through oxide or hydroxide bridges. After binding a second Mn^{2+} and photooxidation, a rate-limiting protein conformation change affords IM_2 that is quickly transformed into the OEC with additional Mn^{2+} equivalents in kinetically unresolved steps that must include deprotonation and incorporation of water as oxide donors.[41,42,44–47]

With respect to the mechanism of catalysis, a dependence of O_2 production on the number of short flashes of light on chloroplasts was discovered, as early as the 1960s.[48–50] Dark-adapted chloroplasts gave a spike in O_2 production on the third millisecond flash, followed by shorter spikes every four subsequent flashes until steady-state O_2 production was observed. Kok proposed that each flash corresponded to a photooxidative event, with three oxidizing equivalents stored until the fourth flash, upon which four-electron oxidation of water to O_2 occurs. In this so-called S-state cycle (Scheme 5.1), S_1 is the dark stable state and S_4 is the transiently formed state

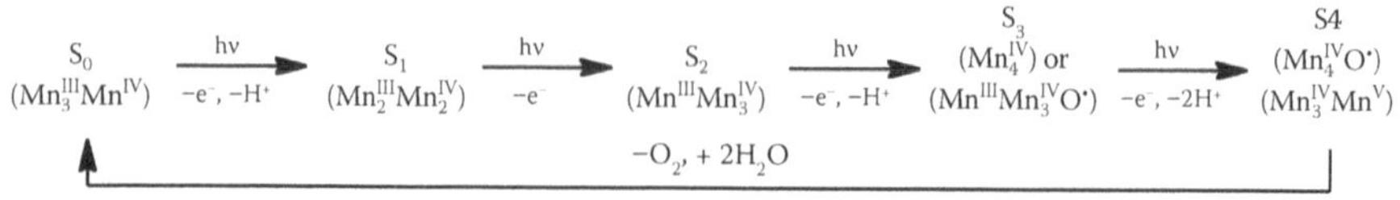

$$S_0 \xrightarrow[-e^-,\,-H^+]{h\nu} S_1 \xrightarrow[-e^-]{h\nu} S_2 \xrightarrow[-e^-,\,-H^+]{h\nu} \begin{array}{c} S_3 \\ (Mn_4^{IV}) \text{ or} \\ (Mn^{III}Mn_3^{IV}O^{\bullet}) \end{array} \xrightarrow[-e^-,\,-2H^+]{h\nu} \begin{array}{c} S4 \\ (Mn_4^{IV}O^{\bullet}) \\ (Mn_3^{IV}Mn^{V}) \end{array}$$

S_0 $(Mn_3^{III}Mn^{IV})$ S_1 $(Mn_2^{III}Mn_2^{IV})$ S_2 $(Mn^{III}Mn_3^{IV})$

$-O_2, +2H_2O$

SCHEME 5.1 High oxidation state pathway for the S-state cycle.

that releases O_2 and relaxes back to S_0. The four oxidations of the OEC have to be negative of $E^{\circ\prime}$ = ca. 0.9 V as necessitated by the potential of the P680$^+$; concurrent deprotonation helps keep the overall OEC charge low and thus levels the potentials of the S-state transition.[51–54] For the S-state cycle, two possibilities have been put forward for the Mn oxidation states, the "high" (Scheme 5.1) and the "low" pathways. The high-oxidation-state pathway has been supported by EPR,[55] [55]Mn electron nuclear double resonance (ENDOR),[56] X-ray absorption spectroscopy (XAS),[57] and Kβ X-ray emission spectroscopy (XES).[58–60] However, biochemical and spectroscopic data have also been interpreted to support the lower oxidation state cycle with an $Mn^{II}Mn_2^{II}Mn^{IV}$ or Mn_4^{III} S_1 state.[61–63]

Utilizing time-resolved mass spectrometry, Ollinger and Radmer[64] and then Messinger, Wydrzynski, and coworkers[65–72] studied the kinetics of substrate water binding to the OEC throughout the S-state cycle. These studies found that water is exchangeable through S_3, suggesting that no intermediate oxidations of water occur[64]; in all of the S-states, there are kinetically distinct, fast (40 s^{-1} for S_3 to ≥120 s^{-1} for S_0 and S_1) and slow (0.02 s^{-1} for S_1 to 10 s^{-1} for S_0) exchanging substrate waters, consistent with two separate sites of water coordination to the OEC[65,67,73]; both substrate waters are bound by the S_2 state[69]; and Sr^{2+} substitution of Ca^{2+} gives an increase in rate for the slow-exchanging water, suggesting it is bound to Ca^{2+}.[70] There are a number of possible ways to explain the slow- and fast-exchanging waters, including protonation state, Mn^{III,IV} or Ca^{2+} coordination, and terminal or bridging ligation mode. Thus, these studies are an important consideration for many mechanistic proposals for O–O bond formation (see Section 5.5).

5.3 STRUCTURAL PROPOSALS FOR THE OEC: A HISTORICAL PERSPECTIVE

Structural understanding of the OEC has gradually developed during the last 30 years, with many methods across multiple disciplines being paramount. Although multiple XRD structures are now known and provide the location and amino acid ligands of the OEC,[9,10,35,74–79] changes in the structure of the OEC due to reductive X-ray damage has been a concern.[80] EPR[81–84] and XAS[85–89]—techniques used to study the OEC since the early 1980s[90–93]—complement the XRD data to afford more complete structural information. Crucial for these two methods was the parallel growth in the synthetic inorganic coordination chemistry of manganese, particularly of multinuclear cluster chemistry (Section 5.6).[94–100] The synthetic systems not only showed what was chemically reasonable to propose for the OEC based on precedent, but, equally important, they acted as spectroscopic benchmarks, providing starting points for hypotheses on how the OEC's spectra relate to structure. Simple synthetic complexes were also key in benchmarking quantum mechanical (QM) computation,[101–110] which has emerged as a crucial method in the study of the OEC. QM methods have improved drastically in the last decade to allow for structural hypotheses for every S-state and mechanisms for substrate water incorporation and O_2 formation (Section 5.4).[107,111,112]

With all of these different methods, and the improvements to each over time, numerous OEC structures have been proposed with a significant amount of disagreement and controversy over the years. The main models discussed during the last 25 years are shown schematically in Figure 5.2.[113] One of the earliest models for the OEC with a specified geometry for the four manganese centers was the Mn_4O_4 cubane/Mn_4O_6 adamantane model proposed by Brudvig and Crabtree in 1986

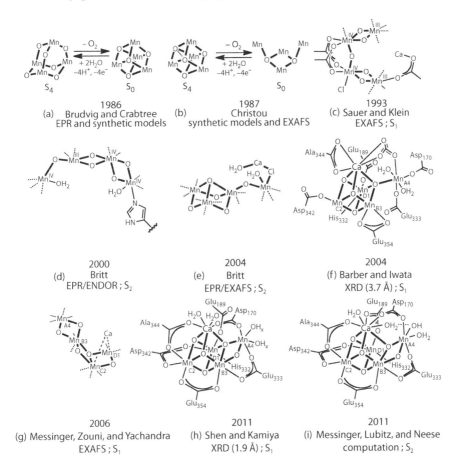

FIGURE 5.2 Key structural models of the OEC. Research group, year, main spectroscopic support, and S-state are included below each structure. Crystal structure resolutions are in parentheses. Dashed lines represent generic coordination sites and could represent amino acids or water. In f through i, Mn numbering combines EXAFS nomenclature ($Mn_{A–D}$)[31] with that of the 2005 and 2011 crystal structures ($Mn_{1–4}$).[10,76] (In the style of Rapatskiy, L. et al., *J. Am. Chem. Soc.* 134, 16619, 2012.)

(Figure 5.2a).[114] They proposed that a *pseudo*-Jahn–Teller distorted Mn_4O_4 cubane could explain their recent EPR data on the S_2 state that suggested two antiferromagnetically coupled dimers ferromagnetically coupled to the other.[115] They also posited that a large structural change in the S_2 to S_3 transition was consistent with X-ray absorption near-edge spectroscopy (XANES) K-edge data of the time that showed a decrease in edge energy between S_2 and S_3.[116] However, the high symmetry of the proposed cubane and adamantane geometries did not prove consistent with extended X-ray absorption fine structure (EXAFS) reported subsequently.[117,118]

Another OEC model based on an Mn_4O_4 cubane was put forward soon after and was dubbed the "double-pivot" mechanism by Vincent and Christou (Figure 5.2b).[119,120] Here, an Mn_4O_2 butterfly structure in the S_0 to S_2 states was proposed to bind and deprotonate two water molecules to afford an $Mn_4O_2(OH)_2$ cubane structure that upon double deprotonation affords dioxygen and the S_0 butterfly structure. Key to the proposal was a synthetically characterized Mn_4O_2 structure (Figure 5.4c) that contained Mn–Mn vectors at ca. 2.7 and 3.3 Å, similar to those found in past EXAFS studies.[93,121] Although further EXAFS studies[122] were not consistent with this proposal, synthetic work by the groups of Christou and Dismukes detailed a variety of structural motifs and properties of clusters of these types (Section 5.6).

A structural model based on oriented-membrane EXAFS was proposed in 1993 and is generally referred to as the "dimer-of-dimers" model.[123] The basic structure is two $Mn_2(\mu_2\text{-}O)_2$ dimers connected through a mono-μ-O and/or κ^2-carboxylates (Figure 5.2c). The Cl^- and Ca^{2+} cofactors were originally proposed to bind Mn and to bridge to the end of one Mn_2O_2 dimer unit through a carboxylate, respectively. The dimer-of-dimers served as the basis for a number of mechanistic proposals, including a metalloradical mechanism[124] and a nucleophilic attack by calcium-ligated hydroxide/water on an electrophilic Mn^V=O.[125]

A different structure, the "trimer/monomer," "3+1," or "dangling Mn" model (Figure 5.2d,e), was proposed on the basis of EPR experiments. Britt and coworkers posited that the magnetic interaction of the Mn in the dimer-of-dimers model could not explain the high-spin $g = 4.1$ signal and the changes to the $g = 2$ multiline signal upon addition of methanol and ammonia.[126] The 3+1 motif, which had been included as a possible structure based on EXAFS data on S_2 (i.e., Figure 9 of Ref. 127), could explain the EPR data as a strongly antiferromagnetically coupled III,IV,IV trimer only weakly coupled to the fourth, "dangling" Mn^{IV}. A handful of trimer/monomer arrangements were proposed that fit the EPR and EXAFS data of the time.[55] Soon after this, the first crystal structures of PSII were reported,[9,74] and although the resolution was only 3.8 Å (2001) or 3.7 Å (2003), the manganese electron density was consistent with a 3+1 arrangement. In 2004, based on a higher resolution of 3.5 Å, Barber and Iwata proposed a more specific structure: an Mn_3CaO_4 cubane with a fourth manganese connected by a cubane oxygen (Figure 5.2f).[75] This was consistent with the EPR proposal and also Ca K-edge XAS data that suggested a Ca–Mn distance of 3.4 Å.[30] In 2005, a higher-resolution structure of 3.0 Å was published that reported the OEC in a 3+1 arrangement more distorted and elongated than a cubane motif, without proposing the position for the bridging oxides (the same group published a 2.9 Å PSII structure in 2009 with no change to the OEC geometry).[35,76]

XAS studies in 2005 showed that the X-ray dose used in the XRD analysis of PSII caused reductive damage to the OEC, shedding doubt on the accuracy of the proposed OEC arrangement as based on crystal structures.[80] Polarized EXAFS studies at much lower X-ray dosage on PSII single crystals were used to provide an updated structure of the OEC (Figure 5.2g) with an asymmetric dimer of Mn_2O_2 diamond cores.[31] A different interpretation of the EXAFS data invoked a cubane with a dangler motif.[54] In 2011, a significantly higher-resolution (1.9 Å) crystal structure was published[10] with purported X-ray dosage below the damage level reported in 2005. Here, a "chair" geometry of the Mn_4CaO_5 was observed at atomic resolution, similar to that proposed in the 2004 crystal structure but with an extra μ_2-O between the dangling Mn and the cubane (Figures 5.2h and 5.3).

There has been controversy over the OEC assignment in this recent XRD study because some of the Mn–O bond lengths are not consistent with a supposed S_1 oxidation state of $Mn_2^{III}Mn_2^{IV}$. Three explanations have been given in the literature, all based on computational modeling: the OEC structure is accurate and supports the low-oxidation-state Kok cycle, with an Mn_4^{III} S_1 state[63,128]; X-ray damage has produced a mixture of reduced oxidation states including S_{-n} states[129–132]; and the observed electron density is a superposition of two S_1 substates in equilibrium by a μ-O migration and proton transfer.[133] Another computational study suggested a similar substate equilibrium for S_2, claiming to explain the two S_2 EPR signals through changes in the magnetic coupling caused by μ-O migration from the cubane unit to form a diamond core with the dangling Mn (Figure 5.2i).[130,134,135] Although some

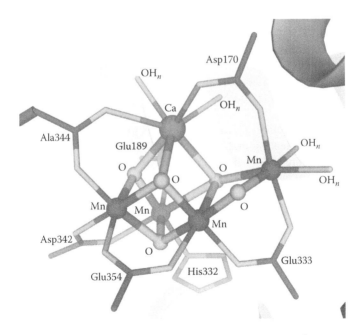

FIGURE 5.3 Oxygen-evolving complex as described by the 1.9 Å resolution crystal structure.[10]

controversy still remains about the structural assignment of the OEC from the 1.9 Å PSII structure, the present data converge toward a cluster with a CaMn$_3$ site (part of a cubane or distorted cubane) and a dangler Mn center, with bridging oxido moieties.

5.4 COMPUTATIONAL INVESTIGATIONS OF THE OEC

The OEC and its coordination sphere contain hundreds of atoms, which make computational treatments challenging. XAS, EPR, and XRD have provided information about the relative positions of various moieties during several states of the cluster. Using these starting points, computational chemists have employed a variety of methods to model the large OEC system. Density functional theory (DFT) combines reasonable scaling, computation time, and sufficient accuracy when hybrid functionals are used.[136,137] The most popular hybrid functional due to its success addressing different chemical reactions is Becke's three-parameter hybrid functional known as B3LYP.[138,139] This has been the main method used to investigate chemical reactions of large molecules and has thus been applied to the oxygen-evolving reaction at the OEC of PSII.[140]

The first computational studies of the OEC were published near the turn of the century,[101–103] with the dimer-of-dimers as structural model. At that time, highly simplified models were used at the B3LYP level to keep the computational cost reasonable. In 1997,[101] a ferromagnetically coupled Mn$_2$O$_2$ model was explored with the coordination shell completed with H$_2$O (or OH$^-$). This model was slightly modified in 1999 to include Ca^{2+} and Cl$^-$,[102] and in 2000, this model was further improved by including a larger Mn$_3$CaClO$_{15}$H$_n$ cluster, with the oxygen radical mechanism investigated through the S-state cycle.[103] This method included the effective core potentials for Mn but all electrons for the Ca, O, H, and Cl atoms. For the single-point energy, a set of polarization functions were added to all atoms—higher angular momentum functions used to describe electron correlations and bonds more accurately—and one set of diffuse functions on Mn—necessary to describe properly any loosely bound electrons.[141–144]

The first crystal structures by Kamiya and Shen in 2003 (3.7 Å), Barber and Iwata in 2004 (3.5 Å), and Saenger and Zouni in 2005 (3.0 Å)[74–76] allowed computational chemists to test a more restricted set of connectivities. At the same time, advances in computational power and theory[145] made it possible to treat a larger number of atoms, making the computational models more realistic. The combination of these factors gave computation an important role in helping determine the oxygen-evolution mechanism of the OEC. Multiple groups proposed different models based on these crystal structures, including those of Siegbahn,[146,147] Brudvig/Batista,[106,148,149] and Kusunoki[150]—who investigated the "high" oxidation path mechanism (Table 5.1, Scheme 5.1)—and Pace/Stranger[151–153] and Isobe[154]—who studied both the "high" and "low" oxidation path (Table 5.2). The different computational method, chemical connectivity, and ligands used for the computational models have many implications on the mechanism proposed, particularly in the oxidation pattern of the Mn through the Kok cycle (Tables 5.1 and 5.2).[112] We discuss briefly the different proposals.

The first proposed mechanism for the Kok cycle based on computational calculations was from Siegbahn, who used the coordinates of the Barber and Iwata

TABLE 5.1
"High" Oxidation Path Mechanism

	Siegbahn		Batista		Pace/Stranger ("High")			Kusunoki	Neese
XRD	**B&I**	**S&Z**	**B&I (a)**	**B&I (b)**	**B&I**	**S&Z**	**K&S'03**	**S&Z**	**K&S'11**
Ref. No.	146	155[a]	156[b]	156	152	152	152	150	129
S0	3433	3343	3433	3433	4432	3433	3433	3334	–
S1	3443	3344	4433	3434	4442	4433	4433	4334	–
S2	4443	3444	4443	4434	4443	4443	4434	4434	3444
S3	4444	4444	4444	4444	–	–	–	4444	–

Note: The table shows a comparison of the different Mn oxidation states assigned for the S-state cycle by the various computation models. The XRD coordinates for the OEC were based on the Kamiya and Shen 2003 (K&S'03),[74] Barber and Iwata 2004 (B&I),[75] Saenger and Zouni 2005 (S&Z),[76] or Kamiya and Shen 2011 (K&S'11)[10] structure. The oxidation states are given in the order Mn_{D1}, Mn_{C2}, Mn_{B3}, Mn_{A4}, following the numbering convention of Figure 5.2.

[a] Siegbahn's more recent publications are consistent with the oxidation states proposed in 2008.

[b] Batista's later work is consistent with the 2006 structure (a).

TABLE 5.2
"Low" Oxidation Path Mechanism

	Isobe	P/S ("Low")			
XRD	**B&I**	**B&I**	**S&Z**	**K&S'03**	**K&S'03**
Ref. No.	154	152	152	152	153
S0	[333]2[a]	3332	3332	2432	3332
S1	[333]3	3432	3432	3432	3432
S2	[334]3	4432	3433	3433	3433
S3	[334]4	4442	3443	3443	3434

Note: The table shows a comparison of the different Mn oxidation states assigned for the S-state cycle by the various computation models: Pace/Stranger (P/S)[151–153] and Isobe.[154] Nomenclature is the same as Table 5.1.

[a] Brackets correspond to an unspecific assignment of an oxidation state to three of the Mn centers.

2004 crystal structure to create simplified computational models.[157] These models included the Mn ion position joined by oxo-bridges and a simplified coordination sphere (aspartate and glutamate side chains represented by formate and histidine by imidazole) that was combined with XAS data of the time. From this approach, five $CaMn_4$ cluster orientations were tested because at that time the oxo-bridges arrangement was not fully resolved by the XRD. The computational method used for this model was B3LYP/lacv3p*//B3LYP/lacvp. This model was further refined in 2005 by the same authors by including polarization functions on the hydrogen atoms (i.e., B3LYP/lacv3p**//B3LYP/lacvp) and slight differences in ligand set, affording a slightly different cluster connectivity and Mn–Mn distances in closer agreement with available EXAFS measurements (Table 5.3).[146] In both models, the Mn atoms were ferromagnetically coupled, although it has been shown that the optimized geometrical parameters are not so dependent on the spin configurations.[129] In 2008, Siegbahn[147] published his more definitive model but this time based on the Saenger and Zouni 2005 crystal structure. In this model, most of the carboxylate ligands are coordinated in bidentate fashion. The amino acids coordinating the metallic cluster were simplified by replacing the α-carbon with a methyl group, except the His332 and Glu333 that were left intact because they form a bridge between the dangler Mn and the Mn_3Ca unit. This structural model is fairly consistent with the EXAFS data (Table 5.3).[88,158] For this model, the procedure used was of higher quality than the previous models, B3LYP/lacv3p+/cc-pvtz(−f)//B3LYP/lacvp*. For the optimization, the polarization functions were included for all the atoms except H; for the single-point calculations, the diffuse functions were included for metals and the correlation consistent basis set without f functions, cc-pvtz(−f), was used for all other atoms. The diffuse functions added are essential in describing properly loosely bound electrons such as anions or excited states, which is somewhat expected in the OER mechanism. However, diffuse functions are not commonly used in large molecules because of computational constraints, i.e., larger number of basis sets and longer convergence time. The inclusion of polarization functions for the optimization and diffuse functions for the single-point energy gave a high-quality oxo–oxyl model/mechanism that has only been slightly refined since.[111,132,159]

Following the Barber and Iwata 2004 crystal structure, the Batista group generated a model in 2005 that includes all amino acid residues with α-carbons <15 Å from the OEC.[148] The optimized structure contains a D1-Glu333 forming a carboxylate bridge between two Mn ions, and waters bound to Ca^{2+} (two), the dangling Mn (three), and to the Mn (two) of the core. This model contained >2000 atoms, which was too large to be handled by QM methods; thus, Batista and coworkers used a molecular mechanics (MM) approach. In the MM approach, balls and springs represent the atoms and bonds, and Newtonian equations are applied with the potential dictated by the Amber force field (FF).[160] The advantage of being able to treat thousands to millions of atoms with MM comes with the cost of not being able to model the electronic properties of the system, the QM regime. Accordingly, in 2006, Batista and coworkers used a hybrid QM/MM method where the reaction center (cluster and first coordination shell: Glu189, Glu354, Ala344, Glu333, Asp170, Asp342, and the imidazole ring of His332) was treated by QM methods while the distal moieties (beyond the first coordination shell) are treated by MM.[156] This model includes all the amino acid residues within 15–20 Å of the OEC. This approach produced two sets of structures, a and b (Tables 5.1 and 5.3), with the only difference being that model a has a pentacoordinated Mn_{A4} and model b has an hexacoordinated Mn_{A4} with an extra water, which changes the oxidation state pattern for the Kok cycle (Table 5.1). The Mn–Mn distances for model a are consistent with experimental EXAFS data (Table 5.3)[31]; however, in the Batista models, the Cl^- ion is bound to Ca^{2+}, which was later proved to be incorrect.[10,35] The computational method used for the QM shell consists of B3LYP with basis set lacvp for Mn, and 6-31G basis for the rest of the atoms; also, only the bridging oxo ions have two d- and one f-type polarization functions, i.e., 6-31G(2df). This was used for both the optimization and single-point energy.

TABLE 5.3
Mn–Mn Distance in S_1 for Different Computational Models Compared with Experiments

S_1	XRD	Siegbahn		Batista	P/S ("Low")			Kusunoki	Neese[a]	Yano	Dau	K&S'11
XRD	B&I	B&I	S&Z	B&I (a)	B&I	S&Z	K&S'03	S&Z	K&S'11	Exp.	Exp.	Exp.
Ref. No.	157	146	147	156	152	152	152	150	129	31	88	10
Mn_{D1}–Mn_{C2}	3.03	2.82	2.76	2.74	2.74	2.74	2.70	2.76	2.79	2.8	3.2	2.8
Mn_{D1}–Mn_{B3}	2.92	3.07	3.12	2.63	3.24	3.48	3.48	3.16	3.33	3.3	3.2	3.3
Mn_{C2}–Mn_{B3}	3.02	2.92	2.78	2.77	2.86	2.82	2.78	2.76	2.77	2.7	2.7	2.9
Mn_{B3}–Mn_{A4}	3.50	3.52	2.86	3.68	3.24	2.93	3.17	3.37	2.69	2.7	2.7	3.0

Note: The distance between Mn centers is in Å. Experiments from Yano,[31] Dau,[88] and Kamiya and Shen.[10] Nomenclature is the same as Tables 5.1 and 5.2.

[a] The Neese computations were for S_2 rather than S_1.

The MM used the Amber FF.[160] Using the same QM/MM computational method, Batista and coworkers calculated a complete S-state cycle in 2008[149] that supported a nucleophilic attack mechanism (Section 5.5) and reviewed their QM/MM work later that year.[107]

A direct comparison of the total energy between structures generated by the QM/MM approach[149] and a pure QM approach in the S_1 state was performed in 2009.[161] This comparison took the coordinates of the models and treated them with the same QM approach; B3LYP/lacv3p+/cc-pvtz(–f)//B3LYP/lacvp* and dielectric correction ($\varepsilon = 6$ and probe radius = 1.40 Å). This allows direct comparison of the total energy of the models. It was found that the proposed structure for the OEC obtained by QM/MM is a local minimum almost 70 kcal/mol less stable than a similar model obtained through pure QM.[111,147,162] This study ruled out the reported QM/MM models as possible candidates. In the same study, other models based on EXAFS[163] were ruled out on the same ground (34–63 kcal/mol less stable than the pure QM model).

Kusunoki developed another model for the Kok cycle based on the Saenger and Zouni 2005 crystal structure that produced Mn–Mn distances similar to the experimental structures (Table 5.3) with a different pattern for the Mn oxidation states (Table 5.2).[150] A comparative study based on models derived from three crystal structures[74–76] was performed by Pace, Stranger, and coworkers.[151] The model based on the Saenger and Zouni 2005 crystal structure produced computationally stable results, similar to the findings of Siegbahn.[147] The model generally matched the EXAFS data (Table 5.3), with the models based on Kamiya and Shen 2003 and Saenger and Zouni 2005 giving a similar oxidation pattern (Table 5.1). In 2011, the groups of Neese, Lubitz, and Messinger[129] calculated various protonation scenarios in the S_2 state. The coordinates of the OEC were based on the Kamiya and Shen 2011 structure, and a newer computational method was utilized, including third-generation van der Waals corrections and relativistic corrections. The computed structure agreed with earlier studies by Siegbahn[111,147] and Kusunoki.[133]

5.5 MECHANISM OF O–O BOND FORMATION

Paralleling the wide assortment of OEC structures put forward, several mechanisms for O–O bond formation have been proposed.[7,54,83,111,164–167] Consistency with XAS, EPR, XRD, and substrate water exchange studies are required for advancing any mechanism, and developments in these fields have disproved many past proposals, such as the adamantane and double-pivot mechanisms discussed previously.[113,116,117,119,122,168] As with the structural hypotheses, some mechanisms have their basis in the chemistry of synthetic transition metal complexes. Copper has been shown to break and form the O–O bond of O_2 in an equilibrium between a $Cu_2^{III}(\mu_2-O)_2$ diamond core and a $Cu_2^{II}(\mu_2-\eta^2:\eta^2-O_2)$ bridging peroxide.[169] Similar proposals for diamond core O–O bond formation in the OEC exist (Scheme 5.2a)[4,119,170]; however, the fast and slow water exchange kinetics are difficult to explain by such mechanisms.[168] Dinuclear ruthenium water oxidation catalysts have been shown to function through an H-bonding, nucleophilic water attacking an Ru^V-oxo intermediate,[166,171] and an Mn_2O_2 O_2-evolving catalyst has been proposed to act similarly through an Mn^V-oxo[172,173] or Mn^{IV}-oxyl radical.[103] Water attack on an

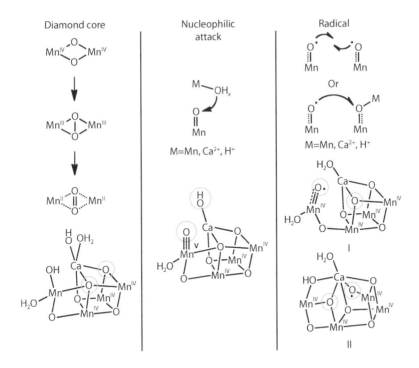

SCHEME 5.2 Proposed mechanisms for O–O bond formation depicted minimally with metal-oxo species (top) and as part of the most recent structural models of the OEC (bottom).

electrophilic $Mn^{IV/V}$-oxo has likewise been proposed for the OEC, with the attacking water in a number of different states, both terminal and bridging: as an H-bonding water/hydroxide,[65] Mn^{n+}-bound water/hydroxide,[124] or Ca^{2+}-bound water/hydroxide (Scheme 5.2b).[125,149,172,174,175] Brudvig and Batista proposed a Ca^{2+}-OH_2 nucleophilic attack mechanism supported by QM/MM calculations[106,107,149]; however, another QM approach used for the OEC and the first coordination shell implicated a different ground-state structure.[161]

The computational work of Siegbahn has supported an oxyl radical (Mn^{IV}–$O^{\bullet}$) in O–O bond formation at the OEC (Scheme 5.2c).[102–104,111,147,161,176] Others have proposed mechanisms including radical intermediates as well.[177–179] Both terminal[147] (Scheme 5.2c-I) and bridging[111] (Scheme 5.2c-II) oxyl radical intermediates have been discussed, with recent computational work supporting the coupling of an $Mn/Mn/Ca$-μ_3-oxo and an Mn/Ca-μ_2-oxyl in the S_4 state (Scheme 5.2c-II).[159] Recent ^{17}O-ENDOR studies mapped the substrate water exchange kinetics onto both the nucleophilic attack or bridging oxyl-coupling mechanisms.[113] Overall, a truly interdisciplinary approach of combining spectroscopy, structural characterization, computation, and comparisons with synthetic models has funneled the mechanistic proposals for water oxidation to only a few candidates. Further detailing the mechanism of O–O bond formation is very desirable for both fundamental reasons and application toward the development of practical artificial catalysts. Additional studies from multiple perspectives are necessary to achieve this goal.

5.6 SYNTHETIC OEC MODEL COORDINATION COMPLEXES

Synthetic manganese coordination clusters have played an important role in our understanding of the OEC, both inspiring the structural and mechanistic hypotheses of their time, and also being targeted because of the OEC structural motifs proposed on the basis of other various analytical techniques. Model complexes, detailed in a number of reviews,[94–100] have been an instrumental benchmarking tool for XAS, EPR, water exchange rates, and computation. This historically collaborative effort has been highlighted by numerous examples during the past 40 years, from the original manganese–bipyridine dimer and tetramanganese dimer-of-dimer models, to the more recent manganese–calcium heterometallic models.

In 1972, Plaksin and coworkers published the X-ray crystal structure of di-μ-oxo-tetrakis(2,2'-bipyridine)dimanganese(III,IV), showing an Mn–Mn distance of 2.716 Å and finding strong antiferromagnetic coupling (Figure 5.4a).[180] These two observations on a model complex were utilized to conclude that the Mn_2O_2 diamond core was a key structural motif within the OEC: comparison of the S_2-state multiline EPR signal to that of the complex[90,91,181] supported the idea of an antiferromagnetic $Mn^{III}Mn^{IV}$ pair within the OEC, and the original OEC EXAFS studies of 1981[93] found Mn–Mn distances of 2.7 Å, consistent with the Mn_2O_2 core characterized by crystallography. More recently, a similar Mn_2O_2 dimer using terpyridine rather than bipyridine was reported to oxidize water using hypochlorite (NaOCl) or oxone (H_2SO_5) as the stoichiometric oxidant (Figure 5.4h).[172,173] Relevant to mechanistic interpretations for the OEC, the water exchange rates of the Mn^{III}–O–Mn^{IV} units of the bi- and terpyridine manganese dimers were measured by a time-resolved mass spectrometry technique.[168,182] The exchange rates were much slower (10^{-3}–10^{-4} s^{-1}) than those of the OEC (ca. 1 s^{-1}), not consistent with mechanisms that invoked bridging oxo units as substrate water in the OEC.

Wieghardt and coworkers synthesized the first Mn_4^{IV} complex, an $Mn_4O_6^{4+}$ adamantane stabilized by three chelating 1,4,7-triazacyclononane ligands (Figure 5.4b).[183,184] With this precedent, the adamantane/cubane mechanistic proposal for the OEC invoked access to such a high oxidation state cluster.[113] Each Mn^{IV} displays a *pseudo*-octahedral coordination environment with three μ_2-oxido and three terminal N donors. Armstrong and coworkers further studied the Mn_4O_6 adamantane core structure, evaluating the effect of altering the chelating N_3 ligand on the basicity of the μ_2-oxido moiety and on the pH-dependent reduction potential.[185–187] More recently, the Mn^{IV}–O–Mn^{IV} water exchange rate of the adamantane geometry was measured to be ≤10^{-8} s^{-1}.[168]

The Mn_4O_4 cubane geometry was common to both the adamantane/cubane (1986) and the double-pivot (1987) mechanisms.[113,120] Tetramanganese complexes had been isolated in such a geometry[188,189]; however, these were low-oxidation-state Mn_4^{II} structures with μ_3-alkoxides rather than oxides. Bashkin and coworkers synthesized the first cubane complex with μ_3-O bridges: $[Mn_3^{III}Mn^{IV}O_3Cl_6(ImH)(OAc)_3]^{2-}$ (ImH = neutral imidazole; Figure 5.4d).[190] During the following decade, the $Mn_3^{III}Mn^{IV}O_3X$ cubanes/partial cubanes were studied in great detail, with variation of terminal ligands (Cl$^-$, pyridines, acetylacetonates, etc.) and the anionic μ_3-X position (X = Cl$^-$, Br$^-$, I$^-$, F$^-$, N$_3^-$, O$_2$CR$^-$, OMe$^-$, and OH$^-$).[190–203] For example, they were able to

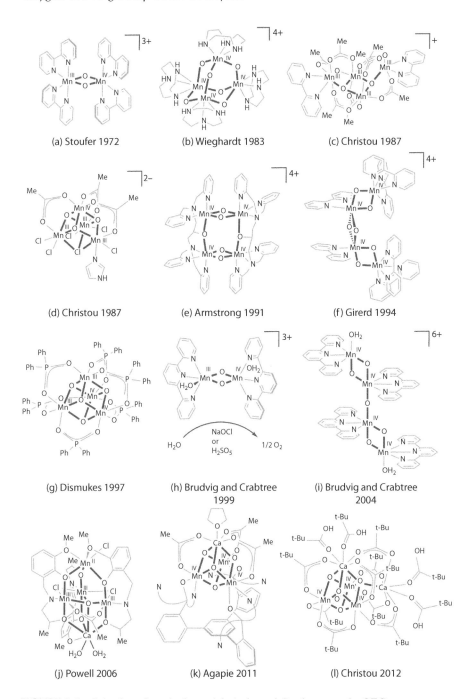

FIGURE 5.4   Selection of synthetic models (a through l) relevant to the OEC.

synthetically model the S_1 to S_2 step of the proposed double-pivot mechanism,[191] utilizing an $Mn_4^{III}O_2$ butterfly complex (Figure 5.4c) to form an $Mn_3^{III}Mn^{IV}O_3Cl$ cubane by addition of chloride and disproportionation. In another reactivity study, water was selectively deprotonated and incorporated into the μ_3-X position, modeling a key functional step in OEC photoassembly and turnover.[199]

Extensive magnetism[192,194,197,198,201,203] and XAS[200] studies were performed on the Mn_4O_3X cubanes to test the hypothesis that the OEC was not a high-symmetry cubane structure.[118,121] Although the K-edge XANES and EXAFS spectra looked superficially similar to those of the OEC in the S_1 state, detailed analysis indicated that the structural motif contained in these synthetic clusters did not match the data from the biological system. Also of note, the K-edge energy varied by >3 eV for a series of cubanes in the same oxidation state and similar geometry, supporting the notion that in addition to the formal metal oxidation state, the nature of the ligands strongly affects the edge energy. This convolution of effects complicates the interpretation of the edge energies of various clusters and continues to cause disagreement over the oxidation state of the OEC.[62]

Other systems that gave some support for the double-pivot mechanism were the diarylphosphonate-stabilized $Mn_2^{III}Mn_2^{IV}O_4^{6+}$ /$Mn^{III}Mn_3^{IV}O_4^{7+}$ cubanes synthesized by Dismukes and coworkers (Figure 5.4g).[204,205] They found that these cubane complexes lose one phosphinate ligand and a molecule of O_2 upon UV photolysis in the gas phase,[206] indicating the ability of an Mn_4O_4 cubane to form dioxygen as previously proposed for PSII.[119] This system was later found to electrochemically oxidize water if embedded in Nafion[207–209]; however, further study showed that decomposition to an amorphous manganese oxide provided the active catalyst.[210] The Mn_3–μ_3–O water exchange rates measured in organic solvent (10^{-5} s^{-1}) were one to two orders of magnitude slower than the synthetic complex Mn^{III}–μ_2–O–Mn^{IV} rate (10^{-3}–10^{-4} s^{-1}) and thus much slower than those found for the OEC.[211]

As spectroscopic[15,115,212,213] and biochemical[16,17] support for a tetramanganese OEC grew, Mn_4 complexes were targeted that contained the 2.7 and 3.3 Å Mn–Mn vectors reported for the OEC.[94,97] For example, alongside the butterfly systems discussed previously,[214] Armstrong's group reported a series of dimer-of-dimer geometries.[215–217] They contained two 2.7 Å Mn–Mn vectors each, and the EPR of the highest oxidation state dimer—with two Mn^{III}–$(\mu$-O$)_2$–Mn^{IV} diamond cores (Figure 5.4e)—modeled that of the S_1 state. Toward modeling the EXAFS dimer-of-dimers proposal in 1993,[123] complexes such as the Mn_4^{IV} diamond core chain structure by Girerd and coworkers (Figure 5.4f)[218] and the $[Mn^{IV}$–$(\mu_2$-O$)_2$–$Mn^{IV}]_2O$ dimer-of-dimers by Brudvig and coworkers (Figure 5.4i)[219] were reported.

On the basis of Ca K-edge XAS data, the calcium ion was proposed to be closely associated with the tetramanganese motif of the OEC, with an Mn–Ca vector of 3.4 Å.[30] In agreement, the 2004 crystal structure proposed an OEC structure displaying an Mn_3CaO_4 cubane moiety. Calcium is necessary for photoactivation (cluster assembly from Mn^{2+} in solution under light) and turnover of the OEC. Synthetic Mn/Ca complexes were targeted to understand the effect of the redox-inactive metal on the chemistry of manganese clusters. The first high-oxidation-state Mn/Ca cluster was isolated in 2005 and contained an Mn_4CaO_4 motif quite similar to the 2004 crystal structure as part of a high-nuclearity $Mn_{13}Ca_2O_{10}$ cluster coordinated by benzoates.[220] A Ca K-edge XAS study on this cluster showed an Mn–Ca vector of ca. 3.5 Å similar to the one in the OEC.[221] Two Mn/Ca complexes have been synthesized with the correct Mn_4Ca metal stoichiometry, although in low oxidation state and with low oxide content. The first contains a trigonal bipyramidal arrangement of metals with an Mn^{II} and Ca^{2+} at the two vertices and one μ_4-oxide, with a low $Mn_3^{III}Mn^{II}$ oxidation state (Figure 5.4j).[222] Similar complexes isolated later by the same group showed O_2 evolution in the presence of O-atom transfer agents and water.[223] A more recent cluster displays an Mn_4^{III} metallocrown moiety with a Ca^{2+} center coordinated to one side of the crown and chelated by carboxylates; this cluster contains no bridging oxido ligands.[224] Other Mn/Ca structures—a low-oxidation-state $Mn_4^{II}Ca_2$ cluster[225] and a high-nuclearity $Mn_6^{III}Ca_2O_2$ complex[226]— have also been reported.

5.7 SITE-DIFFERENTIATED OEC MODEL COMPLEXES

Most of the multinuclear complexes discussed previously were synthesized by a self-assembly method that offers only low control over the geometry and nuclearity of the final complex. Although this manganese cluster chemistry has been invaluable to understanding the OEC, new methods for the controlled synthesis of Mn/Ca complexes are important, especially with the structure of the OEC emerging as a low-symmetry Mn_4CaO_5 cubane/open cubane. In related bioinorganic studies, Holm and coworkers pioneered a synthetic protocol, termed "subsite-specific functionalization," to study the properties of ubiquitous Fe_4S_4 biological clusters. A wide array of ligand-differentiated $Fe_4S_4X_3X'$ and metal-differentiated Fe_3MS_4 complexes were accessible using this method.[227–232] The basis of this synthetic strategy is a semirigid tridentate ligand design that can accommodate binding three metals of the Fe_4S_4 core, leaving the fourth metal center open to ligand substitution or replacement by a heterometal. In the context of metal-oxide clusters, a series of complexes prepared by Christou afforded site differentiation, although not by ligand design, in $Mn_3^{III}Mn^{IV}O_3X$ cubanes with respect to substitution of a μ_3-X anion to Cl^-, Br^-, I^-, F^-, N_3^-, $RCOO^-$, MeO^-, and HO^-.[190–203] This allowed a controlled study of the magnetic properties and XAS K-edge energies as affected by a single change in the cubane structure.

Toward well-defined syntheses of heterometallic metal-oxide clusters, the site-differentiated functionalization method based on ligand design was employed recently to access a series of Mn_3MO_n OEC model complexes.[233–237] The ligand framework was designed to bind three metal centers in close proximity, to accommodate multiple coordination modes, and to be oxidatively robust. These design criteria led to 1,3,5-tris(2-di(2'-pyridyl)hydroxymethylphenyl)benzene (H_3L or L^{3-}), a 1,3,5-triarylbenzene framework appended with dipyridyl-alcohols in one of the *ortho* positions of each of the three arenes on the periphery (Scheme 5.3). The variability in the binding mode of dipyridyl ketone and the corresponding hemiacetal or *gem*-diol is well documented,[238] and indeed plays an important role in the chemistry of this multinucleating ligand L^{3-}. A trimetallic Mn_3^{II} (1) species was isolated upon treatment with $Mn(OAc)_2$ and based under anaerobic conditions. The three alkoxide moieties bridge between metal centers, forming a chair-shaped

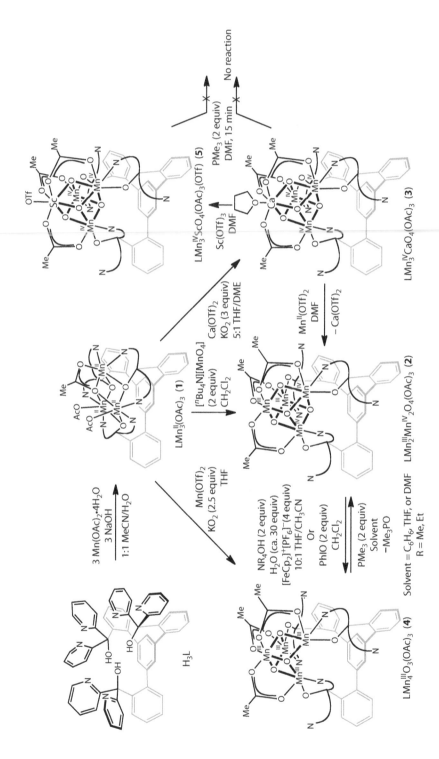

SCHEME 5.3 Synthesis and reactivity of site-differentiated tetrametallic complexes.

M_3O_3 ring, and the two pyridines of each aryl arm bind to two separate metals, resulting in a structure with *pseudo*-C_3 symmetry.[234] The three acetates complete six coordination at manganese. Trinuclear complex **1** was employed as precursor to tetrametallic clusters. Homometallic clusters such as $Mn_2^{III}Mn_2^{IV}O_4$ and $Mn_4^{III}O_3$ were accessed from **1**, oxidant, and a source of Mn for the fourth metal (Scheme 5.3; Figure 5.5). A heterometallic cubane, $Mn_3^{IV}CaO_4$ (**3**), was obtained by using calcium instead of manganese salts as the source of the fourth metal (Figure 5.4k). Another synthetic route to heterometallic clusters involves metal substitution, by treatment of the calcium cubane with metal salts, e.g., scandium triflate to generate $Mn_3^{IV}ScO_4$ (**5**). For all clusters, three manganese centers are coordinated to the trinucleating ligand and the fourth metal is supported by bridging oxide and acetate moieties.

Heterometallic cubane **3** is an accurate synthetic model for the cubane subsite of the current OEC structural model and proves the chemical feasibility of such a structure (Figure 5.5c). The Mn–Mn vectors are ca. 2.83 Å, and the Mn–Ca vector is ca. 3.23 Å, close to the 2.7 and 3.4 Å vectors of the OEC. The Mn–O distances

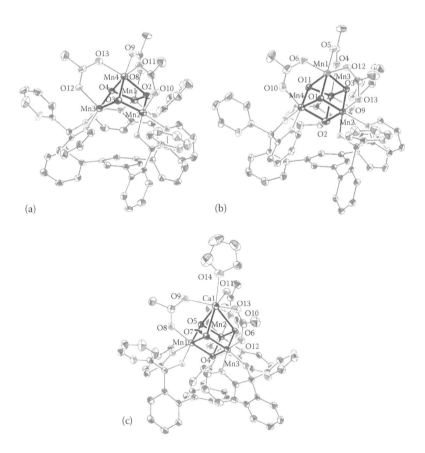

FIGURE 5.5 Solid-state structures of (a) complex **4**, (b) complex **2**, and (c) complex **3**.

are consistent with an Mn_3^{IV} oxidation state, paralleling S_2–S_4. More recently, a second $Mn_3^{IV}CaO_4$ cubane coordination complex with similar structural parameters was synthesized by self-assembly.[239] It contains a second calcium bridging to the cubane unit, similar to the dangling Mn, which is ligated by bridging and terminal carboxylates, and has been studied by EPR and XAS (Figure 5.4l).

The series of complexes 2–5 shows systematic variation within the Mn_3MO_x cluster: 2, 3, and 5 have the same overall cluster geometry but vary the apical metal center, and 2 and 4 only differ by an oxygen atom. Comparison of clusters with such controlled structural changes have offered insight relevant to the OEC regarding the effect of Ca^{2+} on the chemistry of the manganese cluster,[233] the feasibility of μ-oxo migrations proposed as equilibria in the S_1 and S_2 states, the oxidative water incorporation during catalysis and photoactivation, and the rates of exchange of bridging-oxido moieties with water.[235]

Comparison of the electrochemistry of 2, 3, and 5 (Figure 5.6) provides information about the effect of the redox-inactive metal in the cluster. The reduction

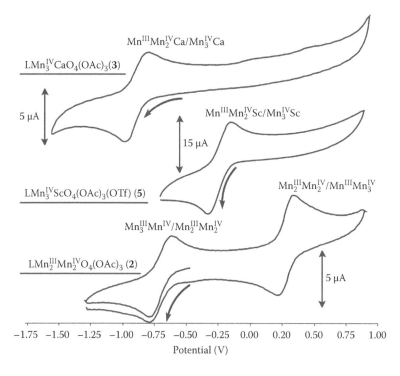

FIGURE 5.6 Cyclic voltammograms of complexes 2, 3, and 5 referenced to $FeCp_2/FeCp_2^+$. The scan rate was 50 mV/s for 2 and 100 mV/s for 3 and 5 at an analyte concentration of 1 mM and electrolyte of 0.1 M $^nBu_4NPF_6$ in dimethylacetamide (2 and 3) and dimethylformamide (5). Arrows depict the direction and beginning of the scan. $E_{1/2}$ values: for 2, −0.70 V ($Mn_3^{III}Mn^{IV}/Mn_2^{III}Mn_2^{IV}$) and +0.29 V ($Mn_2^{III}Mn_2^{IV}/Mn^{III}Mn_3^{IV}$); for 5, −0.24 V ($Mn^{III}Mn_2^{IV}Sc/Mn_3^{IV}Sc$); for 3, −0.94 V ($Mn^{III}Mn_2^{IV}Ca/Mn_3^{IV}Ca$).

potential of the $Mn_3^{IV}/Mn^{III}Mn_2^{IV}$ couple is affected significantly by the nature of the apical metal. As the Lewis acidity of this metal increases, the reduction potential of the cluster becomes more positive. The Mn_3^{IV} oxidation level in **3** is accessed at a potential remarkably more negative (>1 V) than the same oxidation state in **2**. This large difference is notable considering that high oxidation states ($\geq Mn_4^{IV}$) have been proposed for the S-states just before the step of O–O bond formation and that an increasing potential for cluster oxidation has to be accommodated by the same biological oxidant, P680+. Therefore, a role of calcium in the OEC may be to lower the potential at which the cluster can access the high oxidation states necessary for O_2 generation. Consistent with this, if Ca^{2+} is removed from active PSII, the OEC cannot proceed past S_2.[240]

Relevant to the mechanism of water incorporation into the OEC during photo-activation and turnover and to the reactivity of the μ-oxido ligands once within the OEC, the exchange of **2–5** with water and the mechanism of interconversion of **2** and **4** were studied.[235] The oxido moieties of homo- and heterometallic cubanes **2–5** show no exchange with $^{18}OH_2$ over hours, consistent with water exchange studies of other synthetic systems.[168,211,241] The lack of exchange is inconsistent with mechanisms for the OEC that invoke O–O bond formation with a bridging oxido substrate,[111,159] although recent studies suggest μ-oxido equilibria that may affect the exchange rates.[113,242]

Insight into the O-atom transfer abilities of these high-oxidation-state metal-oxo complexes has been obtained from their reactivity with phosphines. Complex **2** was converted to **4** by reaction with PMe_3 in quantitative yield within minutes, generating phosphine oxide as a by-product. Interestingly, unlike the rapid O-atom transfer from **2**, the heterometallic Ca^{2+} and Sc^{3+} cubanes showed no reactivity with PMe_3. Complex **5** is more oxidizing than **2** (Figure 5.6) suggesting that a kinetic barrier causes the difference in reactivity. Further experimentation supported this: PEt_3 reacted much slower than PMe_3 with **2**, and acetate scrambling was complete with **2** after 1 min while significantly slower in **3** and **5** (ca. 1 h). A mechanistic proposal consistent with these results invokes acetate dissociation to facilitate O-atom transfer; as complexes **3** and **5** display only inert Mn^{IV}-OAc, the reaction is slower in comparison with **2** that has labile Mn^{III} centers.

The above mechanistic hypothesis was interrogated by computation. The optimized geometries agree well with the experimental structures, with a root mean square difference for bonds of 0.007 Å and 0.012 Å for Mn_3CaO_4 (**3**) and Mn_4O_4 (**2**), respectively. The reduction potentials obtained by M06[243] vs. B3LYP show lower absolute differences compared with experiments (0.1 V vs. 0.2 V, respectively).[235] With these calibrations of the method, the computed mechanism supports a mechanism in which an Mn^{III}–OAc bond breaks to allow PMe_3 to attack one of the three "top" oxide moieties (i.e., the oxygens further from the triarylbenzene framework). The dissociation energy for the carboxylate bound to the Mn^{III} is 13.2 kcal/mol, while for Mn^{IV} it is 18.6 kcal/mol in the $Mn_2^{III}Mn_2^{IV}O_4$ cubane. The dissociation of the carboxylate bound to an Mn^{IV} in the $Mn_3^{IV}CaO_4$ cluster is 27.1 kcal/mol. These energetic differences are consistent with the fast reactivity for **2** and lack of reactivity of **3**. The generation of **4** by O-atom transfer was computed to involve oxide migration between bottom and top positions in the cluster. Overall, these experimental and computational studies highlight that the reactivity of the oxido moieties can be modulated indirectly by the metal oxidation state affecting the lability of the ancillary ligands.

During OEC turnover and assembly, the substrate and supporting oxygenous ligands are provided by water. Access to partial cubane **4** offered the possibility of investigating the mechanism of water incorporation into a complicated manganese cluster structurally related to the OEC. Indeed, **4** converts to full cubane **2** upon treatment with excess water, two equivalents of base (NR_4OH; R = Me, Et), and excess one-electron oxidant ($FeCp_2PF_6$). This oxidative oxide incorporation from water conceptually models photoactivation of the OEC. Isotopic labeling experiments were designed by taking advantage of the synthetic cycle between partial cubane **4** and cubane **2**. Importantly, the site differentiation of the cluster provided by the trinucleating ligand allows for a mechanistic interpretation of the labeling experiments. Oxidative incorporation of $^{18}OH_2$ into **4** could afford two isotopologues of singly labeled $Mn_4^{16}O_3^{18}O$ (**2***): with the label either in one of the three top positions (**2^{T*}**) or in the bottom position closest to the ligand framework (**2^{B*}**; Scheme 5.4). Reaction of resulting **2*** with PMe_3 could give **4**, **4***, or a mixture of the isotopomers depending on the sequence of mechanisms for water incorporation and O-atom transfer. Experimentally, we observe a mixture of **4** and **4*** (Scheme 5.4). Top-selective incorporation and O-atom transfer is consistent with the labeling observed, although we cannot rule out unselective incorporation. Notwithstanding, the pathways consistent with our labeling study and the computational studies all require μ-oxido migration steps (Scheme 5.5), supporting the possibility of μ-oxido migration equilibria discussed for the S_1 and S_2 states.[133,134]

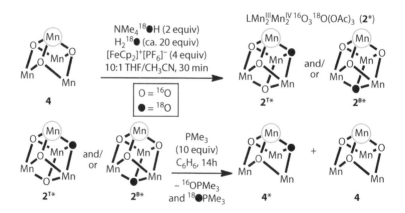

SCHEME 5.4 ^{18}O-labeling study design and outcome. Ligand framework **L** is below Mn_4O_n units as drawn in Scheme 5.3; the site-differentiated Mn center is circled.

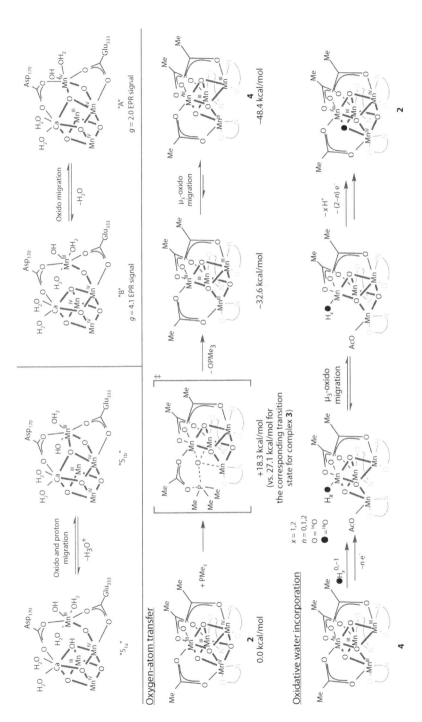

SCHEME 5.5 Comparison of proposed substrate equilibria in S_1 (top left)[134] and S_2 (top right)[135] to μ-oxido migration steps proposed for synthetic model complexes.[235]

5.8   CONCLUSIONS

New synthetic systems, spectroscopic methods, computations, and the concurrent collaborations have produced ever-refined structures of the OEC and more accurate mechanisms for OEC action. Biochemical and spectroscopic experimental results on PSII provided the motivation for synthetic experiments key to benchmarking and supporting various proposals. These synthetic models, in turn, inspired new structural and mechanistic proposals and were crucial for testing computational methods as these became powerful enough to study metalloenzymes. More recently, crystal structures have provided atomic coordinates for more powerful computational work. The recent high-resolution crystal structure of the PSII has prompted spectroscopic, synthetic, and computational developments. Site-differentiated Mn_3MO_n complexes, including an $Mn_3^{IV}CaO_4$ cubane, have been instrumental in studying the reactivity and properties of complicated clusters structurally related to the OEC. These studies indicate that a potential role of the redox-inactive metal, Ca^{2+}, is to tune the reduction potential of the cluster. Additionally, μ-oxido migration was shown to occur within high-oxidation-state synthetic manganese clusters, supporting recent proposals for equilibria between different structures of the OEC dependent on oxide migration. Water exchange was measured to be slow for μ_3-oxos of homo- and heterometallic cubanes. Overall, the interplay of synthetic, structural, spectroscopic, mechanistic, and computational work has led to tremendous insight into the chemistry and properties of manganese clusters relevant to the OEC. Despite these advances, the mechanism of water oxidation remains open for debate. The development of more accurate models, including of the full OEC, is a direction that will likely provide exciting new insight toward understanding not only the function of the biological system, but also toward delineating the design elements for improved catalysts for artificial photosynthesis.

Note: Several articles relevant to material in this chapter have been published while the book was in press. Please see References 244–248.

REFERENCES

1. Joliot, Pierre, and Bessel Kok. "Oxygen evolution in photosynthesis." In *Bioenergetics of Photosynthesis*, edited by Govindjee, 387–411. New York: Academic Press, 1975.
2. Pecoraro, Vincent L. (ed.). *Manganese Redox Enzymes*. New York: VCH Publishers Inc., 1992.
3. Debus, Richard J. "The manganese and calcium-ions of photosynthetic oxygen evolution." *Biochim. Biophys. Acta* 1102, no. 3 (1992): 269–352.
4. Yachandra, Vittal K., Kenneth Sauer, and Melvin P. Klein. "Manganese cluster in photosynthesis: Where plants oxidize water to dioxygen." *Chem. Rev.* 96, no. 7 (1996): 2927–50.
5. Ort, Donald R., and Charles F. Yocum (eds.). *Oxygenic Photosynthesis: The Light Reactions*. Dordrecht: Kluwer Academic Publishers, 1996.
6. Wydrzynski, Tom, and Kazuhiko Satoh (eds.). *The Light-Driven Water: Plastoquinone Oxidoreductase*, Vol. 22. Dordrecht: Springer, 2005.
7. McEvoy, James P., and Gary W. Brudvig. "Water-splitting chemistry of photosystem II." *Chem. Rev.* 106, no. 11 (2006): 4455–83.

8. Babcock, Gerald T., and Charles Yocum. "Dioxygen production: Photosystem II." In *Biological Inorganic Chemistry, Structure and Reactivity*, edited by Ivano Bertini, Edward Stiefel, Joan Selverstone Valentine, and Harry Gray, 302–18. Sausalito, CA: University Science Books, 2007.

9. Zouni, Athina, Horst-Tobias Witt, Jan Kern et al. "Crystal structure of photosystem II from *Synechococcus elongatus* at 3.8 angstrom resolution." *Nature* 409, no. 6821 (2001): 739–43.

10. Umena, Yasufumi, Keisuke Kawakami, Jian-Ren Shen, and Nobuo Kamiya. "Crystal structure of oxygen-evolving photosystem II at a resolution of 1.9 angstrom." *Nature* 473, no. 7345 (2011): U55–U65.

11. Yocum, Charles F. "Calcium activation of photosynthetic water oxidation." *Biochim. Biophys. Acta* 1059, no. 1 (1991): 1–15.

12. Kessler, Erich. "On the role of manganese in the oxygen-evolving system of photosynthesis." *Arch. Biochem. Biophys.* 59, no. 2 (1955): 527–9.

13. Cheniae, George M., and Iris F. Martin. "Photoreactivation of manganese catalyst in photosynthetic oxygen evolution." *Biochem. Biophys. Res. Commun.* 28, no. 1 (1967): 89–95.

14. Emerson, Robert, and Charlton M. Lewis. "Factors influencing the efficiency of photosynthesis." *Am. J. Bot.* 26, no. 10 (1939): 808–22.

15. Yocum, Charles F., Cristine T. Yerkes, Robert E. Blankenship, Robert R. Sharp, and Gerald T. Babcock. "Stoichiometry, inhibitor sensitivity, and organization of manganese associated with photosynthetic oxygen evolution." *Proc. Natl. Acad. Sci. U.S.A.* 78, no. 12 (1981): 7507–11.

16. Murata, Norio, Masanobu Miyao, Tatsuo Omata, Hiroaki Matsunami, and Tomohiko Kuwabara. "Stoichiometry of components in the photosynthetic oxygen evolution system of photosystem-II particles prepared with Triton X-100 from spinach-chloroplasts." *Biochim. Biophys. Acta* 765, no. 3 (1984): 363–9.

17. Ohno, Takashi, Kazuhiko Satoh, and Sakae Katoh. "Chemical-composition of purified oxygen-evolving complexes from the thermophilic cyanobacterium *Synechococcus* sp." *Biochim. Biophys. Acta* 852, no. 1 (1986): 1–8.

18. Piccioni, Richard G., and David C. Mauzerall. "Increase effected by calcium-ion in rate of oxygen evolution from preparations of phormidium-luridum." *Biochim. Biophys. Acta* 423, no. 3 (1976): 605–9.

19. Piccioni, Richard G., and David C. Mauzerall. "Calcium and photosynthetic oxygen evolution in cyanobacteria." *Biochim. Biophys. Acta* 504, no. 3 (1978): 384–97.

20. Ghanotakis, Demetrios F., James N. Topper, Gerald T. Babcock, and Charles F. Yocum. "Water-soluble 17-kDa and 23-kDa polypeptides restore oxygen evolution activity by creating a high-affinity binding-site for Ca^{2+} on the oxidizing side of photosystem-II." *FEBS Lett.* 170, no. 1 (1984): 169–73.

21. Yocum, Charles F. "The calcium and chloride requirements of the O_2 evolving complex." *Coord. Chem. Rev.* 252, no. 3–4 (2008): 296–305.

22. Ghanotakis, Demetrios F., Gerald T. Babcock, and Charles F. Yocum. "Calcium reconstitutes high-rates of oxygen evolution in polypeptide depleted photosystem-II preparations." *FEBS Lett.* 167, no. 1 (1984): 127–30.

23. Ädelroth, Pia, Katrin Lindberg, and Lars-Erik Andreasson. "Studies of Ca^{2+} binding in spinach photosystem-II using $^{45}Ca^{2+}$." *Biochemistry* 34, no. 28 (1995): 9021–7.

24. Boussac, Alain, and A. William Rutherford. "Nature of the inhibition of the oxygen-evolving enzyme of photosystem-II induced by NaCl washing and reversed by the addition of Ca^{2+} or Sr^{2+}." *Biochemistry* 27, no. 9 (1988): 3476–83.

25. Boussac, Alain, Jean-Luc Zimmermann, and A. William Rutherford. "EPR signals from modified charge accumulation states of the oxygen evolving enzyme in Ca^{2+}-deficient photosystem-II." *Biochemistry* 28, no. 23 (1989): 8984–9.

26. Sivaraja, Mohanram, J. Tso, and G. Charles Dismukes. "A calcium-specific site influences the structure and activity of the manganese cluster responsible for photosynthetic water oxidation." *Biochemistry* 28, no. 24 (1989): 9459–64.

27. Ono, Taka-aki, and Yorinao Inoue. "Abnormal redox reactions in photosynthetic O_2-evolving centers in NaCl/EDTA-washed PS II—A dark-stable EPR multiline signal and an unknown positive charge accumulator." *Biochim. Biophys. Acta* 1020, no. 3 (1990): 269–77.

28. Boussac, Alain, Pierre Sétif, and A. William Rutherford. "Inhibition of tyrosine-Z photooxidation after formation of the S_3 state in Ca^{2+}-depleted and (Cl^-)-depleted photosystem-II." *Biochemistry* 31, no. 4 (1992): 1224–34.

29. Kim, Sun Hee, Wolfgang Gregor, Jeffrey M. Peloquin, Marcin Brynda, and R. David Britt. "Investigation of the calcium-binding site of the oxygen evolving complex of photosystem II using [87]Sr ESEEM spectroscopy." *J. Am. Chem. Soc.* 126, no. 23 (2004): 7228–37.

30. Cinco, Roehl M., Karen L. McFarlane Holman, John Robblee et al. "Calcium EXAFS establishes the Mn–Ca cluster in the oxygen–evolving complex of photosystem II." *Biochemistry* 41, no. 43 (2002): 12928–33.

31. Yano, Junko, Jan Kern, Kenneth Sauer et al. "Where water is oxidized to dioxygen: Structure of the photosynthetic Mn_4Ca cluster." *Science* 314, no. 5800 (2006): 821–5.

32. Arnon, Daniel I., and F. R. Whatley. "Is chloride a coenzyme of photosynthesis." *Science* 110, no. 2865 (1949): 554–6.

33. Izawa, Seikichi, Robert L. Heath, and Geoffrey Hind. "Role of chloride ion in photosynthesis. 3. Effect of artificial electron donors upon electron transport." *Biochim. Biophys. Acta* 180, no. 2 (1969): 388–98.

34. Lindberg, Katrin, Tore Vanngard, and Lars-Erik Andreasson. "Studies of the slowly exchanging chloride in photosystem-II of higher-plants." *Photosynth. Res.* 38, no. 3 (1993): 401–8.

35. Guskov, Albert, Jan Kern, Azat Gabdulkhakov, Matthias Broser, Athina Zouni, and Wolfram Saenger. "Cyanobacterial photosystem II at 2.9-angstrom resolution and the role of quinones, lipids, channels and chloride." *Nat. Struct. Mol. Biol.* 16, no. 3 (2009): 334–42.

36. Chow, Wah Soon, and Eva-Mari Aro. "Photoinactivation and mechanisms of recovery." In *The Light-Driven Water: Plastoquinone Oxidoreductase*, edited by Tom J. Wydrzynski and Kazuhiko Satoh, 627–48. Dordrecht: Springer, 2005.

37. Cheniae, George M., and Iris F. Martin. "Photoactivation of manganese catalyst of O_2 evolution. 1. Biochemical and kinetic aspects." *Biochim. Biophys. Acta* 253, no. 1 (1971): 167–81.

38. Miller, Anne-Frances, and Gary W. Brudvig. "Manganese and calcium requirements for reconstitution of oxygen-evolution activity in manganese-depleted photosystem-II membranes." *Biochemistry* 28, no. 20 (1989): 8181–90.

39. Miller, Anne-Frances, and Gary W. Brudvig. "Electron-transfer events leading to reconstitution of oxygen-evolution activity in manganese-depleted photosystem-II membranes." *Biochemistry* 29, no. 6 (1990): 1385–92.

40. Burnap, Robert L. "D1 protein processing and Mn cluster assembly in light of the emerging photosystem II structure." *Phys. Chem. Chem. Phys.* 6, no. 20 (2004): 4803–9.

41. Bartlett, John E., Sergei V. Baranov, Gennady M. Ananyev, and G. Charles Dismukes. "Calcium controls the assembly of the photosynthetic water-oxidizing complex: A cadmium(II) inorganic mutant of the Mn_4Ca core." *Philos. Trans. R. Soc. B Biol. Sci.* 363, no. 1494 (2008): 1253–61.

42. Dasgupta, Jyotishman, Gennady M. Ananyev, and G. Charles Dismukes. "Photoassembly of the water-oxidizing complex in photosystem II." *Coord. Chem. Rev.* 252, no. 3–4 (2008): 347–60.

43. Campbell, Kristy A., Dee Ann Force, Peter J. Nixon, François Dole, Bruce A. Diner, and R. David Britt. "Dual-mode EPR detects the initial intermediate in photoassembly of the photosystem II Mn cluster: The influence of amino acid residue 170 of the D1 polypeptide on Mn coordination." *J. Am. Chem. Soc.* 122, no. 15 (2000): 3754–61.

44. Tyryshkin, Alexei M., Richard K. Watt, Sergei V. Baranov, Jyotishman Dasgupta, Michael P. Hendrich, and G. Charles Dismukes. "Spectroscopic evidence for Ca^{2+} involvement in the assembly of the Mn_4Ca cluster in the photosynthetic water-oxidizing complex." *Biochemistry* 45, no. 42 (2006): 12876–89.

45. Ananyev, Gennady M., and G. Charles Dismukes. "Calcium induces binding and formation of a spin-coupled dimanganese(II,II) center in the apo-water oxidation complex of photosystem II as precursor to the functional tetra-Mn/Ca cluster." *Biochemistry* 36, no. 38 (1997): 11342–50.

46. Zaltsman, Lyudmila, Gennady M. Ananyev, Edward Bruntrager, and G. Charles Dismukes. "Quantitative kinetic model for photoassembly of the photosynthetic water oxidase from its inorganic constituents: Requirements for manganese and calcium in the kinetically resolved steps." *Biochemistry* 36, no. 29 (1997): 8914–22.

47. Dasgupta, Jyotishman, Alexei M. Tyryshkin, Sergei V. Baranov, and G. Charles Dismukes. "Bicarbonate coordinates to Mn^{3+} during photo-assembly of the catalytic Mn_4Ca core of photosynthetic water oxidation: EPR characterization." *Appl. Magn. Reson.* 37, no. 1–4 (2010): 137–50.

48. Joliot, Pierre. "Cinetiques des reactions liees a lemission doxygene photosynthetique." *Biochim. Biophys. Acta* 102, no. 1 (1965): 116–34.

49. Joliot, Pierre, and Anne Joliot. "A polarographic method for detection of oxygen production and reduction of Hill Reagent by isolated chloroplasts." *Biochim. Biophys. Acta* 153, no. 3 (1968): 625–34.

50. Kok, Bessel, Bliss Forbush, and Marion McGloin. "Cooperation of charges in photosynthetic O_2 evolution. 1. A linear four step mechanism." *Photochem. Photobiol.* 11, no. 6 (1970): 457–75.

51. Förster, Verena, and Wolfgang Junge. "Stoichiometry and kinetics of proton release upon photosynthetic water oxidation." *Photochem. Photobiol.* 41, no. 2 (1985): 183–90.

52. Caudle, M. Tyler, and Vincent L. Pecoraro. "Thermodynamic viability of hydrogen atom transfer from water coordinated to the oxygen-evolving complex of photosystem II." *J. Am. Chem. Soc.* 119, no. 14 (1997): 3415–6.

53. Schlodder, Eberhard, and Horst-Tobias Witt. "Stoichiometry of proton release from the catalytic center in photosynthetic water oxidation—Reexamination by a glass electrode study at pH 5.5–7.2." *J. Biol. Chem.* 274, no. 43 (1999): 30387–92.

54. Dau, Holger, and Michael Haumann. "The manganese complex of photosystem II in its reaction cycle—Basic framework and possible realization at the atomic level." *Coord. Chem. Rev.* 252, no. 3–4 (2008): 273–95.

55. Carrell, Thomas G., Alexei M. Tyryshkin, and G. Charles Dismukes. "An evaluation of structural models for the photosynthetic water-oxidizing complex derived from spectroscopic and X-ray diffraction signatures." *J. Biol. Inorg. Chem.* 7, no. 1–2 (2002): 2–22.

56. Kulik, Leonid, Boris Epel, Wolfgang Lubitz, and Johannes Messinger. "Electronic structure of the Mn_4O_xCa cluster in the S_0 and S_2 states of the oxygen-evolving complex of photosystem II based on pulse ^{55}Mn-ENDOR and EPR spectroscopy." *J. Am. Chem. Soc.* 129, no. 44 (2007): 13421–35.

57. Roelofs, Theo A., Wenchuan C. Liang, Matthew J. Latimer et al. "Oxidation states of the manganese cluster during the flash-induced S-state cycle of the photosynthetic oxygen-evolving complex." *Proc. Natl. Acad. Sci. U.S.A.* 93, no. 8 (1996): 3335–40.

58. Bergmann, Uwe, Melissa M. Grush, Craig R. Horne et al. "Characterization of the Mn oxidation states in photosystem II by K beta X-ray fluorescence spectroscopy." *J. Phys. Chem. B* 102, no. 42 (1998): 8350–2.

59. Visser, Hendrik, Elodie Anxolabéhère-Mallart, Uwe Bergmann et al. "Mn K-edge XANES and Kβ XES studies of two Mn-Oxo binuclear complexes: Investigation of three different oxidation states relevant to the oxygen-evolving complex of photosystem II." *J. Am. Chem. Soc.* 123, no. 29 (2001): 7031–9.

60. Pizarro, Shelly A., Pieter Glatzel, Hendrik Visser et al. "Mn oxidation states in Tri- and tetra-nuclear Mn compounds structurally relevant to photosystem II: Mn K-edge X-ray absorption and Kβ X-ray emission spectroscopy studies." *Phys. Chem. Chem. Phys.* 6, no. 20 (2004): 4864–70.

61. Kolling, Derrick R. J., Nicholas Cox, Gennady M. Ananyev, Ron J. Pace, and G. Charles Dismukes. "What are the oxidation states of manganese required to catalyze photosynthetic water oxidation?" *Biophys. J.* 103, no. 2 (2012): 313–22.

62. Pace, Ron J., Lu Jin, and Rob Stranger. "What spectroscopy reveals concerning the Mn oxidation levels in the oxygen evolving complex of photosystem II: X-ray to near infrared." *Dalton Trans.* 41, no. 36 (2012): 11145–60.

63. Gatt, Phillip, Simon Petrie, Rob Stranger, and Ron J. Pace. "Rationalizing the 1.9 angstrom crystal structure of photosystem II—A remarkable Jahn–Teller balancing act induced by a single proton transfer." *Angew. Chem. Int. Ed.* 51, no. 48 (2012): 12025–8.

64. Radmer, Richard, and Otto Ollinger. "Do the higher oxidation-states of the photosynthetic O_2-evolving system contain bound H_2O." *FEBS Lett.* 195, no. 1–2 (1986): 285–9.

65. Messinger, Johannes, Murray Badger, and Tom Wydrzynski. "Detection of one slowly exchanging substrate water molecule in the S_3 state of photosystem-II." *Proc. Natl. Acad. Sci. U.S.A.* 92, no. 8 (1995): 3209–13.

66. Messinger, Johannes, Warwick Hillier, Murray Badger, and Tom Wydrzynski. Heterogeneity in Substrate Water Binding to Photosystem II. In *Photosynthesis: From Light to Biosphere*, Vol. II. (1995): 283–6. The Netherlands: Kluwer Academic Publishers.

67. Hillier, Warwick, Johannes Messinger, and Tom Wydrzynski. "Kinetic determination of the fast exchanging substrate water molecule in the S_3 state of photosystem II." *Biochemistry* 37, no. 48 (1998): 16908–14.

68. Hillier, Warwick, Johannes Messinger, and Tom Wydrzynski. "Substrate water [18]O exchange kinetics in the S_2 state of photosystem II." In *Photosynthesis: Mechanisms and Effects*, Vols. I–V. (1998): 1307–10. The Netherlands: Kluwer Academic Publishers.

69. Hendry, Garth, and Tom Wydrzynski. "The two substrate-water molecules are already bound to the oxygen-evolving complex in the S_2 state of photosystem II." *Biochemistry* 41, no. 44 (2002): 13328–34.

70. Hendry, Garth, and Tom Wydrzynski. "[18]O isotope exchange measurements reveal that calcium is involved in the binding of one substrate-water molecule to the oxygen-evolving complex in photosystem II." *Biochemistry* 42, no. 20 (2003): 6209–17.

71. Hillier, Warwick, and Tom Wydrzynski. "[18]O-water exchange in photosystem II: Substrate binding and intermediates of the water splitting cycle." *Coord. Chem. Rev.* 252, no. 3–4 (2008): 306–17.

72. Singh, Sonita, Richard J. Debus, Tom Wydrzynski, and Warwick Hillier. "Investigation of substrate water interactions at the high-affinity Mn site in the photosystem II oxygen-evolving complex." *Philos. Trans. R. Soc. B Biol. Sci.* 363, no. 1494 (2008): 1229–34.

73. Hillier, Warwick, and Tom Wydrzynski. "The affinities for the two substrate water binding sites in the O_2 evolving complex of photosystem II vary independently during S-state turnover." *Biochemistry* 39, no. 15 (2000): 4399–405.

74. Kamiya, Nobuo, and Jian-Ren Shen. "Crystal structure of oxygen-evolving photosystem II from *Thermosynechococcus vulcanus* at 3.7-angstrom resolution." *Proc. Natl. Acad. Sci. U.S.A.* 100, no. 1 (2003): 98–103.

75. Ferreira, Kristina N., Tina M. Iverson, Karim Maghlaoui, James Barber, and So Iwata. "Architecture of the photosynthetic oxygen-evolving center." *Science* 303, no. 5665 (2004): 1831–8.

76. Loll, Bernhard, Jan Kern, Wolfram Saenger, Athina Zouni, and Jacek Biesiadka. "Towards complete cofactor arrangement in the 3.0 angstrom resolution structure of photosystem II." *Nature* 438, no. 7070 (2005): 1040–4.

77. Barber, James, and James W. Murray. "Revealing the structure of the Mn-cluster of photosystem II by X-ray crystallography." *Coord. Chem. Rev.* 252, no. 3–4 (2008): 233–43.

78. Barber, James, and James W. Murray. "The structure of the Mn_4Ca^{2+} cluster of photosystem II and its protein environment as revealed by X-ray crystallography." *Philos. Trans. R. Soc. B Biol. Sci.* 363, no. 1494 (2008): 1129–37.

79. Koua, Faisal H. M., Yasufumi Umena, Keisuke Kawakami, and Jian-Ren Shen. "Structure of Sr-substituted photosystem II at 2.1 angstrom resolution and its implications in the mechanism of water oxidation." *Proc. Natl. Acad. Sci. U.S.A.* 110, no. 10 (2013): 3889–94.

80. Yano, Junko, Jan Kern, Klaus-Dieter Irrgang et al. "X-ray damage to the Mn_4Ca complex in single crystals of photosystem II: A case study for metalloprotein crystallography." *Proc. Natl. Acad. Sci. U.S.A.* 102, no. 34 (2005): 12047–52.

81. Zheng, Ming, and G. Charles Dismukes. "Orbital configuration of the valence electrons, ligand field symmetry, and manganese oxidation states of the photosynthetic water oxidizing complex: Analysis of the S_2 state multiline EPR signals." *Inorg. Chem.* 35, no. 11 (1996): 3307–19.

82. Peloquin, Jeffrey M., and R. David Britt. "EPR/ENDOR characterization of the physical and electronic structure of the OEC Mn cluster." *Biochim. Biophys. Acta, Bioenerg.* 1503, no. 1–2 (2001): 96–111.

83. Britt, R. David, Kristy A. Campbell, Jeffrey M. Peloquin et al. "Recent pulsed EPR studies of the photosystem II oxygen-evolving complex: Implications as to water oxidation mechanisms." *Biochim. Biophys. Acta, Bioenerg.* 1655, no. 1–3 (2004): 158–71.

84. Haddy, Alice. "EPR spectroscopy of the manganese cluster of photosystem II." *Photosynth. Res.* 92, no. 3 (2007): 357–68.

85. Ono, Taka-aki, Takumi Noguchi, Yorinao Inoue, Masami Kusunoki, Tadashi Matsushita, and Hiroyuki Oyanagi. "X-ray-detection of the period-4 cycling of the manganese cluster in photosynthetic water oxidizing enzyme." *Science* 258, no. 5086 (1992): 1335–7.

86. Robblee, John, Roehl M. Cinco, and Vittal K. Yachandra. "X-ray spectroscopy-based structure of the Mn cluster and mechanism of photosynthetic oxygen evolution." *Biochim. Biophys. Acta, Bioenerg.* 1503, no. 1–2 (2001): 7–23.

87. Sauer, Kenneth, Junko Yano, and Vittal K. Yachandra. "X-ray spectroscopy of the photosynthetic oxygen-evolving complex." *Coord. Chem. Rev.* 252, no. 3–4 (2008): 318–35.

88. Dau, Holger, Alexander Grundmeier, Paola Loja, and Michael Haumann. "On the structure of the manganese complex of photosystem II: Extended-range EXAFS data and specific atomic-resolution models for four S-states." *Philos. Trans. R. Soc. B Biol. Sci.* 363, no. 1494 (2008): 1237–43.

89. Yano, Junko, Jan Kern, Yulia Pushkar et al. "High-resolution structure of the photosynthetic Mn_4Ca catalyst from X-ray spectroscopy." *Philos. Trans. R. Soc. B Biol. Sci.* 363, no. 1494 (2008): 1139–47.

90. Dismukes, G. Charles, and Yona Siderer. "EPR spectroscopic observations of a manganese center associated with water oxidation in spinach-chloroplasts." *FEBS Lett.* 121, no. 1 (1980): 78–80.

91. Dismukes, G. Charles, and Yona Siderer. "Intermediates of a polynuclear manganese center involved in photosynthetic oxidation of water." *Proc. Natl. Acad. Sci. U.S.A.* 78, no. 1 (1981): 274–8.

92. Kirby, Jon A., David B. Goodin, Tom Wydrzynski, A. S. Robertson, and Melvin P. Klein. "State of manganese in the photosynthetic apparatus. 2. X-ray absorption-edge studies on manganese in photosynthetic membranes." *J. Am. Chem. Soc.* 103, no. 18 (1981): 5537–42.

93. Kirby, Jon A., A. S. Robertson, J. P. Smith, Albert C. Thompson, Stephen R. Cooper, and Melvin P. Klein. "State of manganese in the photosynthetic apparatus. 1. Extended X-ray absorption fine-structure studies on chloroplasts and di-μ-oxo-bridged dimanganese model compounds." *J. Am. Chem. Soc.* 103, no. 18 (1981): 5529–37.

94. Christou, George. "Manganese carboxylate chemistry and its biological relevance." *Acc. Chem. Res.* 22, no. 9 (1989): 328–35.

95. Wieghardt, Karl. "The active-sites in manganese-containing metalloproteins and inorganic model complexes." *Angew. Chem. Int. Ed.* 28, no. 9 (1989): 1153–72.

96. Limburg, Julian, Veronika A. Szalai, and Gary W. Brudvig. "A mechanistic and structural model for the formation and reactivity of a Mn^V=O species in photosynthetic water oxidation." *J. Chem. Soc. Dalton Trans.* no. 9 (1999): 1353–61.

97. Mukhopadhyay, Sumitra, Sanjay K. Mandal, Sumit Bhaduri, and William H. Armstrong. "Manganese clusters with relevance to photosystem II." *Chem. Rev.* 104, no. 9 (2004): 3981–4026.

98. Cady, Clyde W., Robert H. Crabtree, and Gary W. Brudvig. "Functional models for the oxygen-evolving complex of photosystem II." *Coord. Chem. Rev.* 252, no. 3–4 (2008): 444–55.

99. Meelich, Kristof, Curtis M. Zaleski, and Vincent L. Pecoraro. "Using small molecule complexes to elucidate features of photosynthetic water oxidation." *Philos. Trans. R. Soc. B Biol. Sci.* 363, no. 1494 (2008): 1271–9.

100. Mullins, Christopher S., and Vincent L. Pecoraro. "Reflections on small molecule manganese models that seek to mimic photosynthetic water oxidation chemistry." *Coord. Chem. Rev.* 252, no. 3–4 (2008): 416–43.

101. Blomberg, Margareta R. A., Per E. M. Siegbahn, Stenbjörn Styring, Gerald T. Babcock, Björn Akermark, and Peter Korall. "A quantum chemical study of hydrogen abstraction from manganese-coordinated water by a tyrosyl radical: A model for water oxidation in photosystem II." *J. Am. Chem. Soc.* 119, no. 35 (1997): 8285–92.

102. Siegbahn, Per E. M., and Robert H. Crabtree. "Manganese oxyl radical intermediates and O–O bond formation in photosynthetic oxygen evolution and a proposed role for the calcium cofactor in photosystem II." *J. Am. Chem. Soc.* 121, no. 1 (1999): 117–27.

103. Siegbahn, Per E. M. "Theoretical models for the oxygen radical mechanism of water oxidation and of the water oxidizing complex of photosystem II." *Inorg. Chem.* 39, no. 13 (2000): 2923–35.

104. Lundberg, Marcus, Margareta R. A. Blomberg, and Per E. M. Siegbahn. "Oxyl radical required for O–O bond formation in synthetic Mn-catalyst." *Inorg. Chem.* 43, no. 1 (2004): 264–74.

105. Lundberg, Marcus, and Per E. M. Siegbahn. "Agreement between experiment and hybrid DFT calculations for O–H bond dissociation enthalpies in manganese complexes." *J. Comput. Chem.* 26, no. 7 (2005): 661–7.

106. Sproviero, Eduardo M., José A. Gascón, James P. McEvoy, Gary W. Brudvig, and Victor S. Batista. "Characterization of synthetic oxomanganese complexes and the inorganic core of the O$_2$-evolving complex in photosystem-II: Evaluation of the DFT/B3LYP level of theory." *J. Inorg. Biochem.* 100, no. 4 (2006): 786–800.

107. Sproviero, Eduardo M., José A. Gascón, James P. McEvoy, Gary W. Brudvig, and Victor S. Batista. "Computational studies of the O-2-evolving complex of photosystem II and biomimetic oxomanganese complexes." *Coord. Chem. Rev.* 252, no. 3–4 (2008): 395–415.

108. Orio, Maylis, Dimitrios A. Pantazis, Taras Petrenko, and Frank Neese. "Magnetic and spectroscopic properties of mixed valence manganese(III,IV) dimers: A systematic study using broken symmetry density functional theory." *Inorg. Chem.* 48, no. 15 (2009): 7251–60.

109. Batista, Victor S., Ting Wang, and Gary W. Brudvig. "Characterization of proton cou-pled electron transfer in a biomimetic oxomanganese complex: Evaluation of the DFT B3LYP level of theory." *J. Chem. Theory Comput.* 6, no. 3 (2010): 755–60.

110. Batista, Victor S., Ting Wang, and Gary W. Brudvig. "Study of proton coupled elec-tron transfer in a biomimetic dimanganese water oxidation catalyst with terminal water ligands." *J. Chem. Theory Comput.* 6, no. 8 (2010): 2395–401.

111. Siegbahn, Per E. M. "Structures and energetics for O_2 formation in photosystem II." *Acc. Chem. Res.* 42, no. 12 (2009): 1871–80.

112. Gatt, Phillip, Rob Stranger, and Ron J. Pace. "Application of computational chemistry to understanding the structure and mechanism of the Mn catalytic site in photosystem II—A review." *J. Photochem. Photobiol. B* 104, no. 1–2 (2011): 80–93.

113. Rapatskiy, Leonid, Nicholas Cox, Anton Savitsky et al. "Detection of the water-binding sites of the oxygen-evolving complex of photosystem II using W-band ^{17}O electron–electron double resonance-detected NMR spectroscopy." *J. Am. Chem. Soc.* 134, no. 40 (2012): 16619–34.

114. Brudvig, Gary W., and Robert H. Crabtree. "Mechanism for photosynthetic O_2 evolu-tion." *Proc. Natl. Acad. Sci. U.S.A.* 83, no. 13 (1986): 4586–8.

115. Depaula, Julio C., Warren F. Beck, and Gary W. Brudvig. "Magnetic-properties of man-ganese in the photosynthetic O_2-evolving complex. 2. Evidence for a manganese tetra-mer." *J. Am. Chem. Soc.* 108, no. 14 (1986): 4002–9.

116. Goodin, David B., Vittal K. Yachandra, R. David Britt, Kenneth Sauer, and Melvin P. Klein. "The state of manganese in the photosynthetic apparatus. 3. Light-induced-changes in X-ray absorption (K-edge) energies of manganese in photosynthetic mem-branes." *Biochim. Biophys. Acta* 767, no. 2 (1984): 209–16.

117. McDermott, Ann E., Vittal K. Yachandra, R. D. Guiles et al. "The state of manganese in the photosynthetic apparatus. 9. Characterization of the manganese O_2-evolving com-plex and the iron quinone acceptor complex in photosystem-II from a thermophilic cya-nobacterium by electron-paramagnetic resonance and X-ray absorption-spectroscopy." *Biochemistry* 27, no. 11 (1988): 4021–31.

118. Penner-Hahn, James, Richard M. Fronko, Vincent L. Pecoraro, Charles F. Yocum, Scott D. Betts, and Neil R. Bowlby. "Structural characterization of the manganese sites in the photosynthetic oxygen-evolving complex using X-ray absorption-spectroscopy." *J. Am. Chem. Soc.* 112, no. 7 (1990): 2549–57.

119. Christou, George, and John B. Vincent. "The molecular double-pivot mechanism for water oxidation." *Biochim. Biophys. Acta* 895, no. 3 (1987): 259–74.

120. Vincent, John B., and George Christou. "A molecular double-pivot mechanism for water oxidation." *Inorg. Chim. Acta Bioinorg.* 136, no. 3 (1987): L41–3.

121. Yachandra, Vittal K., R. D. Guiles, Ann E. McDermott et al. "The state of manganese in the photosynthetic apparatus. 4. Structure of the manganese complex in photosystem-II studied using EXAFS spectroscopy—The S_1 state of the O_2-evolving photosystem-II complex from spinach." *Biochim. Biophys. Acta* 850, no. 2 (1986): 324–32.

122. George, Graham N., Roger C. Prince, and Stephen P. Cramer. "The manganese site of the photosynthetic water-splitting enzyme." *Science* 243, no. 4892 (1989): 789–91.

123. Yachandra, Vittal K., Victoria J. DeRose, Matthew J. Latimer, Ishita Mukerji, Kenneth Sauer, and Melvin P. Klein. "Where plants make oxygen—A structural model for the photosynthetic oxygen-evolving manganese cluster." *Science* 260, no. 5108 (1993): 675–9.

124. Hoganson, Curtis W., and Gerald T. Babcock. "A metalloradical mechanism for the gen-eration of oxygen from water in photosynthesis." *Science* 277, no. 5334 (1997): 1953–6.

125. Pecoraro, Vincent L., Michael J. Baldwin, M. Tyler Caudle, Wen-Yuan Hsieh, and Neil A. Law. "A proposal for water oxidation in photosystem II." *Pure Appl. Chem.* 70, no. 4 (1998): 925–9.

126. Peloquin, Jeffrey M., Kristy A. Campbell, David W. Randall et al. "Mn-55 ENDOR of the S$_2$-state multiline EPR signal of photosystem II: Implications on the structure of the tetranuclear Mn cluster." *J. Am. Chem. Soc.* 122, no. 44 (2000): 10926–42.

127. DeRose, Victoria J., Ishita Mukerji, Matthew J. Latimer, Vittal K. Yachandra, Kenneth Sauer, and Melvin P. Klein. "Comparison of the manganese oxygen-evolving complex in photosystem-II of spinach and *Synechococcus* sp. with multinuclear manganese model compounds by X-ray-absorption spectroscopy." *J. Am. Chem. Soc.* 116, no. 12 (1994): 5239–49.

128. Petrie, Simon, Phillip Gatt, Rob Stranger, and Ron J. Pace. "Modelling the metal atom positions of the photosystem II water oxidising complex: A density functional theory appraisal of the 1.9 angstrom resolution crystal structure." *Phys. Chem. Chem. Phys.* 14, no. 32 (2012): 11333–43.

129. Ames, William, Dimitrios A. Pantazis, Vera Krewald et al. "Theoretical evaluation of structural models of the S$_2$ state in the oxygen evolving complex of photosystem II: Protonation states and magnetic interactions." *J. Am. Chem. Soc.* 133, no. 49 (2011): 19743–57.

130. Luber, Sandra, Ivan Rivalta, Yasufumi Umena et al. "S$_1$-state model of the O$_2$-evolving complex of photosystem II." *Biochemistry* 50, no. 29 (2011): 6308–11.

131. Siegbahn, Per E. M. "The effect of backbone constraints: The case of water oxidation by the oxygen-evolving complex in PSII." *ChemPhysChem* 12, no. 17 (2011): 3274–80.

132. Galstyan, Artur, Arturo Robertazzi, and Ernst Walter Knapp. "Oxygen-evolving Mn cluster in photosystem II: The protonation pattern and oxidation state in the high-resolution crystal structure." *J. Am. Chem. Soc.* 134, no. 17 (2012): 7442–9.

133. Kusunoki, Masami. "S-1-state Mn$_4$Ca complex of photosystem ii exists in equilibrium between the two most-stable isomeric substates: XRD and EXAFS evidence." *J. Photochem. Photobiol. B* 104, no. 1–2 (2011): 100–10.

134. Pantazis, Dimitrios A., William Ames, Nicholas Cox, Wolfgang Lubitz, and Frank Neese. "Two interconvertible structures that explain the spectroscopic properties of the oxygen-evolving complex of photosystem II in the S$_2$ state." *Angew. Chem. Int. Ed.* 51, no. 39 (2012): 9935–40.

135. Isobe, Hiroshi, Mitsuo Shoji, Shusuke Yamanaka et al. "Theoretical illumination of water-inserted structures of the CaMn$_4$O$_5$ cluster in the S$_2$ and S$_3$ states of oxygen-evolving complex of photosystem II: Full geometry optimizations by B3LYP hybrid density functional." *Dalton Trans.* 41, no. 44 (2012): 13727–40.

136. Sodupe, Mariona, Juan Bertran, Luis Rodríguez-Santiago, and E. J. Baerends. "Ground state of the (H$_2$O)$_2^+$ radical cation: DFT versus post-Hartree–Fock methods." *J. Phys. Chem. A* 103, no. 1 (1999): 166–70.

137. Sousa, Sérgio F., Pedro A. Fernandes, and Maria J. Ramos. "General performance of density functionals." *J. Phys. Chem. A* 111, no. 42 (2007): 10439–52.

138. Becke, Axel D. "Density-functional exchange-energy approximation with correct asymptotic-behavior." *Phys. Rev. A* 38, no. 6 (1988): 3098–100.

139. Becke, Axel D. "Density-functional thermochemistry. 3. The role of exact exchange." *J. Chem. Phys.* 98, no. 7 (1993): 5648–52.

140. Siegbahn, Per E. M. "Modeling aspects of mechanisms for reactions catalyzed by metalloenzymes." *J. Comput. Chem.* 22, no. 14 (2001): 1634–45.

141. Hay, P. Jeffrey, Willard R. Wadt, and Thom H. Dunning, Jr. "Theoretical studies of molecular electronic-transition lasers." *Annu. Rev. Phys. Chem.* 30 (1979): 311–46.

142. Hay, P. Jeffrey, and Willard R. Wadt. "*Ab initio* effective core potentials for molecular calculations—Potentials for K to Au including the outermost core orbitals." *J. Chem. Phys.* 82, no. 1 (1985): 299–310.

143. Hay, P. Jeffrey, and Willard R. Wadt. "*Ab initio* effective core potentials for molecular calculations—Potentials for the transition-metal atoms Sc to Hg." *J. Chem. Phys.* 82, no. 1 (1985): 270–83.

144. Wadt, Willard R., and P. Jeffrey Hay. "*Ab initio* effective core potentials for molecular calculations—Potentials for main group elements Na to Bi." *J. Chem. Phys.* 82, no. 1 (1985): 284–98.

145. Perczel, András, Ödön Farkas, Imre Jákli, Igor A. Topol, and Imre G. Csizmadia. "Peptide models. XXXIII. Extrapolation of low-level Hartree–Fock data of peptide conformation to large basis set SCF, MP2, DFT, and CCSD(T) results. The Ramachandran surface of alanine dipeptide computed at various levels of theory." *J. Comput. Chem.* 24, no. 9 (2003): 1026–42.

146. Siegbahn, Per E. M., and Marcus Lundberg. "The mechanism for dioxygen formation in PSII studied by quantum chemical methods." *Photochem. Photobiol. Sci.* 4, no. 12 (2005): 1035–43.

147. Siegbahn, Per E. M. "A structure-consistent mechanism for dioxygen formation in photosystem II." *Chem. Eur. J.* 14, no. 27 (2008): 8290–302.

148. McEvoy, James P., José A. Gascón, Victor S. Batista, and Gary W. Brudvig. "The mechanism of photosynthetic water splitting." *Photochem. Photobiol. Sci.* 4, no. 12 (2005): 940–9.

149. Sproviero, Eduardo M., José A. Gascón, James P. McEvoy, Gary W. Brudvig, and Victor S. Batista. "Quantum mechanics/molecular mechanics study of the catalytic cycle of water splitting in photosystem II." *J. Am. Chem. Soc.* 130, no. 11 (2008): 3428–42.

150. Kusunoki, Masami. "Mono-manganese mechanism of the photosystem II water splitting reaction by a unique Mn_4Ca cluster." *Biochim. Biophys. Acta, Bioenerg.* 1767, no. 6 (2007): 484–92.

151. Petrie, Simon, Rob Stranger, Phillip Gatt, and Ron J. Pace. "Bridge over troubled water: Resolving the competing photosystem II crystal structures." *Chem. Eur. J.* 13, no. 18 (2007): 5082–9.

152. Petrie, Simon, Rob Stranger, and Ron J. Pace. "Structural, magnetic coupling and oxidation state trends in models of the $CaMn_4$ cluster in photosystem II." *Chem. Eur. J.* 14, no. 18 (2008): 5482–94.

153. Petrie, Simon, Rob Stranger, and Ron J. Pace. "Location of potential substrate water binding sites in the water oxidizing complex of photosystem II." *Angew. Chem. Int. Ed.* 49, no. 25 (2010): 4233–6.

154. Isobe, Hiroshi, Mitsuo Shoji, Kenichi Koizumi et al. "Electronic and spin structures of manganese clusters in the photosynthesis II system." *Polyhedron* 24, no. 16–17 (2005): 2767–77.

155. Siegbahn, Per E. M. "Mechanism and energy diagram for O–O bond formation in the oxygen-evolving complex in photosystem II." *Philos. Trans. R. Soc. B Biol. Sci.* 363, no. 1494 (2008): 1221–8.

156. Sproviero, Eduardo M., José A. Gascón, James P. McEvoy, Gary W. Brudvig, and Victor S. Batista. "QM/MM models of the O_2-evolving complex of photosystem II." *J. Chem. Theory Comput.* 2, no. 4 (2006): 1119–34.

157. Lundberg, Marcus, and Per E. M. Siegbahn. "Theoretical investigations of structure and mechanism of the oxygen-evolving complex in PSII." *Phys. Chem. Chem. Phys.* 6, no. 20 (2004): 4772–80.

158. Yano, Junko, and Vittal K. Yachandra. "Where water is oxidized to dioxygen: Structure of the photosynthetic Mn_4Ca cluster from X-ray spectroscopy." *Inorg. Chem.* 47, no. 6 (2008): 1711–26.

159. Siegbahn, Per E. M. "Mechanisms for proton release during water oxidation in the S_2 to S_3 and S_3 to S_4 transitions in photosystem II." *Phys. Chem. Chem. Phys.* 14, no. 14 (2012): 4849–56.

160. Wang, Junmei, Romain M. Wolf, James W. Caldwell, Peter A. Kollman, and David A. Case. "Development and testing of a general Amber force field." *J. Comput. Chem.* 25, no. 9 (2004): 1157–74.

161. Siegbahn, Per E. M. "An energetic comparison of different models for the oxygen evolving complex of photosystem II." *J. Am. Chem. Soc.* 131, no. 51 (2009): 18238–9.

162. Siegbahn, Per E. M. "Water oxidation in photosystem II: Oxygen release, proton release and the effect of chloride." *Dalton Trans.* 2009, no. 45 (2009): 10063–8.

163. Pantazis, Dimitrios A., Maylis Orio, Taras Petrenko et al. "Structure of the oxygen-evolving complex of photosystem II: Information on the S_2 state through quantum chemical calculation of its magnetic properties." *Phys. Chem. Chem. Phys.* 11, no. 31 (2009): 6788–98.

164. Volkov, Alexander G. "Oxygen evolution in the course of photosynthesis—Molecular mechanisms." *Bioelectrochem. Bioenerg.* 21, no. 1 (1989): 3–24.

165. Tommos, Cecilia, and Gerald T. Babcock. "Oxygen production in nature: A light-driven metalloradical enzyme process." *Acc. Chem. Res.* 31, no. 1 (1998): 18–25.

166. Liu, Feng, Javier J. Concepcion, Jonah W. Jurss, Thomas Cardolaccia, Joseph L. Templeton, and Thomas J. Meyer. "Mechanisms of water oxidation from the blue dimer to photosystem II." *Inorg. Chem.* 47, no. 6 (2008): 1727–52.

167. Brudvig, G. W. "Water oxidation chemistry of photosystem II." *Philos. Trans. R. Soc. B Biol. Sci.* 363, no. 1494 (2008): 1211–8.

168. Tagore, Ranitendranath, Hongyu Chen, Robert H. Crabtree, and Gary W. Brudvig. "Determination of μ-oxo exchange rates in di-μ-oxo dimanganese complexes by electrospray ionization mass spectrometry." *J. Am. Chem. Soc.* 128, no. 29 (2006): 9457–65.

169. Halfen, Jason A., Samiran Mahapatra, Elizabeth C. Wilkinson et al. "Reversible cleavage and formation of the dioxygen O–O band within a dicopper complex." *Science* 271, no. 5254 (1996): 1397–400.

170. Dasgupta, Jyotishman, Rogier T. van Willigen, and G. Charles Dismukes. "Consequences of structural and biophysical studies for the molecular mechanism of photosynthetic oxygen evolution: Functional roles for calcium and bicarbonate." *Phys. Chem. Chem. Phys.* 6, no. 20 (2004): 4793–802.

171. Yang, Xiaofan, and Mu-Hyun H. Baik. "*cis,cis*-[(bpy)$_2$Ru^{V}O]$_2$O^{4+} catalyzes water oxidation formally via *in situ* generation of radicaloid Ru^{IV}-O•." *J. Am. Chem. Soc.* 128, no. 23 (2006): 7476–85.

172. Limburg, Julian, John S. Vrettos, Louise M. Liable-Sands, Arnold L. Rheingold, Robert H. Crabtree, and Gary W. Brudvig. "A functional model for O–O bond formation by the O$_2$-evolving complex in photosystem II." *Science* 283, no. 5407 (1999): 1524–7.

173. Limburg, Julian, John S. Vrettos, Hongyu Chen, Julio C. de Paula, Robert H. Crabtree, and Gary W. Brudvig. "Characterization of the O$_2$-evolving reaction catalyzed by [(terpy)(H$_2$O)Mn^{III}(O)$_2$Mn^{IV}(OH$_2$)(terpy)](NO$_3$)$_3$ (terpy = 2,2':6,2"-terpyridine)." *J. Am. Chem. Soc.* 123, no. 3 (2001): 423–30.

174. Vrettos, John S., Julian Limburg, and Gary W. Brudvig. "Mechanism of photosynthetic water oxidation: Combining biophysical studies of photosystem II with inorganic model chemistry." *Biochim. Biophys. Acta Bioenerg.* 1503, no. 1–2 (2001): 229–45.

175. McEvoy, James P., and Gary W. Brudvig. "Structure-based mechanism of photosynthetic water oxidation." *Phys. Chem. Chem. Phys.* 6, no. 20 (2004): 4754–63.

176. Siegbahn, Per E. M. "O–O bond formation in the S-4 state of the oxygen-evolving complex in photosystem II." *Chem. Eur. J.* 12, no. 36 (2006): 9217–27.

177. Haumann, Michael, and Wolfgang Junge. "Photosynthetic water oxidation: A simplex-scheme of its partial reactions." *Biochim. Biophys. Acta Bioenerg.* 1411, no. 1 (1999): 86–91.

178. Liang, Wenchuan C., Theo A. Roelofs, Roehl M. Cinco et al. "Structural change of the Mn cluster during the $S_2 \rightarrow S_3$ state transition of the oxygen-evolving complex of photosystem II. Does it reflect the onset of water/substrate oxidation? Determination by Mn X-ray absorption spectroscopy." *J. Am. Chem. Soc.* 122, no. 14 (2000): 3399–412.

179. Messinger, Johannes. "Evaluation of different mechanistic proposals for water oxidation in photosynthesis on the basis of Mn_4O_xCa structures for the catalytic site and spectroscopic data." *Phys. Chem. Chem. Phys.* 6, no. 20 (2004): 4764–71.

180. Plaksin, P. M., Gus J. Palenik, R. Carl Stoufer, and M. Mathew. "Novel antiferromagnetic oxo-bridged manganese complex." *J. Am. Chem. Soc.* 94, no. 6 (1972): 2121–2.

181. Cooper, Stephen R., G. Charles Dismukes, Melvin P. Klein, and Melvin Calvin. "Mixed-valence interactions in di-μ-oxo bridged manganese complexes—Electron-paramagnetic resonance and magnetic-susceptibility studies." *J. Am. Chem. Soc.* 100, no. 23 (1978): 7248–52.

182. Tagore, Ranitendranath, Robert H. Crabtree, and Gary W. Brudvig. "Distinct mechanisms of bridging-oxo exchange in di-μ-O dimanganese complexes with and without water-binding sites: Implications for water binding in the O_2-evolving complex of photosystem II." *Inorg. Chem.* 46, no. 6 (2007): 2193–203.

183. Wieghardt, Karl, Ursula Bossek, and Walter Gebert. "Synthesis of a tetranuclear manganese(IV) cluster with adamantane skeleton—$[(C_6H_{15}N_3)_4Mn_4O_6]^{4+}$." *Angew. Chem. Int. Ed.* 22, no. 4 (1983): 328–9.

184. Wieghardt, Karl, Ursula Bossek, Bernhard Nuber et al. "Synthesis, crystal structures, reactivity, and magnetochemistry of a series of binuclear complexes of manganese(II), -(III), and -(IV) of biological relevance. The crystal structure of $[L'Mn^{IV}(\mu-O)_3Mn^{IV}L']$ $(PF_6)_2•H_2O$ containing an unprecedented short Mn...Mn distance of 2.296 Å." *J. Am. Chem. Soc.* 110, no. 22 (1988): 7398–411.

185. Hagen, Karl S., T. David Westmoreland, Michael J. Scott, and William H. Armstrong. "Structural and electronic consequences of protonation in $(Mn_4O_6)^{4+}$ cores—pH dependent properties of oxo-bridged manganese complexes." *J. Am. Chem. Soc.* 111, no. 5 (1989): 1907–9.

186. Dubé, Christopher E., David W. Wright, and William H. Armstrong. "Multiple reversible protonations of the adamantane-shaped $\{Mn_4O_6\}^{4+}$ core: Detection of protonation stereoisomers at the $\{Mn_4O_4(OH)_2\}^{6+}$ level." *J. Am. Chem. Soc.* 118, no. 44 (1996): 10910–1.

187. Dubé, Christopher E., David W. Wright, Samudranil Pal, Peter J. Bonitatebus, Jr., and William H. Armstrong. "Tetranuclear manganese–ore aggregates relevant to the photosynthetic water oxidation center. Crystal structure, spectroscopic properties and reactivity of adamantane-shaped $[Mn_4O_6(bpea)_4]^{4+}$ and the reduced mixed-valence analog $[Mn_4O_6(bpea)_4]^{3+}$." *J. Am. Chem. Soc.* 120, no. 15 (1998): 3704–16.

188. McKee, Vickie, and William B. Shepard. "X-ray structural-analysis of a tetra-manganese(II) complex of a new (4×4) Schiff-base macrocycle incorporating a cubane-like $Mn_4(alkoxy)_4$ core." *J. Chem. Soc. Chem. Commun.* no. 3 (1985): 158–9.

189. Brooker, Sally, Vickie McKee, William B. Shepard, and Lewis K. Pannell. "Formation of a (4+4) Schiff-base macrocyclic ligand by a template rearrangement—Crystal and molecular-structures of two tetranuclear manganese(II) complexes." *J. Chem. Soc. Dalton Trans.* no. 11 (1987): 2555–62.

190. Bashkin, John S., Hsiu-Rong Chang, William E. Streib, John C. Huffman, David N. Hendrickson, and George Christou. "Modeling the photosynthetic water oxidation center—Preparation and physical properties of a tetranuclear oxide bridged Mn complex corresponding to the native S_2 state." *J. Am. Chem. Soc.* 109, no. 21 (1987): 6502–4.

191. Wang, She Yi, Kirsten Folting, William E. Streib et al. "Chloride-induced conversion of $Mn_4O_2(OAc)_6(py)_2(dbm)_2$ to $Mn_4O_3Cl(OAc)_3(dbm)_3$—Potential relevance to photosynthetic water oxidation." *Angew. Chem. Int. Ed.* 30, no. 3 (1991): 305–6.

192. Hendrickson, David N., George Christou, Edward A. Schmitt et al. "Photosynthetic water oxidation center—Spin frustration in distorted cubane $Mn^{IV}Mn_3^{III}$ model complexes." *J. Am. Chem. Soc.* 114, no. 7 (1992): 2455–71.

193. Wang, She Yi, Hui-Lien Tsai, William E. Streib, George Christou, and David N. Hendrickson. "Bromide incorporation into a high-oxidation-state manganese aggregate, and reversible redox processes for the $Mn_4O_3X(OAc)_3(dbm)_3$ (X = Cl, Br) complexes." *J. Chem. Soc. Chem. Commun.* no. 19 (1992): 1427–9.

194. Wemple, Michael W., Hui-Lien Tsai, Kirsten Folting, David N. Hendrickson, and George Christou. "Distorted cubane $Mn_4O_3Cl^{6+}$ complexes with arenecarboxylate ligation—Crystallographic, magnetochemical, and spectroscopic characterization." *Inorg. Chem.* 32, no. 10 (1993): 2025–31.

195. Wang, She Yi, Hui-Lien Tsai, Karl S. Hagen, David N. Hendrickson, and George Christou. "New structural type in manganese carboxylate chemistry via coupled oxidation/oxide incorporation—Potential insights into photosynthetic water oxidation." *J. Am. Chem. Soc.* 116, no. 18 (1994): 8376–7.

196. Wemple, Michael W., David M. Adams, Kirsten Folting, David N. Hendrickson, and George Christou. "Incorporation of fluoride into a tetranuclear Mn/O/RCO$_2$ aggregate—Potential relevance to inhibition by fluoride of photosynthetic water oxidation." *J. Am. Chem. Soc.* 117, no. 27 (1995): 7275–6.

197. Wemple, Michael W., David M. Adams, Karl S. Hagen, Kirsten Folting, David N. Hendrickson, and George Christou. "Site-specific ligand variation in manganese-oxide cubane complexes, and unusual magnetic-relaxation effects in $Mn_4O_3X(OAc)_3(Dbm)_3$ (X = N_3^-), OCN$^-$ Hdbm = dibenzoylmethane)." *J. Chem. Soc. Chem. Commun.* no. 15 (1995): 1591–3.

198. Aubin, Sheila J., Michael W. Wemple, David M. Adams, Hui-Lien Tsai, George Christou, and David N. Hendrickson. "Distorted $Mn^{IV}Mn_3III$ cubane complexes as single-molecule magnets." *J. Am. Chem. Soc.* 118, no. 33 (1996): 7746–54.

199. Aromí, Guillem, Michael W. Wemple, Sheila J. Aubin, Kirsten Folting, David N. Hendrickson, and George Christou. "Modeling the photosynthetic water oxidation complex: Activation of water by controlled deprotonation and incorporation into a tetranuclear manganese complex." *J. Am. Chem. Soc.* 120, no. 23 (1998): 5850–1.

200. Cinco, Roehl M., Annette Rompel, Hendrik Visser et al. "Comparison of the manganese cluster in oxygen-evolving photosystem II with distorted cubane manganese compounds through X-ray absorption spectroscopy." *Inorg. Chem.* 38, no. 26 (1999): 5988–98.

201. Andres, Hanspeter, Reto Basler, Hans-Ulrich Güdel et al. "Inelastic neutron scattering and magnetic susceptibilities of the single-molecule magnets $Mn_4O_3X(OAc)_3(dbm)_3$ (X = Br, Cl, Oac, and F): Variation of the anisotropy along the series." *J. Am. Chem. Soc.* 122, no. 50 (2000): 12469–77.

202. Aromí, Guillem, Sumit Bhaduri, Pau Artús, Kirsten Folting, and George Christou. "Bridging nitrate groups in $[Mn_4O_3(NO_3)(O_2CMe)_3(R_2dbm)_3]$ (R = H, Et) and $[Mn_4O_2(NO_3)(O_2CEt)_6(bpy)_2](ClO_4)$: Acidolysis routes to tetranuclear manganese carboxylate complexes." *Inorg. Chem.* 41, no. 4 (2002): 805–17.

203. Aliaga-Alcalde, Núria, Rachel S. Edwards, Stephen O. Hill, Wolfgang Wernsdorfer, Kirsten Folting, and George Christou. "Single-molecule magnets: Preparation and properties of low symmetry $[Mn_4O_3(O_2CPh-R)_4(dbm)_3]$ complexes with S = 9/2." *J. Am. Chem. Soc.* 126, no. 39 (2004): 12503–16.

204. Ruettinger, Wolfgang F., Charles Campana, and G. Charles Dismukes. "Synthesis and characterization of $Mn_4O_4L_6$ complexes with cubane-like core structure: A new class of models of the active site of the photosynthetic water oxidase." *J. Am. Chem. Soc.* 119, no. 28 (1997): 6670–1.

205. Ruettinger, Wolfgang F., Douglas M. Ho, and G. Charles Dismukes. "Protonation and dehydration reactions of the $Mn_4O_4L_6$ cubane and synthesis and crystal structure of the oxidized cubane $[Mn_4O_4L_6]^+$: A model for the photosynthetic water oxidizing complex." *Inorg. Chem.* 38, no. 6 (1999): 1036–7.

206. Ruettinger, Wolfgang, Masayuki Yagi, Kurt Wolf, Steven Bernasek, and G. Charles Dismukes. "O_2 evolution from the manganese-oxo cubane core $Mn_4O_4^{6+}$: A molecular mimic of the photosynthetic water oxidation enzyme?" *J. Am. Chem. Soc.* 112, no. 42 (2000): 10353–7.

207. Brimblecombe, Robin, Gerhard F. Swiegers, G. Charles Dismukes, and Leone Spiccia. "Sustained water oxidation photocatalysis by a bioinspired manganese cluster." *Angew. Chem. Int. Ed.* 47, no. 38 (2008): 7335–8.

208. Brimblecombe, Robin, Derrick R. J. Kolling, Alan M. Bond, G. Charles Dismukes, Gerhard F. Swiegers, and Leone Spiccia. "Sustained water oxidation by $Mn_4O_4^{7+}$ core complexes inspired by oxygenic photosynthesis." *Inorg. Chem.* 48, no. 15 (2009): 7269–79.

209. Brimblecombe, Robin, Annette Koo, G. Charles Dismukes, Gerhard F. Swiegers, and Leone Spiccia. "Solar driven water oxidation by a bioinspired manganese molecular catalyst." *J. Am. Chem. Soc.* 132, no. 9 (2010): 2892–4.

210. Hocking, Rosalie K., Robin Brimblecombe, Lan-Yun Chang et al. "Water-oxidation catalysis by manganese in a geochemical-like cycle." *Nat. Chem.* 3, no. 6 (2011): 461–6.

211. Ohlin, C. André, Robin Brimblecombe, Leone Spiccia, and William H. Casey. "Oxygen isotopic exchange in an $Mn^{III}Mn_3^{IV}$ -oxo cubane." *Dalton Trans.* no. 27 (2009): 5278–80.

212. Guiles, R. D., Jean-Luc Zimmermann, Ann E. McDermott et al. "The S3 state of photosystem-II—Differences between the structure of the manganese complex in the S_2 and S_3 states determined by X-ray absorption-spectroscopy." *Biochemistry* 29, no. 2 (1990): 471–85.

213. Kim, Dennis H., R. David Britt, Melvin P. Klein, and Kenneth Sauer. "The g = 4.1 EPR signal of the S_2 state of the photosynthetic oxygen-evolving complex arises from a multinuclear Mn cluster." *J. Am. Chem. Soc.* 112, no. 25 (1990): 9389–91.

214. Vincent, John B., Cheryl Christmas, John C. Huffman, George Christou, Hsiu-Rong Chang, and David N. Hendrickson. "Modeling the photosynthetic water oxidation center—Synthesis, structure, and magnetic-properties of $[Mn_4O_2(OAc)_7(bipy)_2]$ $(ClO_4)\cdot 3H_2O$ (bipy = 2,2′-bipyridine)." *J. Chem. Soc. Chem. Commun.* no. 4 (1987): 236–8.

215. Chan, Michael K., and William H. Armstrong. "A novel tetranuclear manganese complex that displays multiple high-potential redox processes—Synthesis, structure, and properties of $\{[Mn_2(TPHPN)(O_2CCH_3)(H_2O)]_2O\}(ClO_4)_4\cdot 2CH_3OH^1$." *J. Am. Chem. Soc.* 111, no. 25 (1989): 9121–2.

216. Chan, Michael K., and William H. Armstrong. "Tetranuclear manganese oxo complex with a 2.7-A Mn = Mn separation and intramolecular H_2O = (μ-O) hydrogen-bonded contacts: $[Mn_4O_2(TPHPN)_2(H_2O)_2(CF_3SO_3)_2](CF_3SO_3)_3$—Possible mode for binding of water at the active site of the oxygen-evolving complex in photosystem-II." *J. Am. Chem. Soc.* 112, no. 12 (1990): 4985–6.

217. Chan, Michael K., and William H. Armstrong. "Support for a dimer of di-mu-oxo dimers model for the photosystem-ii manganese aggregate—Synthesis and properties of $[(Mn_2O_2)_2(TPHPN)_2](ClO_4)_4$." *J. Am. Chem. Soc.* 113, no. 13 (1991): 5055–7.

218. Philouze, Christian, Geneviève Blondin, Jean-Jacques Girerd, Jean Guilhem, Claudine Pascard, and Doris Lexa. "Aqueous chemistry of high-valent manganese—Structure, magnetic, and redox properties of a new-type of Mn-oxo cluster, $[Mn_4^{IV}O_6(bpy)_6]^{4+}$: Relevance to the oxygen-evolving center in plants." *J. Am. Chem. Soc.* 116, no. 19 (1994): 8557–65.

219. Chen, Hongyu, J. W. Faller, Robert H. Crabtree, and Gary W. Brudvig. "Dimer-of-dimers model for the oxygen-evolving complex of photosystem II. Synthesis and properties of $[Mn_4^{IV}O_5(terpy)_4(H_2O)_2](ClO_4)_6$." *J. Am. Chem. Soc.* 126, no. 23 (2004): 7345–9.

220. Mishra, Abhudaya, Wolfgang Wernsdorfer, Khalil A. Abboud, and George Christou. "The first high oxidation state manganese-calcium cluster: Relevance to the water oxidizing complex of photosynthesis." *Chem. Commun.* no. 1 (2005): 54–6.

221. Mishra, Abhudaya, Junko Yano, Yulia Pushkar, Vittal K. Yachandra, Khalil. A. Abboud, and George Christou. "Heteronuclear Mn–Ca/Sr complexes, and Ca/Sr EXAFS spectral comparisons with the oxygen-evolving complex of photosystem II." *Chem. Commun.* no. 15 (2007): 1538–40.

222. Hewitt, Ian J., Jin-Kui Tang, N. T. Madhu et al. "A series of new structural models for the OEC in photosystem II." *Chem. Commun.* no. 25 (2006): 2650–2.

223. Nayak, Sanjit, Hari Pada Nayek, Stefanie Dehnen, Annie K. Powell, and Jan Reedijk. "Trigonal propeller-shaped $[Mn_3^{III}M^{II}Na]$ complexes (M = Mn, Ca): Structural and functional models for the dioxygen evolving centre of PSII." *Dalton Trans.* 40, no. 12 (2011): 2699–702.

224. Koumousi, Evangelia S., Shreya Mukherjee, Christine M. Beavers, Simon J. Teat, George Christou, and Theocharis C. Stamatatos. "Towards models of the oxygen-evolving complex (OEC) of photosystem II: A Mn_4Ca cluster of relevance to low oxidation states of the OEC." *Chem. Commun.* 47, no. 39 (2011): 11128–30.

225. Jerzykiewicz, Lucjan B., Józef Utko, Marek Duczmal, and Piotr Sobota. "Syntheses, structure, and properties of a manganese-calcium cluster containing a Mn_4Ca_2 core." *Dalton Trans.* no. 8 (2007): 825–6.

226. Kotzabasaki, Vasiliki, Milosz Siczek, Tadeusz Lis, and Constantinos J. Milios. "The first heterometallic Mn–Ca cluster containing exclusively Mn(III) centers." *Inorg. Chem. Commun.* 14, no. 1 (2011): 213–6.

227. Stack, T. Daniel P., and Richard H. Holm. "Subsite-specific functionalization of the $[4Fe–4S]^{2+}$ analog of iron sulfur protein clusters." *J. Am. Chem. Soc.* 109, no. 8 (1987): 2546–7.

228. Stack, T. Daniel P., and Richard H. Holm. "Subsite-differentiated analogs of biological $[4Fe–4S]^{2+}$ clusters—Synthesis, solution and solid-state structures, and subsite-specific reactions." *J. Am. Chem. Soc.* 110, no. 8 (1988): 2484–94.

229. Ciurli, Stefano, Michel Carrie, John A. Weigel et al. "Subsite-differentiated analogs of native $[4Fe–4S]^{2+}$ clusters—Preparation of clusters with five- and six-coordinate subsites and modulation of redox potentials and charge-distributions." *J. Am. Chem. Soc.* 112, no. 7 (1990): 2654–64.

230. Ciurli, Stefano, and Richard H. Holm. "Heterometal cubane-type clusters—A $ReFe_3S_4$ single-cubane cluster by cleavage of an iron-bridged double cubane and the site-voided cubane $[Fe_3S_4]$ as a cluster ligand." *Inorg. Chem.* 30, no. 4 (1991): 743–50.

231. Zhou, Jian, and Richard H. Holm. "Synthesis and metal-ion incorporation reactions of the cuboidal Fe_3S_4 cluster." *J. Am. Chem. Soc.* 117, no. 45 (1995): 11353–4.

232. Zhou, Jian, James W. Raebiger, Charles A. Crawford, and Richard H. Holm. "Metal ion incorporation reactions of the cluster $[Fe_3S_4(LS_3)]^{3-}$, containing the cuboidal $[Fe_3S_4]^0$ core." *J. Am. Chem. Soc.* 119, no. 27 (1997): 6242–50.

233. Kanady, Jacob S., Emily Y. Tsui, Michael W. Day, and Theodor Agapie. "A synthetic model of the Mn_3Ca subsite of the oxygen-evolving complex in photosystem II." *Science* 333, no. 6043 (2011): 733–6.

234. Tsui, Emily Y., Jacob S. Kanady, Michael W. Day, and Theodor Agapie. "Trinuclear first row transition metal complexes of a hexapyridyl, trialkoxy 1,3,5-triarylbenzene ligand." *Chem. Commun.* 47, no. 14 (2011): 4189–91.

235. Kanady, Jacob S., Jose L. Mendoza-Cortes, Emily Y. Tsui, Robert J. Nielsen, William A. Goddard III, and Theodor Agapie. "Oxygen atom transfer and oxidative water incorporation in cuboidal Mn_3MO_n complexes based on synthetic, isotopic labeling, and computational studies." *J. Am. Chem. Soc.* 135, no. 3 (2013): 1073–82.

236. Tsui, Emily Y., Rosalie Tran, Junko Yano, and Theodor Agapie. "Redox-inactive metals modulate the reduction potential in heterometallic manganese-oxido clusters." *Nat. Chem.* 5, no. 4 (2013): 293–9.

237. Tsui, Emily Y., and Theodor Agapie. "Reduction potentials of heterometallic manganese-oxido cubane complexes modulated by redox-inactive metals." *Proc. Natl. Acad. Sci. U.S.A.* 110, no. 25 (2013): 10084–8.

238. Stamatatos, Theocharis C., Constantinos G. Efthymiou, Constantinos C. Stoumpos, and Spyros P. Perlepes. "Adventures in the coordination chemistry of di-2-pyridyl ketone and related ligands: From high-spin molecules and single-molecule magnets to coordination polymers, and from structural aesthetics to an exciting new reactivity chemistry of coordinated ligands." *Eur. J. Inorg. Chem.* no. 23 (2009): 3361–91.

239. Mukherjee, Shreya, Jamie A. Stull, Junko Yano et al. "Synthetic model of the asymmetric $[Mn_3CaO_4]$ cubane core of the oxygen-evolving complex of photosystem II." *Proc. Natl. Acad. Sci. U.S.A.* 109, no. 7 (2012): 2257–62.

240. de Paula, Julio C., Peter Mark Li, Anne-Frances Miller, Brian W. Wu, and Gary W. Brudvig. "Effect of the 17- and 23-kilodalton polypeptides, calcium, and chloride on electron-transfer in photosystem-II." *Biochemistry* 25, no. 21 (1986): 6487–94.

241. Tagore, Ranitendranath, Hongyu Chen, Hong Zhang, Robert H. Crabtree, and Gary W. Brudvig. "Homogeneous water oxidation by a di-μ-oxo dimanganese complex in the presence of Ce^{4+}." *Inorg. Chim. Acta* 360, no. 9 (2007): 2983–89.

242. Siegbahn, Per E. M. "Substrate water exchange for the oxygen evolving complex in PSII in the S_1, S_2, and S_3 states." *J. Am. Chem. Soc.* 135, no. 25 (2013): 9442–49.

243. Zhao, Yan, and Donald G. Truhlar. "The M06 suite of density functionals for main group thermochemistry, thermochemical kinetics, noncovalent interactions, excited states, and transition elements: Two new functionals and systematic testing of four m06-class functionals and 12 other functionals." *Theor. Chem. Acc.* 120, no. 1–3 (2008): 215–41.

244. Cox, Nicholas, Dimitrios A. Pantazis, Frank Neese, and Wolfgang Lubitz. "Biological Water Oxidation." *Acc. Chem. Res.* 46, no. 7 (2013): 1588–96.

245. Siegbahn, Per E. M. "Substrate water exchange for the oxygen evolving complex in PSII in the S_1, S_2, and S_3 states." *J. Am. Chem. Soc.* 135, no. 25 (2013): 9442–49.

246. Yano, Junko, and Vittal Yachandra. "Mn_4Ca cluster in photosynthesis: Where and how water is oxidized to dioxygen." *Chem. Rev.* 114, no. 8 (2014): 4175–205.

247. Kanady, Jacob S., Po-Heng Lin , Kurtis M. Carsch et al. "Toward models for the full oxygen-evolving complex of photosystem II by ligand coordination to lower the symmetry of the Mn_3CaO_4 Cubane: Demonstration that electronic effects facilitate binding of a fifth metal." *J. Am. Chem. Soc.* 136, no. 41 (2014): 14373–6.

248. Cox, Nicholas, Marius Retegan, Frank Neese, Dimitrios A. Pantazis, Alain Boussac, and Wolfgang Lubitz. "Electronic structure of the oxygen-evolving complex in photosystem II prior to O-O bond formation." *Science* 345, no. 6198 (2014): 804–8.

6 Electrifying Metalloenzymes

Fraser A. Armstrong

CONTENTS

6.1 HISTORIC CONTEXT

Twenty-five years ago, few people associated enzymes with dynamic (amperometric) electrochemistry in any context other than biosensor research. There has been a handful of articles describing catalytic turnover by enzymes (structurally ill defined at that time) adsorbed on electrodes; however, the results were not interpreted beyond comments that the enzyme established an electrode potential that was appropriate for the substrate redox couple being studied.[1] Much more has been published about (noncatalytic) the electrochemistry of small electron carrier proteins in solution: the important breakthrough being the revelation, independently by the groups of Hill[2] and Kuwana,[3] that a solution of mitochondrial cytochrome *c* exhibits reversible cyclic voltammetry (CV) at certain electrodes. Cytochrome *c* is more than 60 times as massive as ferrocene yet the CV in each case was of textbook shape, with peak currents determined by the rate of diffusion of protein to the electrode surface. Given the three-dimensional (3D) structure of cytochrome *c* known at that time and our maturing understanding of long-range electron transfer, the results meant that there must be good transient contact between the electrode and the region of the protein close to the heme edge.

Dynamic electrochemical techniques promised much for unraveling the properties of much larger redox enzymes, particularly with regard to studying catalytic electron transport or resolving and quantifying coupled chemical reactions. This prediction turned out to be true, to such an extent that it can now be argued that the electrochemistry of enzymes is teaching us more about basic electrochemistry and electrocatalysis—in a sense, a "reversal of role." Despite their size, enzymes are very likely the best electrocatalysts identified to date.[4] The main trick has been to realize that with a molecule as large as an enzyme (far more massive than cytochrome c and even exceeding 0.5 MDa), the enzyme must be strongly adsorbed—completely overruling an old prejudice (anonymous) that "enzymes do not show direct electrochemistry because they adsorb too strongly." Protein film electrochemistry, PFE as we now abbreviate it, has benefited from the information provided by so many enzyme crystal structures and spectroscopic investigations.[5] In turn, PFE has complemented structural and spectroscopic information and established links between enzyme mechanism, electrocatalysis, and semiconductor-based solar-fuel catalysis: how is this so?

Now bearing little resemblance to the solution electrochemistry of ferrocene or cytochrome c, PFE has produced its own special rules: it links electrocatalysis, genetic engineering, fundamental enzyme chemistry, and artificial photosynthesis. This chapter is intended as a broad guide: primary results are limited to a few examples needed to convey some experimental insight, while the main focus is on summarizing some interesting systems that can be studied and resolved.

6.2   PRINCIPLES

6.2.1   Enzyme–Electrode Interface

Electrochemical instrumentation is relatively inexpensive, and investigators can be creative with experiments and equipment. The basic principles defining the relationship between the enzyme and the electrochemical platform (the electrode and the electrochemical workstation controlling the electrode) have been described at length[6,7] and are shown in Figure 6.1a.

The enzyme molecule is strongly adsorbed (attached) at an electrode surface in such a way that electrons can easily enter or leave the enzyme's internal electron transfer system, under the strict potential control provided by a potentiostat. The entire enzyme is attached to the electrode surface, and any unintended stray molecules in solution do not contribute. The current that flows due to catalytic turnover is not only a direct reporter of activity at a given potential but also a parameter, equivalent to a chromophore, that is used to measure the rates of coupled reactions, such as activation and inactivation. The system resembles a series of resistors, with each stage having a conductance expressed by a rate constant[6]: k_E refers to the interfacial electron transfer between electrode and the first relay center (the electron entry–exit center) in the enzyme, whereas k_{cat} refers to the catalytic turnover rate of the enzyme. It is important to mention the "orientation problem," noting that many enzymes are excellent at electron transfer but in a highly anisotropic way. An electron-transport

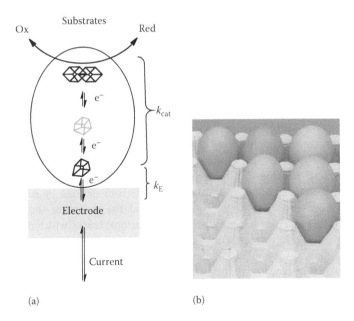

Substrates

Ox   Red

$-k_{cat}$

$\}\, k_E$

Electrode

Current

(a)   (b)

FIGURE 6.1 (a) Cartoon showing an enzyme attached to an electrode for a PFE study. Electrons produced by catalysis enter or leave the enzyme at an electron relay center close to the enzyme surface. The net direction and rate constant k_E for this process depend on the electrode potential. Within the enzyme, electrons flow at a rate proportional to the turnover rate k_{cat} of the enzyme. Either of these processes could be rate (current) limiting. (b) Analogy between the binding of enzyme molecules at a rough surface and the stabilization and extended surface contact provided to eggs in an egg tray.

enzyme has no difficulty in receiving or delivering electrons in its natural environment, and it might be expected that an electrode, to be successful, should present a very similar environment to that of the enzyme's natural partner. An important concept here is that of an egg tray (Figure 6.1b), with each egg having the letter "X" marked on its surface to represent the position of the electron entry–exit center. On a flat table, an egg rolls around—offering little chance that it will settle in such a way that the "X" contacts the surface. In contrast, an egg tray not only stabilizes the egg but also fits snugly around at least part of the egg, the numerous contact points stabilizing the binding and making it highly likely that "X" will lie at an interfacial contact point. An electrode should therefore present a rough surface, with many weak binding interactions rather than one strong one. The question of orientation is therefore misguided, as it conveys the idea (perpetuated in cartoons such as Figure 6.1a) that the surface is flat and offers few contact points.

The surface of the electrode should be valence satisfied, i.e., there should be no reactive surface atoms that can attack protein groups aggressively and indiscriminately to cause denaturation. A pyrolytic graphite "edge" (PGE) electrode surface abraded with coarse grit is rough and contains numerous oxidized carbon functionalities to link with the enzyme surface. The enzyme film is usually prepared by dabbing a tiny amount of enzyme solution onto the freshly polished electrode and then removing the excess liquid. Although additions of agents such as poly ions can greatly enhance binding, more elaborate methods are often highly desirable. The electrode can be enhanced for enzyme attachment by immobilizing carbon nanostructures, resulting in a "3D" electrode—a network of interaction points. The enzyme can be covalently attached to these extended nanostructures, for example using carbodiimide coupling to form bonds between lysines $(-NH_3^+)$ on the enzyme surface and carboxylate groups introduced on the material (use of the hydrophobic pyrene carboxylic acid is one option). The higher current densities and stabilities resulting from these procedures are particularly useful in applications such as fuel cells.[8]

The interfacial electron-transfer rate constant k_E varies with electrode potential, and we regard transfer to occur across a tunnel barrier defined by reorganization energy, electrode–cofactor orbital separation, and the nature of the intervening medium. The enzyme may adopt numerous orientations, few of which may be really productive for fast electron transfer. A model that allows for a dispersion of interfacial rate constants is very helpful in explaining why enzymes tend to give drawn-out current–potential traces that rarely reach a limiting value.[9]

6.2.2 SPECIAL FEATURES OF THE ELECTROCHEMICAL CELL

A glovebox, fully equipped with ports for electronics and gas transfer, is necessary for studying air-sensitive enzymes and (if constructed from metal) also acts as a Faraday cage to minimize noise. The electrochemical cell is designed to provide an exact fit for the electrode rotator shaft that enters from above and is thus gastight. Gases are introduced either by injecting a small aliquot of saturated solution into the cell electrolyte or by constant flow of a gas mixture (produced by precise mass flow controllers) into the headspace using inlet and outlet ports. In either case, mixing into the 1–2 mL of electrolyte solution is accelerated by the electrode rotation. For most purposes, a PGE electrode of diameter 3 mm is appropriate, and the entire enzyme under investigation is attached to the electrode surface (there is no enzyme in solution). Electrode rotation is important because catalytic turnover is often so fast that the current and voltammetric waveshape are affected by substrate depletion rather than by properties of the enzyme; also, a product that may inhibit the enzyme can be swept away. The rotation rate (ω) dependence of the catalytic current (i_{cat}) should always be checked, using the Koutecky ($i_{cat} \propto \omega^{1/2}$) or Koutecky–Levich ($1/i_{cat} \propto 1/\omega^{1/2}$) relationships.[10]

6.3 WHAT WE LEARN ABOUT ENZYMES

Figure 6.2 shows the different inputs that PFE can make during the investigation of an enzyme. The task box *Reconnaissance* exploits the fact that just tiny amounts of material are required to build a useful picture in a short time, whereas *Measurement* exploits the ability to measure catalysis and rates of active-site interconversions under excellent potential control. By adding the potential dimension, PFE complements structural and spectroscopic investigations and pinpoints the conditions under which particular species exist.

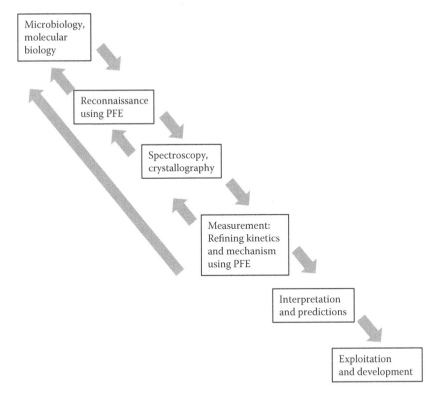

FIGURE 6.2 Inputs and feedbacks made by PFE during the investigation of an electron–transport enzyme.

The primary experimental method is CV, which is often carried out in the presence of both oxidized and reduced forms of the substrate for which interconversion is catalyzed. Ideally, the electrode is rotated at sufficient speed that the current is independent of rotation rate, although this condition may not be achievable with very active enzymes and low substrate concentrations. CV gives immediate insight into the catalytic bias of an enzyme (relative rates for catalyzing in the oxidizing or reducing direction), overpotential requirements, and interconversion between active and inactive states that is often marked by hysteresis. Once the system is reasonably well understood, the kinetics of selective processes are studied at constant potential—initiating the reaction by a potential step or by injection of a reagent, and monitoring the rate of change of the catalytic current.

6.3.1 ENZYMES AS SUPERB ELECTROCATALYSTS

By defining an electrocatalyst as a catalyst for a half-cell reaction,[11] we are adopting a broad interpretation extendable to particulate systems that involve neither electrode nor controlled potential. It is noteworthy that many enzymes that are amenable to study by PFE are also associated with membrane-based electron-transfer systems in living organisms.

A cyclic voltammogram shows two measures of catalytic efficiency—the rate (current) and the energy requirement (overpotential relative to the equilibrium solution potential) that can be represented as two rulers (Figure 6.3). Customarily, an enzyme's efficiency is judged by a rate constant; however, in the PFE experiment, we also measure the driving force that is required to achieve that rate. An efficient electrocatalyst is one that provides a high catalytic rate with minimum of overpotential (equating to energy cost). The ideal electrocatalyst thus behaves "reversibly," i.e., the current cuts cleanly across the potential axis, converting immediately from reduction to oxidation with minuscule overpotential requirement. Very few electrocatalysts work under conditions close to reversibility, although an obvious example is H^+/H_2 interconversion at the platinum electrode.

Enzymes thus far established to work reversibly are hydrogenases, carbon monoxide dehydrogenase (CODH), NADH dehydrogenase (complex I in mitochondria), fumarate reductase, succinate dehydrogenase (complex II in mitochondria), and a W-containing formate dehydrogenase.[4] Enzymes not quite operating reversibly include the blue copper oxidases, fungal laccase, or bilirubin oxidase, which still perform exceptionally well in terms of requiring only a small overpotential to carry out four-electron reduction of O_2.[12]

Enzymes could eventually lead to a unified theory of electrocatalysis. The enzyme is significant because each part of it can be engineered, and we know the 3D structures of enzymes over large size scales. The buried active site is usually well shielded from the electrode or active sites in neighboring enzymes. For any enzyme, we can now ask: what parts control (i) rate, (ii) overpotential requirement, or (iii) stability? By engineering the relay centers, we determine the importance of electron transfers, both across the electrode–enzyme interface and through the enzyme molecule. By engineering the active site, we can identify the fine details governing coupled proton–electron transfer and ligand binding, sensitivity to inhibitors, and redox-linked interconversions between states. Other parts of the enzyme can also be engineered, such as tunnels that carry small molecules and ions to and from the buried active site.

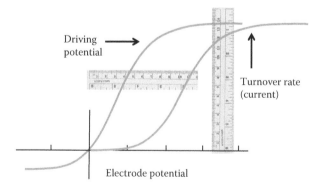

FIGURE 6.3 Two measures of the efficiency of an enzyme. The catalytic rate (current) is the more conventional parameter. The overpotential requirement (represented by the horizontal rule) that indicates the free energy needed to drive catalysis at a reasonable rate is not revealed in conventional studies.

Enzymes may be the best electrocatalysts, but they also project a very large footprint and it is therefore important to compare rates on a per-active-site basis. It is usually very difficult to see nonturnover signals, i.e., reversible signals from discrete redox sites that should appear when turnover is stopped; thus, it is very fortunate that the catalytic turnover rate greatly amplifies the current, with factors of 10^2–10^3 being typical.

6.3.2 REACTIONS OF ACTIVE SITES

Reactions at an active site are observed via changes in the catalytic current that occur as the state of the active site is altered. Different oxidation states of the active site will prevail for each potential region of a voltammogram; thus, inhibitors that bind specifically to a particular oxidation state of the active site will cause marked changes in waveform. Slow binding and release of an inhibitor will appear as hysteresis.

By clearly defining the potentials that control these interconversions, we gain a valuable guide for preparing bulk enzyme samples for spectroscopy, under controlled potential conditions: indeed, it would be difficult to unravel these complexities in other ways. The picture from PFE is unique—and it is no overstatement to make the reference "complexity resolved." It is easiest to explain this point by describing the features of Figure 6.4, which shows the electrocatalytic cyclic voltammograms for two [NiFe]-hydrogenases, Hyd-1 and Hyd-2, each produced by *Escherichia coli*, under an atmosphere of 10% H_2 at pH 6.0.[13] The past 10 years have seen a revolution in our understanding of hydrogenases, and PFE has played an important role.[14]

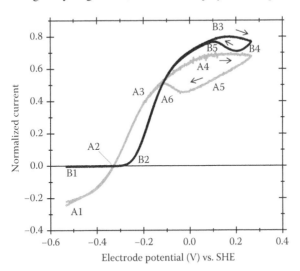

FIGURE 6.4 Cyclic voltammograms comparing the electrocatalytic activities of two [NiFe]-hydrogenases, Hyd-1 (black trace) and Hyd-2 (gray trace), from *E. coli*. Letters and numbers refer to features of the voltammograms explained in the text. Conditions for the experiments are as follows: temperature 30°C, 10% H_2, pH 6.0, scan rate 1 mV s^{-1}, rotation rate of PGE electrode 8500 rpm. (Original experiments described in Lukey, M.J. et al., *J. Biol. Chem.*, 285, 3928, **2010**.)

Dealing first with Hyd-2 (gray trace), the cycle commences at a negative potential (A1). The net current that is initially entirely due to H^+ reduction decreases as the potential is raised, dropping to zero (A2) at the equilibrium potential (−345 mV) as it changes direction in favor of H_2 oxidation (A3). Up to this point, the trace overlays with that of the return scan, signifying steady-state catalysis; however, hysteresis becomes clearly visible as the potential is raised further. A slow inactivation process (A4) continues after the scan direction is reversed (A5), then a sharp and well-defined reactivation process (A6) restores full activity and steady-state catalysis. The inactivation (A4) is due to rate-limiting binding of a hydroxide ion in the active site followed by oxidation of nickel, thereby forming the state known as Ni-B (Ni^{3+}-OH) or "Ready": the reactivation of the enzyme (A6) occurs upon rapid one-electron reduction that induces spontaneous release of OH^-. In electrochemical terms, inactivation is a CE process (a chemical step followed by an electrochemical step), whereas the reactivation process is an EC process. We see that the CV experiment has already described much about how this enzyme works; however, let us first compare the result with the other hydrogenase, Hyd-1. Starting at negative potential as before, no reduction current flows (B1) and H_2 oxidation does not commence until the potential has reached −0.3 V, after which it starts to climb (B2). Like Hyd-2, Hyd-1 also converts to the oxidized inactive state, Ni-B, at high potential and reactivates rapidly as the potential is lowered. Unlike Hyd-2, Hyd-1 does not produce H_2: simulation studies have suggested that the reason Hyd-1 does not produce H_2 is because the reduction potential at which electrons enter the enzyme through the distal Fe–S cluster is too positive.[7] Another difference is that Hyd-1 reactivates from Ni-B (B5) at a much more favorable (positive) potential than Hyd-2. As we will see later, Hyd-1 is O_2 tolerant and the facile reactivation of Ni-B is a major factor contributing to this important property.

A limitation of CV is that the potential and time domains are convoluted, and extraction of quantitative information requires simulation. The kinetics of interconversions between states are best measured directly using chronoamperometry, where the kinetic trace corresponds to the *rate of change* of catalytic rate. Chronoamperometry under strict potential control examines how fast an enzyme responds to a rapid change in circumstances, such as injection of an inhibitor or a potential step from one value to another that causes a change in steady-state catalytic rate. A particularly useful aspect is the resolution of "EC" or "CE" reactions that include ligand exchange or proton transfer driven by (or following) electron-transfer reactions. Rates of true electron transfer-driven processes may conform to Butler–Volmer kinetics, i.e., the rate increases as a logarithmic function of the overpotential applied. The Butler–Volmer equation is analogous to the Marcus theory when the overpotential used to drive the reaction is sufficiently lower than the reorganization energy. Rates of inhibitor binding and release vary from one catalytic state to another, each controllable by the electrode potential.

Hydrogenases are sensitive to oxygen, and the various scenarios have been revealed and examined in detail by using PFE. The most O_2-sensitive enzymes are the [FeFe]-hydrogenases, which have attracted particular interest because they are produced by photosynthetic, O_2-evolving green algae and might be the active components of future hydrogen farms. Studies by PFE have shown how O_2 sensitivity depends strongly on the electrode potential that is applied, which determines the state of the active site (the H-cluster) prevailing at the point of O_2 attack.[15] The mechanism of O_2 inactivation of [FeFe]-hydrogenases has been investigated, exploiting the competition with CO (which binds much more rapidly and tightly) and the fact that whereas O_2 reacts irreversibly with the active site, CO reacts reversibly so that its removal from the cell reinstates activity. The outcome of these studies is the conclusion that O_2 (like CO) targets the most oxidized active intermediate of the H-cluster, known as H_{ox}, which is also the state thought to bind H_2.[15,16] The situation is summarized in Figure 6.5. The binding of O_2 unleashes an irreversible sequence of reactions that results in H-cluster destruction. The more reduced states H_{red} and H_{sred} (sred = superreduced) may resist O_2 binding because considerable ligand reorganization is needed to avoid occupancy of antibonding orbitals.

A particular success of PFE has been in the elucidation of the mechanism of O_2 tolerance of certain [NiFe]-hydrogenases. The principle is shown in Figure 6.6, which also shows aspects of the detective work that led to the mechanism.[17]

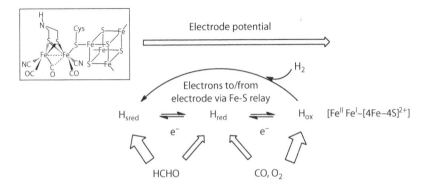

FIGURE 6.5 Selective binding of small-molecule reversible inhibitors and O_2 to different states of [FeFe]-hydrogenases, as determined by PFE studies.

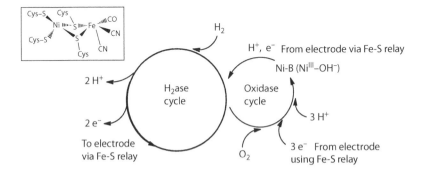

FIGURE 6.6 General mechanism for O_2 tolerance of some [NiFe]-hydrogenases that can also act as four-electron oxidases. The hydrogenase cycle is approximately 3 orders of magnitude faster than the oxidase cycle. The kinetics of individual stages of the oxidase cycle (the kinetics of O_2 attacking the active site and of activation of Ni-B) are resolved by chronoamperometry. (From Cracknell, J.A. et al., *Proc. Natl. Acad. Sci. U. S. A.*, 106, 20681, **2009**.)

The main point is that an O_2 molecule that occasionally attacks the active site is immediately reduced to harmless water, using at least one electron from the active site and up to three electrons from the three FeS clusters of the relay. Through this rapid response mechanism, the active site is able to resume H_2 activation unscathed. The fidelity of this unusual oxidase action of the enzyme is absolutely essential; otherwise, a trapped, reactive oxygen species oxygenates the active site, and the activity of the enzyme sample decreases continually as more sites succumb to attack. Crucial to this fidelity is an unusual [4Fe–3S] cluster, located closest to the active site and (uniquely, thus far) able to transfer two electrons at two closely spaced potentials.[18–20] The second oxidation process is coupled to deprotonation of a peptide-N making it a locally electroneutral (e^-, H^+) process that poses little Coulombic cost. The value of PFE here is its ability to resolve two parts of the protective oxidase co-cycle. The rate at which the enzyme re-enters the normal catalytic cycle is given by the O_2 turnover rate, which depends on (i) the rate at which O_2 attacks, and (ii) the rate of reduction of the resulting Ni-B state by a one-electron reaction.[21] These separate reactions are studied by controlled-potential chronoamperometry. The first reaction, initiated by injection of an O_2-saturated solution, depends on O_2 concentration; the second reaction is the same as shown in Figure 6.4 (top—right and B5). As initiated by a reductive potential step, the rate of activation increases as the potential is made more negative, in accordance with a linear free energy relationship (the rate varies as a logarithmic function of potential). The fraction of H_2 oxidation activity f that is observed for each O_2 level conforms reasonably well to the equation $f = $ [(rate of activation (E))/(rate of inactivation (O_2) + rate of activation (E))].

Because the enzyme is immobilized, substances can be removed after their action has been recorded, a trick that was essential for discovering that formaldehyde is a reversible inhibitor of [FeFe]-hydrogenases: only by removing formaldehyde and seeing that activity is immediately recovered could it be reasoned that it is not damaging the enzyme.[22] The chronoamperometric trace shown in Figure 6.7 is a good example of this kind of experiment. The action of formaldehyde is potential dependent, demonstrating an important attribute of PFE, i.e., the ability to judge which intermediate redox states of an active site are targeted most by different inhibitors (and substrates).[23] A summary of the effect of formaldehyde on the [FeFe]-hydrogenase from *Clostridium acetobutilicum* is included in Figure 6.5. The exact mode of inhibition by formaldehyde remains unclear; however, calculations have suggested that HCHO is well suited to react with an electron-rich Fe-hydrido species, perhaps assisted by the close presence of a pendant N-base.[23]

Perhaps the best example of intermediate-selective inhibitor binding to date involves the Ni-containing CODH.[24] This enzyme is known to interconvert between two EPR-active states, C_{red1} and C_{red2}, which prevail at different regions of potential during CV conducted under a mixture of CO_2 and CO (Figure 6.8). Cyanide, an analogue of CO, inhibits CO oxidation and also CO_2 reduction; however, it is released at a very negative potential. The interpretation is that CN^- binds tightly to C_{red1}, stabilizing it. In contrast, cyanate, an analogue of CO_2, inhibits CO_2 reduction but is released at a potential approximately 0.1 V above the CO_2/CO equilibrium potential. The interpretation is that NCO^- binds specifically to C_{red2}, stabilizing it. Binding of these inhibitors is orders of magnitude slower than catalytic CO_2/CO turnover, and cyclic voltammograms are characterized by marked hysteresis.[24]

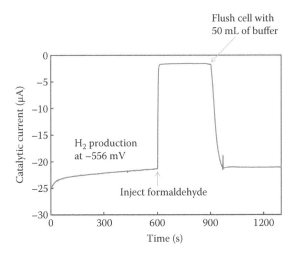

FIGURE 6.7 Experiment demonstrating the reversible inhibition, by formaldehyde, of proton reduction by a [FeFe]-hydrogenase. The enzyme is the [FeFe]-hydrogenase from *C. acetobutylicum*. After proton reduction has been running for 5 min, formaldehyde solution is injected to give a final concentration of 4.5 mM (ca. 2 μM anhydrous HCHO). The current drops rapidly to a very low level, but recovers rapidly and completely when the cell solution is replaced after 3 min. Other conditions: pH 6.0, 10°C, 100% H_2, electrode rotation rate 2500 rpm, electrode potential −558 mV vs. SHE. (Original experiment described in Wait, A.F. et al., *J. Am. Chem. Soc.*, 133, 1282, **2011**.)

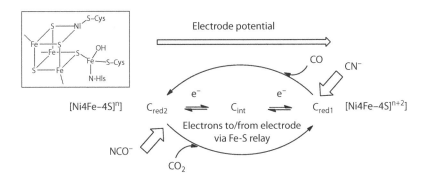

FIGURE 6.8 Selective binding of two inhibitors, cyanide and cyanate, to different redox states of the [Ni4Fe–4S]-containing carbon monoxide dehydrogenase ($CODH_I$) from *Carboxydothermus hydrogenoformans*. (Original experiments described in Wang, V.C.C. et al., *J. Am. Chem. Soc.*, 135, 2198, **2013**.)

6.4 EXTENSIONS AND APPLICATIONS

6.4.1 FUEL-CELL EXPERIMENTS

The performance of a pair of electron-transport enzymes working together is quantified in terms of power—the product of the current flowing between the two enzymes and the potential difference that is established. This interpretation applies in biological situations as it does in a fuel cell. A typical fuel-cell experiment involves applying a variable resistor between anode and cathode; however, a resistor could be replaced by a working load such as a small electronic device—a clock or LED.[25] The limiting cases are high resistance (open circuit), where no current flows and the potential difference is large (ideally that calculated from the Nernst equation and potentials and activities of the reactants), and short circuit, in which maximum current flows and the potential difference collapses. The highest power occurs at an intermediate resistance value. A short circuit may cause an enzyme to become inactive, as explained below.

Enzymes, being specific electrocatalysts, make it possible to construct membraneless fuel cells, in which a mixture of fuel and oxidant is passed into the cell.[26] Fuel is oxidized at the anode and O_2 is reduced at the cathode (Figure 6.9a). Blue Cu oxidases are suitable for the cathode and a variety of enzymes can be used at the anode, depending on the fuel. Uniquely, the O_2-tolerant hydrogenases can generate power from nonexplosive H_2/air mixtures. Recent investigations of fuel cells running on H_2-rich/air mixtures show that it should be possible to achieve power levels well in excess of 1 mW/cm^3 (volume). A study in 2011 described how power densities exceeding 0.1 mW/cm^2 electrode area can be achieved using high-density covalent enzyme attachment to carbon nanorods linked to a graphite surface.[8] The "3D" electrode concept has recently been extended and simplified by adsorbing Hyd-1 and bilirubin oxidase into compacted mesoporous carbon electrodes.[27] Large amounts of enzyme are attached and the fuel cell, producing >1 mW/cm^2 anode area from an 80/20 H_2/air mixture, runs for several days before losing 40% of power. The power-limiting electrode is the cathode, because the O_2 level is very low (4%). Power enhancement is achieved by increasing the area of the cathode relative to the anode. Some results are shown in Figure 6.9b. The open circuit voltage is close to the theoretical maximum.

An interesting scenario occurs when H_2-weak conditions are used.[28] Under 3% H_2 in air, Hyd-1 is overwhelmed by O_2 when the load is decreased to approach the condition of short circuit, and the fuel-cell power collapses. We can visualize what happens by reference to Figure 6.4. Reaction with O_2 converts Hyd-1 into Ni-B, which is normally easily reactivated by lowering the potential; however, as the load/resistance is lowered too far, the voltage drops to such a level that the potential always remains in the inactive region (above B5). Hyd-1 cannot recover because it needs electrons, and no electrons are available because the active site is in the Ni-B state. The fuel cell is instead reactivated by connecting a live anode for a few seconds, equivalent to a driver having a jump start to fix a car with a flat battery.[28]

6.4.2 Light-Driven Reactions by Enzymes Attached to Semiconducting Nanoparticles

As excellent electrocatalysts, which can be engineered to alter and test different characteristics, enzymes offer important insight regarding the design of photoelectrochemical devices producing H_2 or reducing CO_2.[11] Two essential parameters for photoelectrocatalysis are the band gap and the potentials of the valence band and the conduction band (Figure 6.10). The band gap determines the energy of the radiation

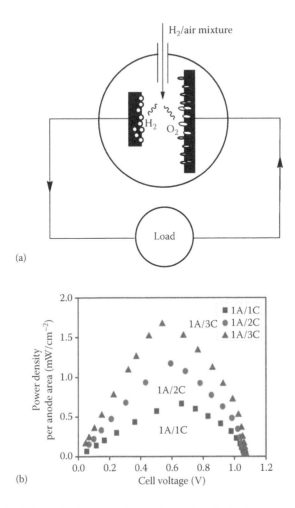

FIGURE 6.9 (a) Schematic diagram of a membrane-less fuel cell running on an H_2-rich/ air mixture. The cathode (to which bilirubin oxidase is attached) is larger than the anode (to which an O_2-tolerant hydrogenase is attached). The load can be a small electronic device, although several fuel cells are required (in series) to provide several volts. (b) Power curves obtained by varying the load resistance. The curves labeled 1A/1C, etc., refer to fuel cells in which the cathode is scaled up relative to the anode, to compensate for the low O_2 levels. The open circuit potential is 1.03–1.05 V; maximum power is drawn at 0.5–0.7 V. (Original experiment described in Xu, L. and Armstrong, F.A., *Energ. Environ. Sci.*, 6, 2166, **2013**.)

that is required to excite an electron. The band potentials determine what chemical reactions can be driven. Thus, for H_2 production, the conduction band potential should be more negative than the standard potential for H^+/H_2, whereas for O_2 evolution, the valence band potential should be more positive than the standard potential for $O_2/2H_2O$.

The electronic properties of the semiconductor can have significant effects on the electrocatalytic behavior, and semiconducting materials are tested by using them

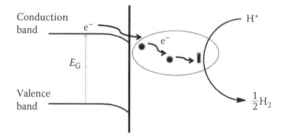

FIGURE 6.10 Schematic diagram of the interface between a semiconductor and a fuel-forming enzyme. The diagram corresponds to an n-type semiconductor held at a potential more negative than the flatband potential, so that electrons accumulate at the surface.

as electrodes. Semiconducting materials differ from most electrodes in that the interface across which electrons transfer includes a space–charge region within the material. The flatband potential E_{fb} of a semiconductor is another parameter that is assessed by using the material as an electrode. For an n-type semiconductor, application of an electrode potential more positive than E_{fb} decreases the carrier density in the space–charge region. It is therefore possible that an electrocatalytic reaction occurring close to the reversible potential could be favorably biased to operate in the desired direction, thus improving efficiency.

Illumination of semiconductor nanoparticles presented as a suspension in a short pathlength container is a standard approach in solar fuel research. Detailed interpretation of results requires consideration that the properties of semiconducting nanoparticles depend not only on chemical identity but also size, as small particles behave differently to large particles.

Unlike electrocatalysis driven by a constant potential, light-driven electrocatalysis has a severe constraint in that excited electrons are only available for very short times before electron–hole recombination occurs. Catalysis therefore has to be able to compete with recombination. An enzyme may trap electrons, using its electron-transfer relay centers, and this may be an important design principle for artificial photosynthesis catalysts because such trapping might compete effectively with electron–hole recombination.[11] A good electrocatalyst may be one that is able to store all the electrons (or holes) required for its reaction, and use the interfacial tunnel barrier to separate these electrons from the semiconductor material. It is less easy for small molecules to do this. The flatband properties of a semiconductor can also provide a gate by which a reaction occurring close to the reversible potential does not run backwards, although small nanoparticles may have dimensions larger than the thickness of the space–charge region.

The conduction band potential of anatase is suitable for H_2 and CO production at neutral pH provided only a small overpotential is required to drive the reaction. However, the band gap ($E_G = 3.1$ eV) is too large for visible light; instead, the conduction band can be populated using electrons supplied by an Ru photosensitizer, similar to the technology developed by Grätzel.[29] The configuration is shown schematically in Figure 6.11.

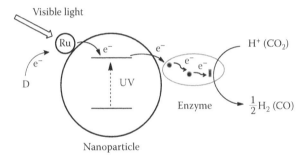

FIGURE 6.11 Schematic diagram of a TiO_2 nanoparticle excited by visible light using an Ru photosensitizer. An electron transferred into the conduction band can transfer to a fuel-forming enzyme. The band gap of anatase is too large to allow direct excitation by visible light. D = sacrificial electron donor.

Alternatively, semiconducting CdX nanoparticles (X = S, Se, Te) have small band gaps and can be used with visible light. Both hydrogenases and CODH have been used to demonstrate H_2 production and CO production at rates far exceeding those of other catalysts.[30,31] This aspect is seen as setting a benchmark for future small-molecule catalysts. A hydrogenase has been measured to give a turnover frequency of 50 s^{-1}, when attached to Ru-sensitized TiO_2 (anatase) nanoparticles and driven by visible light: this is a very high rate.[29] Thus far, only a much slower rate constant has been measured for CO_2 reduction by CODH (0.25 s^{-1} attached to CdS nanoparticles,[31] driven directly by visible light), but this is still higher than has been achieved with other catalysts.

6.4.3 Catalytic Conducting Particles

The scenario of two enzymes co-attached to a common conducting particle forms the basis of a completely new type of catalytic system. The idea behind catalytic conducting particles is to link two reactions together by electron transfer through the supporting particle. Hydrogen is a very useful reducing agent: its direct use is widespread in organic chemistry and industry but not biology. Three systems have been demonstrated, as summarized schematically in Figure 6.12: each example utilizes a hydrogenase and a second enzyme, both attached to a carbon particle.

The first demonstration of this concept was the quantitative reduction of nitrate to nitrite by molecular H_2, using Hyd-1 and nitrate reductase, both from *E. coli*.[32] This system was chosen because nitrite can be assayed easily. The second demonstration was the reduction, by CO, of water (H^+) to H_2, using Hyd-2 and CODH—a reaction equivalent to the water–gas shift reaction.[33] A truly significant aspect of that demonstration, and testimony to the kinetic superiority of enzymes, was that the reaction rate measured at room temperature was higher on a per-site basis than the rate achieved by industrial catalysts at high temperatures; indeed, good rates of conversion were even recorded when the vial was immersed in an ice bath. The third demonstration, and one that has commercial potential, is the continuous regeneration of NADH by H_2, using a hydrogenase and a FAD-containing diaphorase.[34]

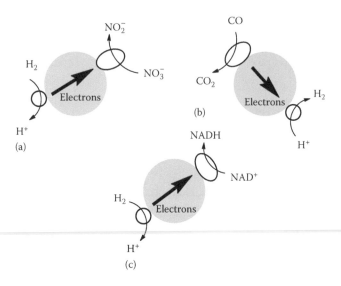

FIGURE 6.12 Cartoon showing catalysts made by combining two enzymes on an electronically conducting carbon particle. (a) Hyd-1 and nitrate reductase; (b) CODH and Hyd-2 (water gas shift reaction); and (c) an O_2-tolerant [NiFe]-hydrogenase and diaphorase (continuous regeneration of NADH).

6.5 OUTLOOK

We have shown how the electrocatalytic waveform for an enzyme is a characteristic property of that enzyme, and serves as a map by which to navigate through all kinds of complex reactivities. With models now under development to simulate the potential–current relationship, an intriguing idea is to predict the waveform for enzymes that currently do not exhibit electrocatalysis, i.e., the properties of the CV we might observe if it were possible to transfer electrons in and out reversibly.[7] Such a hypothetical concept can help us to understand how these enzymes work. Examples that could be revolutionized in this way would be cytochrome c oxidase (a proton pump) and nitrogenase (which requires ATP hydrolysis). What might we expect to see? We could predict that neither enzyme would show reversible catalysis: the coupling to a proton pump and requirement for ATP hydrolysis, respectively, would lead us to expect a sizeable overpotential and one-way catalysis. Even in such a hypothetical situation, prediction of the exact dependence of current on potential, including the deviations that might occur at specific values, would be an interesting exercise.

ACKNOWLEDGMENTS

I am grateful to the many students and postdoctorates who have worked with me over the years, and to the UK Research Councils, Biotechnology and Biological Sciences Research Council, and Engineering and Physical Sciences Research Council, for generous funding.

REFERENCES

1. Armstrong, F. A. *Struct. Bond.* **1990**, *72*, 137–230.
2. Eddowes, M. J.; Hill, H. A. O. *Chem. Commun.* **1977**, 771.
3. Yeh, P.; Kuwana, T. *Chem. Lett.* **1977**, 1145.
4. Armstrong, F. A.; Hirst, J. *Proc. Natl. Acad. Sci. U.S.A.* **2011**, *108*, 14049–14054.
5. Léger, C.; Bertrand, P. *Chem. Rev.* **2008**, *108*, 2379–2438.
6. Léger, C.; Elliott, S. J.; Hoke, K. R.; Jeuken, L. J. C.; Jones, A. K.; Armstrong, F. A. *Biochemistry* **2003**, *42*, 8653–8662.
7. Hexter, S. V.; Grey, F.; Happe, T.; Climent, V.; Armstrong, F. A. *Proc. Natl. Acad. Sci. U.S.A.* **2012**, *109*, 11516–11521.
8. Krishnan, S.; Armstrong, F. A. *Chem. Sci.* **2012**, *3*, 1015–1023.
9. Leger, C.; Jones, A. K.; Albracht, S. P. J.; Armstrong, F. A. *J. Phys. Chem. B* **2002**, *106*, 13058–13063.
10. Bard, A. J.; Faulkner, L. R. *Electrochemical Methods*; Wiley, New York, **2001**.
11. Woolerton, T. W.; Sheard, S.; Chaudhary, Y. S.; Armstrong, F. A. *Energy Environ. Sci.* **2012**, *5*, 7470–7490.
12. dos Santos, L.; Climent, V.; Blanford, C. F.; Armstrong, F. A. *Phys. Chem. Chem. Phys.* **2010**, *12*, 13962–13974.
13. Lukey, M. J.; Parkin, A.; Roessler, M. M.; Murphy, B. J.; Harmer, J.; Palmer, T.; Sargent, F.; Armstrong, F. A. *J. Biol. Chem.* **2010**, *285*, 3928–3938.
14. Vincent, K. A.; Parkin, A.; Armstrong, F. A. *Chem. Rev.* **2007**, *107*, 4366–4413.
15. Stripp, S.; Goldet, G.; Brandmayr, C.; Vincent, K. A.; Armstrong, F. A.; Happe, T. *Proc. Natl. Acad. Sci. U.S.A.* **2009**, *106*, 17331–17336.
16. Goldet, G.; Brandmayr, C.; Stripp, S. T.; Happe, T.; Cavazza, C.; Fontecilla–Camps, J. C.; Armstrong, F. A. *J. Am. Chem. Soc.* **2009**, *131*, 14979–14989.
17. Cracknell, J. A.; Wait, A. F.; Lenz, O.; Friedrich, B.; Armstrong, F. A. *Proc. Natl. Acad. Sci. U.S.A.* **2009**, *106*, 20681–20686.
18. Volbeda, A.; Amara, P.; Darnault, C.; Mouesca, J. M.; Parkin, A.; Roessler, M. M.; Armstrong, F. A.; Fontecilla-Camps, J. C. *Proc. Natl. Acad. Sci. U.S.A.* **2012**, *109*, 5305–5310.
19. Lukey, M. J.; Roessler, M. M.; Parkin, A.; Evans, R. M.; Davies, R. A.; Lenz, O.; Friedrich, B.; Sargent, F.; Armstrong, F. A. *J. Am. Chem. Soc.* **2011**, *133*, 16881–16892.
20. Roessler, M. M.; Evans, R. M.; Davies, R. A.; Harmer, J.; Armstrong, F. A. *J. Am. Chem. Soc.* **2012**, *134*, 15581–15594.
21. Evans, R. M.; Parkin, A.; Roessler, M. M.; Murphy, B. J.; Adamson, H.; Lukey, M. J.; Sargent, F.; Volbeda, A.; Fontecilla-Camps, J. C.; Armstrong, F. A. *J. Am. Chem. Soc.* **2013**, *135*, 2694–2707.
22. Wait, A. F.; Brandmayr, C.; Stripp, S. T.; Cavazza, C.; Fontecilla-Camps, J. C.; Happe, T.; Armstrong, F. A. *J. Am. Chem. Soc.* **2011**, *133*, 1282–1285.
23. Foster, C. E.; Kramer, T.; Wait, A. F.; Parkin, A.; Jennings, D. P.; Happe, T.; McGrady, J. E.; Armstrong, F. A. *J. Am. Chem. Soc.* **2012**, *134*, 7553–7557.
24. Wang, V. C. C.; Can, M.; Pierce, E.; Ragsdae, S. W.; Armstrong, F. A. *J. Am. Chem. Soc.* **2013**, *135*, 2198–2206.
25. Vincent, K. A.; Cracknell, J. A.; Clark, J. R.; Ludwig, M.; Lenz, O.; Friedrich, B.; Armstrong, F. A. *Chem. Commun.* **2006**, 5033–5035.
26. Cracknell, J. A.; Vincent, K. A.; Armstrong, F. A. *Chem. Rev.* **2008**, *108*, 2439–2461.
27. Xu, L.; Armstrong, F. A. *Energy Environ. Sci.* **2013**, *6*, 2166–2171.
28. Wait, A. F.; Parkin, A.; Morley, G. M.; dos Santos, L.; Armstrong, F. A. *J. Phys. Chem. C* **2010**, *114*, 12003–12009.
29. Reisner, E.; Powell, D. J.; Cavazza, C.; Fontecilla-Camps, J. C.; Armstrong, F. A. *J. Am. Chem. Soc.* **2009**, *131*, 18457–18466.

30. Brown, K. A.; Wilker, M. B.; Boehm, M.; Dukovic, G.; King, P. W. *J. Am. Chem. Soc.* **2012**, *134*, 5627–5636.

31. Chaudhary, Y. S.; Woolerton, T. W.; Allen, C. S.; Warner, J. H.; Pierce, E.; Ragsdale, S. W.; Armstrong, F. A. *Chem. Commun.* **2012**, *48*, 58–60.

32. Vincent, K. A.; Li, X.; Blanford, C. F.; Belsey, N. A.; Weiner, J. H.; Armstrong, F. A. *Nat. Chem. Biol.* **2007**, *3*, 761–762.

33. Lazarus, O.; Woolerton, T. W.; Parkin, A.; Lukey, M. J.; Reisner, E.; Seravalli, J.; Pierce, E.; Ragsdale, S. W.; Sargent, F.; Armstrong, F. A. *J. Am. Chem. Soc.* **2009**, *131*, 14154–14155.

34. Reeve, H. A.; Lauterbach, L.; Ash, P. A.; Lenz, O.; Vincent, K. A. *Chem. Commun.* **2012**, *48*, 1589–1591.

7 Iron Uptake Mechanism in Ferritin from *Helicobacter pylori*

Sella Kim and Kyung Hyun Kim

CONTENTS

7.1   IRON AND FERRITIN

Iron has an essential function in living cells as it can gain and lose electrons based on two stable oxidation states, ferrous (Fe^{2+}) and ferric (Fe^{3+}) ions. This property makes iron a useful component of heme-bound and nonheme iron proteins. Excess ferrous ions act as a toxic catalyst to produce free radicals in the presence of H_2O_2 or other reactive oxygen species (ROS),[1] whereas iron deficiency arrests cell growth and leads to cell death. ROS induce stress conditions in cells and damage proteins, nucleotides, and lipids in cell membrane.[2] Cellular iron levels are thus carefully regulated to minimize the pool of otherwise potentially toxic irons. On the other hand, iron lies at the center of a battle for nutritional resource between hosts and microbial pathogens. A host's iron status can influence infection in animals including humans, which affects the pathogenicity of numerous infections, such as malaria, HIV-1, and tuberculosis.[3] A study on iron deficiency in patients, recorded five times the number of infections in those receiving iron supplements relative to a placebo group.[4]

Organisms have many pathways to regulate cellular iron levels by controlling mRNA expression and/or protein levels. There are two specialized iron-binding proteins: extracellular transferrin or siderophore and intracellular ferritin. Iron responsive element (IRE) and IRE-binding protein can regulate transferrin or ferritin at the level of translation through a negative feedback loop.[5,6] Since living organisms have no physiologic regulatory mechanism for excreting iron, iron homeostasis is generally regulated at the level of iron uptake and storage.

Ferritin in living organisms is the principal iron storage protein that stores and releases iron in a controlled manner. Ferritins are observed from prokaryotes to eukaryotes as iron-storage proteins in spite of differences in cellular location and biosynthetic regulation. Ferritin can be a homopolymer or a heteropolymer consisting of subunits designated as H-, M-, and/or L-type subunit in case of eukaryotes. Although the DNA and amino acid sequences of ferritins vary considerably, they share 17%–62% amino acid sequence identity among *Helicobacter pylori* ferritin (Hpf); *Campylobacter jejuni* ferritin (Cjf); *Escherichia coli* ferritin (Ecf); and human H (HuHF), L (HuLF), and M (HuMF) ferritins (Figure 7.1a). The crystal structures of ferritins are also highly conserved, which shows 24 subunits assembled into a hollow shell with a 110–120 Å diameter; 432 point symmetry with 3-fold,

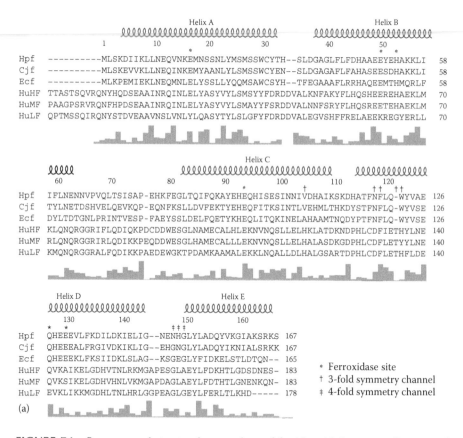

FIGURE 7.1 Sequence and structural comparison of ferritins. (a) Sequence alignment of Hpf, Cjf, Ecf, HuHF, HuMF, and HuLF proteins. Above alignment, amino acid sequences of Hpf are represented in blocks of 10 amino acids and the numbers on the right are the residue number within their own sequences. Helical secondary structure element is indicated at the top, and histograms at the bottom indicate the conservation level. Residues interacting with the ferroxidase center and at the 3-fold and 4-fold channels are marked by asterisk, dagger, and double dagger at the top, respectively.

(Continued)

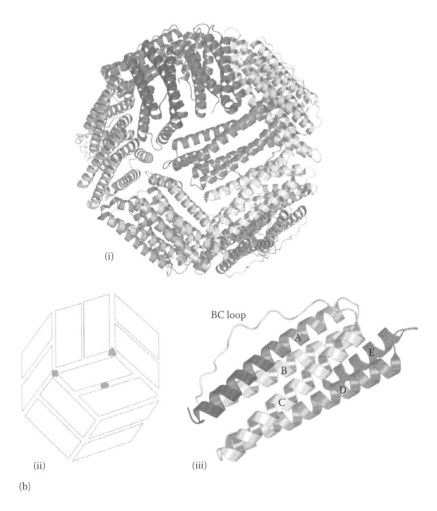

FIGURE 7.1 (CONTINUED)   Sequence and structural comparison of ferritins. (b) Structures of ferritins. (i) Quaternary 24-mer Hpf structure; (ii) 3-fold, 4-fold, and 2-fold symmetry points marked by a triangle, square, and oval, respectively; (iii) monomer structure of Hpf. BC loop and helices A–E are shown.

4-fold, and 2-fold axes (Figure 7.1b). Each subunit is approximately 20 kDa and folded as a four-helix bundle with a fifth short helix, within a dinuclear ferroxidase center catalyzes the oxidation of ferrous ions to ferric irons. The hollow center is a crucial part of protein design since it provides a large cavity for the accumulation of an iron core for storage purposes,[7] which is capable of storing up to 4500 Fe ions as inorganic complexes either of a crystalline mineral ferrihydrite ($5Fe_2O_3 \cdot 9H_2O$) or of an amorphous hydrous ferric phosphate.[5] Fe ions can communicate through hydrophilic channels on the symmetry axes in both eukaryotic and prokaryotic ferritins.[8]

The ferritin family is classified into three subfamilies: the classical ferritin (Ftn), bacterioferritin (Bfr), and DNA-binding protein (Dps). Ftn and Bfr proteins are considered maxi-ferritins, whereas Dps proteins are mini-ferritins. Bfr proteins containing heme have essentially the same architecture as ferritin, which are assembled into a 24-mer cluster to form a hollow construction, but are restricted to the bacterial and archaeal domains of life.[9] The difference between Ftn and Bfr proteins is the presence of heme within the Bfr shell—up to 12 hemes at the interface between 2-fold related subunits. The role of the heme group within Bfr is likely to enable the release of iron from the storage cavity through a reduction mechanism.[10] In comparison, Dps is assembled with only 12 subunits, with a smaller 95 Å diameter and lower iron storage capacity of up to 500 irons. Dps exhibits a 3–3–2 fold symmetry, and each Dps subunit lacks the E helix of the Ftn and Bfr proteins, but contains an additional short helical region within the long BC loop.[11,12] Dps proteins use unique ferroxidase sites at the interface between 2-fold related subunits, unlike Ftn and Bfr.

Ferritin subunits, except for the L-type subunit, incorporate a ferroxidase center that consists of μ-oxo-bridged dinuclear irons and is responsible for iron oxidation.[13–15] Ferrous ions and either dioxygen or hydrogen peroxide are the substrates for each protein subunit catalytic site that initiates biomineral synthesis.[16] Ferritins act as antioxidants by consuming iron and oxygen during biomineralization, which is part of a feedback loop where excess iron and oxygen activate ferritin DNA and mRNA to increase ferritin synthesis. Subsequent protein accumulation, which consumes iron and oxygen, decreases the signals and shuts down ferritin synthesis.[17]

7.2 *Helicobacter pylori* FERRITIN

Helicobacter pylori, a microaerophilic, gram-negative bacterium, is a principal etiological agent of chronic active gastritis, duodenal or gastric ulcers, gastric carcinoma, or gastric lymphoma.[18] At least 50% of the world's human population has *H. pylori* infection.[19] The organism can survive in the acidic environment of the stomach, and its prevalence increases with older age and with lower socioeconomic status during childhood, and thus varies markedly worldwide.[20] The bacterium may have already infected humans around 58,000 years ago when they migrated out of Africa.[21] *Helicobacter pylori* require iron to survive, and proteins involved in iron metabolism are suggested to represent major virulence determinants.[22]

Helicobacter pylori ferritin (Hpf) plays a significant role in iron homeostasis, controlling the intracellular amount of the soluble ferrous ions and protects the bacteria from acid-amplified iron toxicity. An Hpf mutant of *H. pylori* was completely defective for colonization in the gerbil stomach, and rapid iron storage is required for adaptation and survival of the bacterium to the changes in iron availability occurring in the gastric mucosa.[23] Although the amino acid residues at the active site in ferritins are highly conserved (Figure 7.1a), those at the hydrophilic symmetry channels for the uptake of ferrous irons are not well conserved between prokaryotes and eukaryotes. The 24-mer ferritin has two types of channels on the 3-fold and 4-fold symmetry axes, with different functions. It has been known that Fe ions communicate with the solvent via hydrophilic channels on the 3-fold symmetry axes in eukaryotes, whereas other channels along the 4-fold axes are rather tight and hydrophobic, and do not seem to be involved in ion exchange.[24]

7.3   SYMMETRY CHANNELS FOR IRON UPTAKE

The channels in ferritin formed by assembled symmetry are important to the uptake of Fe ions. The crystal structures with metal ions (Cd^{2+}, Zn^{2+}, Ca^{2+}, and Tb^{3+}) bound to the six conserved carboxylates along the 3-fold channels suggested the channels as the likely route of iron entry into the protein shell in animal ferritins,[2,24,25] including insect ferritins.[26] Experimental evidence consistent with several studies showed that substitution of the 3-fold channel carboxylates by other amino acids (alanine, histidine, or leucine) inhibited iron incorporation into animal ferritins.[27,28] Entry by this route means that Fe^{2+} must not only pass through the 10-Å-long hydrophilic 3-fold channel but also traverse a distance of about 20 Å in length along the hydrophilic pathway from the inside of the channel to the catalytic ferroxidase center.[27] The residues that are formed along the 3-fold iron channels in case of human H-type ferritin (HuHF) are hydrophilic: His118, Cys130, Asp131, Glu134, and Thr135. Evidence for the 3-fold route came from the calculations of a molecular surface with electrostatic potential in HuHF, which shows that the negative outer entrance is surrounded by patches of positive potential (Figure 7.2a), and this arrangement leads to electrostatic fields directing cations such as Fe ions toward the channel entrance. The region of negative potential extends through the 3-fold channel to the interior of the molecule.[2,29,30]

In contrast, the negative electrostatic potential at the entrance to the 4-fold symmetry channels suggests that ferrous ions can be pulled inside by an electrostatic gradient with near-positive patch (Figure 7.2a). Whereas, the pore entrance to 3-fold symmetry channels has a diameter of approximately 10 Å but is plugged by hydrophobic amino acid residues Ile109 and Phe117 in Hpf (Figure 7.2b), the amino acid residues along the 4-fold symmetry channel are hydrophilic: Asn146, Glu147, Asn148, and His149. Notably, the 4-fold symmetry channels in prokaryotes were found to have similar chemical properties to those of the 3-fold channels of animal ferritins.

7.4   IRON UPTAKE IN *H. pylori* FERRITIN

The molecular electrostatic potential representation of ferritins illustrates a hydrophilic portion of the surface at the 3-fold axis channel in human ferritin (Figure 7.2a), whereas the same channel is formed mainly by hydrophobic residues in Hpf and *E. coli* ferritin (Ecf), including other prokaryotic and archean ferritins. In contrast, a hydrophilic portion of the surface is found at the 4-fold symmetry axis channel in prokaryotic ferritins, Hpf and Ecf. The pore of the prokaryotic 4-fold channels consists of polar side chains, e.g., Glu147 and His149 in Hpf, and Ser147 and Glu149 in Ecf (Figure 7.2b), including archaeal ferritins. As described previously, metal ions were previously observed at the 4-fold channel in a Bfn.[17,25] A better understanding of the mechanisms of iron uptake by Hpf was gained when Hpf native and mutant proteins were prepared in several iron-bound states: apo, low-iron bound, intermediates, and high-iron bound state. As iron levels in ferritins were increased, significant conformational changes at the entrance to the 4-fold symmetry channel were observed in the intermediate and high-iron bound states, whereas no such

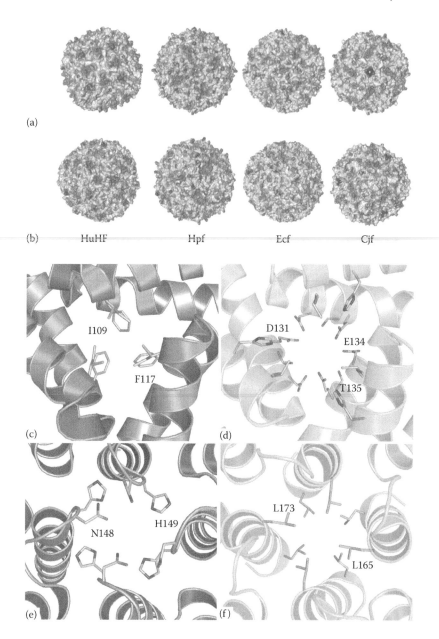

FIGURE 7.2 Overall structures of Hpf. Surface potential representation of the pore entrance at (a) the 4-fold and (b) 3-fold channels, viewed down the 4-fold and 3-fold symmetry axis, respectively. The electrostatic potentials of HuHF, Hpf, Ecf, and Cjf are contoured from −8 (light gray) to +8 kT (dark gray) through 0 (white). Amino acid residues at the entrance to the (c and d) 3-fold and (e and f) 4-fold symmetry channels of Hpf and HuLF, respectively.

(Continued)

FIGURE 7.2 (CONTINUED) Overall structures of Hpf. (g) Pores at the junction of protein subunits at the 4-fold symmetry channel. The conformational changes of Asn148 and His149 occur, depending on the apo, low-iron and high-iron bound states of Hpf. (h) The 3-fold symmetry channel. There is no conformational change of Ile107 and Phe117 depending on iron concentration in contrast with the 4-fold channel.

changes were observed at the 3-fold symmetry channels (Figure 7.2c). In addition, no changes were found in the apo and low-iron bound states. In the high-iron bound state, His149 adopts alternative conformations and the distance between the rings falls into the range of 3.0–3.2 Å. In Hpf, Fe ion should be able to coordinate to four symmetry-related imidazole rings of His149, which are directed inwardly toward the pore, while Asn148 is facing the outside medium. When these Fe sites are spatially superimposed, based on the structures in apo, low-iron bound, intermediates, and high-iron bound states, the Fe ion observed in the electron density map is positioned at 0.3–2.5 Å below the imidazole planes depending on iron-bound forms. Therefore, His149 at the 4-fold channel may act like a swing gate whose conformational changes upon Fe binding are directly coupled to allow the axial translocation of Fe ions through the 4-fold channel.[8]

Kinetic analysis of Hpf, its H149L mutant, Ecf, HuHF, and HuLF revealed that the rate of iron oxidation was monitored by the increase of absorbance at 310 nm at various concentrations of ferrous ammonium sulfate.[31] The K_m and k_{cat} values for the oxidation activity of mature Hpf were 1.84 mM and 8.79 s^{-1} at pH 7.0, respectively (Table 7.1). By contrast, H149L mutants showed markedly reduced iron oxidation activity (59.7 ± 5.2% that of mature Hpf), mainly owing to the lower affinity of Fe ions. In addition, the activity of Hpf was higher than that of Ecf, and the catalytic efficiency of HuHF was markedly high compared with that of Hpf and Ecf. It turned out that the efficiency of Hpf and HuHF was, however, significantly reduced at low pH and the activity of Hpf was significantly decreased at low pH, with approximately a 5-fold decrease at pH 5.5 (data not shown). Stopped-flow experiments were also carried out in which Fe^{2+} ion was mixed against mature Hpf or H149L at different ratios of Fe^{2+}/protein. Although both Hpf and H149L showed very similar curves at low Fe^{2+}/protein ratio, corresponding to the apo and low iron states, Hpf exhibited markedly enhanced curves compared with that of H149L at >500 Fe/proteins.[8] It is very likely that the ferroxidase reaction predominantly occurs in the apo and low-iron states, whereas the mineral surface autoxidation dominates in the intermediate and high-iron states.[8]

Iron thus enters the 4-fold symmetry channel of the shell in Hpf, unlike ferritins of mammals and insects. The 4-fold symmetry channel may serve as a pathway for iron translocation not only in Hpf but also in other prokaryotes, which means that the 4-fold channel can be the earliest path for iron uptake in prokaryotes before the emergence of eukaryotic ferritins. In fact, ferritins in insect and vertebrates contain a variable mixture of H and L subunits facilitating the uptake of iron.[26,32] These ferritins contain varying and characteristic ratios of H and L subunits in different tissues, which reflects tuning of ferritin functions for specific needs.

TABLE 7.1

Kinetic Parameters for Iron Incorporation of Ferritins

Ferritin	K_m (mM)	k_{cat} (s^{-1})	k_{cat}/K_m (mM^{-1} s^{-1})
Mature form	1.84 ± 0.22	8.79 ± 1.96	4.71 ± 0.60
H149L	4.47 ± 0.14	12.53 ± 0.70	2.81 ± 0.24
Ecf	1.56 ± 0.24	2.71 ± 0.55	1.73 ± 0.09
HuHF	0.22 ± 0.01	12.02 ± 0.11	54.46 ± 0.58
HuLF	0.57 ± 0.16	3.88 ± 0.98	6.84 ± 0.16

7.5   IRON UPTAKE MECHANISM IN *H. pylori* FERRITIN

Hpf may have evolved to possess an iron uptake mechanism for iron storage through the 4-fold symmetry channel (Figure 7.3a). First, iron–protein interactions preferentially occur on the external surface at the pore to the 4-fold symmetry channel. His149 at the 4-fold symmetry channel can serve as an iron-sensing gate or an entry route for iron transfer. The ferroxidase center then initiates iron oxidation by binding two ferrous ions that are oxidized to ferric ions that migrate to the inner nucleation site by the cluster of glutamate residues. Second, as the iron nucleation takes place in the core, and ferric oxide clusters are subsequently formed, irons may enter

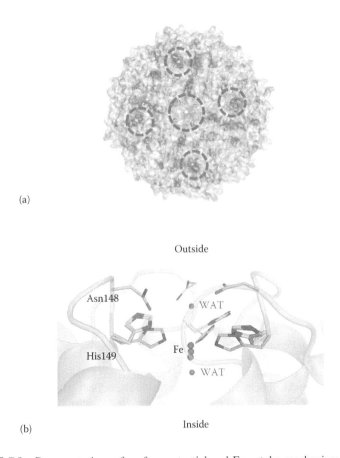

(a)

(b)   Inside

FIGURE 7.3   Representations of surface potential and Fe uptake mechanism. (a) Surface potential representation of the pore entrance to the 4-fold symmetry channel. It is presented with contours from −8 (light gray) to +8 kT (dark gray) through 0 (white). (b) Schematic diagram of proposed iron entry and translocation paths in Hpf. An iron uptake mechanism for iron storage through the 4-fold symmetry channel. Hydrophilic residues are shown in a stick model. Metal ions and water molecules are represented as dark and bright spheres, respectively.

the 4-fold channel directly into the core where oxidation can occur. They are oriented along the axis coordinated to the imidazole rings of His149, and the residues at the channels are able to assist their passage. The progressive movement of irons along the channels is tightly coupled to the alternative conformational changes of His149. It was previously suggested that as the iron core is progressively laid down, the mechanism of iron oxidation may change from a protein-dominated process to a mineral surface-dominated process.[8] Understanding how this uptake mechanism for irons can be controlled in Hpf will be instrumental for developing new therapeutic approaches for treating gastritis and gastric ulcer. Our results also provide a novel target for therapeutic intervention in a broad range of infections caused by pathogenic bacteria including *H. pylori*.

ACKNOWLEDGMENTS

We wish to acknowledge the technical support from the staff at the beamlines of Pohang Light Source (5C) (proposal no. 2012-5C-008) and PF (BL-17A) under the approval of the PF Program Advisory Committee (proposal no. 2012G618). This work was supported by the Basic Science Research Program through NRF (2010-0012853) and the Department of Biotechnology and Bioinformatics at KU under BK21 Plus funded by the Ministry of Education, Korea.

REFERENCES

1. Miller, R. A. and Britigan, B. E. Role of oxidants in microbial pathophysiology. *Clin Microbiol Rev* **10**, 1–18 (1997).
2. Chasteen, N. D. and Harrison, P. M. Mineralization in ferritin: An efficient means of iron storage. *J Struct Biol* **126**, 182–194 (1999).
3. Drakesmith, H. and Prentice, A. M. Hepcidin and the iron-infection axis. *Science* **338**, 768–772 (2012).
4. Murray, M. J., Murray, A. B., Murray, M. B. and Murray, C. J. The adverse effect of iron repletion on the course of certain infections. *Br Med J* **2**, 1113–1115 (1978).
5. Braun, V., Hantke, K. and Koster, W. Bacterial iron transport: Mechanisms, genetics, and regulation. *Met Ions Biol Syst* **35**, 67–145 (1998).
6. Orino, K. et al. Ferritin and the response to oxidative stress. *Biochem J* **357**, 241–247 (2001).
7. Le Brun, N. E., Crow, A., Murphy, M. E., Mauk, A. G. and Moore, G. R. Iron core mineralisation in prokaryotic ferritins. *Biochim Biophys Acta* **1800**, 732–744 (2010).
8. Cho, K. J. et al. The crystal structure of ferritin from *Helicobacter pylori* reveals unusual conformational changes for iron uptake. *J Mol Biol* **390**, 83–98 (2009).
9. Andrews, S. C. The ferritin-like superfamily: Evolution of the biological iron storeman from a rubrerythrin-like ancestor. *Biochim Biophys Acta* **1800**, 691–705 (2010).
10. Weeratunga, S. K. et al. Binding of *Pseudomonas aeruginosa* apobacterioferritin-associated ferredoxin to bacterioferritin B promotes heme mediation of electron delivery and mobilization of core mineral iron. *Biochemistry* **48**, 7420–7431 (2009).
11. Grant, R. A., Filman, D. J., Finkel, S. E., Kolter, R. and Hogle, J. M. The crystal structure of Dps, a ferritin homolog that binds and protects DNA. *Nat Struct Biol* **5**, 294–303, (1998).
12. Ilari, A., Stefanini, S., Chiancone, E. and Tsernoglou, D. The dodecameric ferritin from *Listeria innocua* contains a novel intersubunit iron-binding site. *Nat Struct Biol* **7**, 38–43 (2000).

13. Tatur, J., Hagen, W. R. and Matias, P. M. Crystal structure of the ferritin from the hyperthermophilic archaeal anaerobe *Pyrococcus furiosus*. *J Biol Inorg Chem* **12**, 615–630, (2007).

14. Johnson, E., Cascio, D., Sawaya, M. R., Gingery, M. and Schroder, I. Crystal structures of a tetrahedral open pore ferritin from the hyperthermophilic archaeon *Archaeoglobus fulgidus*. *Structure* **13**, 637–648 (2005).

15. Stillman, T. J. et al. The high-resolution x-ray crystallographic structure of the ferritin (EcFtnA) of *Escherichia coli*; Comparison with human H ferritin (HuHF) and the structures of the Fe(3+) and Zn(2+) derivatives. *J Mol Biol* **307**, 587–603 (2001).

16. Bevers, L. E. and Theil, E. C. Maxi- and mini-ferritins: Minerals and protein nanocages. *Prog Mol Subcell Biol* **52**, 29–47 (2011).

17. Theil, E. C. and Goss, D. J. Living with iron (and oxygen): Questions and answers about iron homeostasis. *Chem Rev* **109**, 4568–4579 (2009).

18. Suerbaum, S. and Michetti, P. *Helicobacter pylori* infection. *N Engl J Med* **347**, 1175–1186 (2002).

19. Everhart, J. E. Recent developments in the epidemiology of *Helicobacter pylori*. *Gastroenterol Clin North Am* **29**, 559–578 (2000).

20. Woodward, M., Morrison, C. and McColl, K. An investigation into factors associated with *Helicobacter pylori* infection. *J Clin Epidemiol* **53**, 175–181 (2000).

21. Linz, B. et al. An African origin for the intimate association between humans and *Helicobacter pylori*. *Nature* **445**, 915–918 (2007).

22. McGee, D. J. and Mobley, H. L. Mechanisms of *Helicobacter pylori* infection: Bacterial factors. *Curr Top Microbiol Immunol* **241**, 155–180 (1999).

23. Arosio, P., Ingrassia, R. and Cavadini, P. Ferritins: A family of molecules for iron storage, antioxidation and more. *Biochim Biophys Acta* **1790**, 589–599 (2009).

24. Hempstead, P. D. et al. Comparison of the three-dimensional structures of recombinant human H and horse L ferritins at high resolution. *J Mol Biol* **268**, 424–448 (1997).

25. Lawson, D. M. et al. Solving the structure of human H ferritin by genetically engineering intermolecular crystal contacts. *Nature* **349**, 541–544 (1991).

26. Nichol, H. and Locke, M. The characterization of ferritin in an insect. *Insect Biochem* **19**, 587–602 (1989).

27. Bauminger, E. R. et al. Iron (II) oxidation and early intermediates of iron–core formation in recombinant human H-chain ferritin. *Biochem J* **296 (Pt 3)**, 709–719 (1993).

28. Levi, S., Santambrogio, P., Corsi, B., Cozzi, A. and Arosio, P. Evidence that residues exposed on the three-fold channels have active roles in the mechanism of ferritin iron incorporation. *Biochem J* **317**, 467–473 (1996).

29. Douglas, T. and Ripoll, D. R. Calculated electrostatic gradients in recombinant human H-chain ferritin. *Prot Sci* **7**, 1083–1091 (1998).

30. Crichton, R. R. and Declercq, J. P. X-ray structures of ferritins and related proteins. *Biochim Biophys Acta* **1800**, 706–718 (2010).

31. Toussaint, L., Bertrand, L., Hue, L., Crichton, R. R. and Declercq, J. P. High-resolution x-ray structures of human apoferritin H-chain mutants correlated with their activity and metal-binding sites. *J Mol Biol* **365**, 440–452 (2007).

32. Arosio, P. and Levi, S. Ferritin, iron homeostasis, and oxidative damage. *Free Radic Biol Med* **33**, 457–463, pii: S0891-5849(02)00842-0 (2002).

8 Multiple-Step Electron Flow in Proteins

Jeffrey J. Warren, Maraia E. Ener,
Jay R. Winkler, and Harry B. Gray

CONTENTS

8.1 BACKGROUND AND INTRODUCTION

The application of the semiclassical electron transfer (ET) theory to metalloprotein redox reactions has been appreciated for 50 years.[1,2] The first qualitative suggestion that electrons could "tunnel" between redox centers was described for photosynthetic reaction centers (PRCs) in 1966.[3] The proposal that "it is unnecessary to postulate any more complicated mechanism than tunneling for moving electrons over a few tens of angstroms" in biomolecules was surprisingly perceptive,[4] considering that structures were not available from which to characterize the nature and arrangement of redox partners.

Work in the 1970s, building on investigations of small-molecule ET reactions during the 1950s and 1960s,[5] demonstrated that small redox proteins could transfer electrons to small-molecule redox reagents.[6] By that time, only a handful of metalloproteins had been characterized by X-ray crystallography, including myoglobin,[7] hemoglobin,[8] cytochrome c,[9] and plastocyanin.[10] In the 1980s, the first generation of artificial transition metal–modified protein systems emerged.[11] These systems employed a small-molecule redox reagent tethered to the exterior of a metalloprotein.

235

ET reactions were initiated with laser pulses and studied using transient spectroscopies. In both the bimolecular and unimolecular protein systems, ET rate constants could be determined and then were rationalized using the semiclassical ET theory.

The semiclassical theory of ET (Equations 8.1 and 8.2) provides the framework for understanding single-step, intramolecular ET.[1,5] The three important parameters are reorganization energy (λ, including inner and outer sphere), driving force ($-\Delta G^\circ$), and electronic coupling between the reactants and products at the transition state (H_{AB}). The salient points from this extensive work on single-step ET reactions of metalloproteins are that many active sites have optimized λ and H_{AB}; the practical limit of separation between redox sites is ~20 Å for ET on biologically relevant timescales (milliseconds or less); and the semiclassical ET theory can provide useful predictions of ET rates in biological systems.[12,13]

$$k_{ET} = \sqrt{\frac{4\pi^3}{h^2 \lambda k_B T}} H_{AB}^2 \exp\left(\frac{-(\Delta G^\circ + \lambda)^2}{4\lambda k_B T}\right) \tag{8.1}$$

$$H_{AB} = H_{AB}^{r_0} \exp(-0.5\beta(r - r_0)) \tag{8.2}$$

A great many proteins rely on net ET reactions over distances much longer than 20 Å. In these cases, nature breaks up a large distance between two cofactors with several shorter ET steps with minimal losses in free energy. This mode of ET is called "hopping." The concept of electrons hopping between separated sites likely grew out of work on solid-state physics in the early 1960s to rationalize electron mobility in extended lattices.[14] In the biological context, and also in the early 1960s, it was proposed that the extended π-stacked structure of DNA could serve as a conducting polymer, with electrons moving through the bases.[15]

The same semiclassical ET theory parameters that govern single-step ET also govern hopping reactions. In this chapter, we describe the development of hopping maps to describe two-step hopping systems. We then use hopping maps to analyze electron flow in a few especially well-characterized systems. Finally, in the context of our predictions from semiclassical theory, we pose important challenges to advanced molecular simulations and experiments in better developing a comprehensive picture of *in vivo* redox enzyme function.

8.2   HOPPING MAPS

We developed hopping maps to rationalize ET rates in artificial two-step donor (D)–intermediate (I)–acceptor (A) hopping systems[16] (Equation 8.3, where X = D$^+$–I–A, Y = D–I$^+$–A, and Z = D–I–A$^+$). We later applied hopping analysis to natural systems.[17,18]

$$X \underset{k_2}{\overset{k_1}{\rightleftharpoons}} Y \underset{k_4}{\overset{k_3}{\rightleftharpoons}} Z \tag{8.3}$$

Note that our hopping maps are derived explicitly for a two-step (three-site) system. However, a similar semiclassical ET theory-based analysis could be adapted to calculate reaction times in systems with additional redox steps/sites. The maps are derived by using the semiclassical ET theory to calculate a tunneling time for hopping ($\tau_{hopping}$, Equation 8.4), which accounts for k_1, k_2, k_3, and k_4, and comparing it to the tunneling time for single-step ET ($\tau_{tunneling}$, Equation 8.5).[17] The expression for C_0 is given in Equation 8.6. Throughout this chapter, we denote the distances between the cofactors as r_1, r_2, and r_T for the D–I, I–A, and D–A distances, respectively. This calculated value is compared with that for single-step ET between the donor and acceptor. The inputs are effectively the same as for a calculation for single-step ET, but with added parameters for the intermediate step (e.g., $-\Delta G°_{DI}$, $-\Delta G°_{IA}$, $-\Delta G°_{total}$).

$$\tau_{hopping} = \frac{\exp\left(\frac{\beta(r_2 - r_0) + (\Delta G°_{IA} + \lambda)^2}{4\lambda RT}\right)\left(1 + \exp\left(\frac{\Delta G°_{DI}}{RT}\right)\right) + \exp\left(\frac{\beta(r_1 - r_0) + (\Delta G°_{DI} + \lambda)^2}{4\lambda RT}\right)\left(1 + \exp\left(\frac{\Delta G°_{IA}}{RT}\right)\right)}{C_0\left(1 + \exp\left(\frac{\Delta G°_{IA}}{RT}\right) + \exp\left(\frac{\Delta G°}{RT}\right)\right)} \tag{8.4}$$

$$\tau_{tunneling} = \frac{1}{C_0 \exp(-\beta(r_T - r_0))\left(\exp\left(\frac{-(\Delta G° + \lambda)^2}{4\lambda RT}\right) + \exp\left(\frac{-(\Delta G° - \lambda)^2}{4\lambda RT}\right)\right)} \tag{8.5}$$

$$C_0 = \sqrt{\frac{4\pi^3}{h^2 \lambda RT}}(H_{AB}(r_0))^2 \tag{8.6}$$

Model hopping maps for hypothetical systems with three different arrangements of cofactors are shown in Figure 8.1. In these maps, the arrangement of cofactors is fixed. The relatively low driving force range shown is relevant to many biological systems, and each map is calculated using our empirical λ and β for reactions of metalloproteins.[12] The shaded areas in hopping maps indicate instances where $\tau_{hopping}$ is faster than $\tau_{tunneling}$. The calculated hopping time at given driving forces is as in the temperature bar at the far right of the figure. The unshaded area at the top of

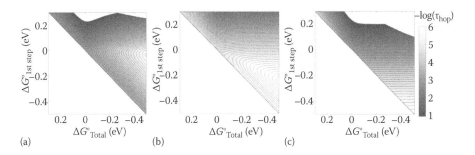

FIGURE 8.1 Hopping maps for a three-site ET system with $r_T = 25$ Å. The cofactor separations are (a) $r_1 = 8$ Å, $r_2 = 17$ Å; (b) $r_1 = r_2 = 12.5$ Å; and (c) $r_1 = 17$ Å, $r_2 = 8$ Å. Each map has $\lambda = 0.8$ eV, $\beta = 1.1$ Å^{-1}, and $H_{AB}^{r0} = 186$ cm^{-1}. The contours are plotted at 0.1 log unit intervals.

some of the maps indicates driving force regimes where single-step ET is faster than hopping, while the unshaded area at the bottom of the map indicates driving forces where the electron/hole is effectively stuck at the intermediate.

We also can make hopping maps where we keep the driving forces and r_T fixed and vary r_1 and r_2, illustrating the effect of the component distance on hopping advantage (Figure 8.2). For these maps, it is more convenient to plot $\log(k_{hop}/k_{ss})$ (equivalently $-\log(\tau_{hopping}/\tau_{tunneling})$) rather than an absolute tunneling time, as in Figure 8.1. The white area at the top and right of these maps is where there is no hopping advantage. The white area at the bottom left represents unphysical arrangements of cofactors. Construction of these maps assumes center-to-center distances between cofactors. The use of these distances versus edge-to-edge distances has been debated[12,19]; however, there probably is no single answer to the ET distance debate. We think that any ET analysis should explicitly consider the origin and destination of electrons/holes.

Analysis of the hopping maps in Figures 8.1 and 8.2 leads to some general conclusions about the factors that are important for hopping function. First, hopping is disfavored, although technically possible, in systems with intermediate or total steps that are endergonic by >0.2 eV. Second, the arrangement of cofactors has a profound effect on the driving force regime where hopping is predicted to occur. The systems with the fastest net ET from donor to acceptor are those where the first step is no larger than the second step. This is more quantitatively shown in Figure 8.2, where the greatest hopping advantage, at any driving force, comes from arrangements where the first step is 0–3 Å less than the second step. Third, there is a critical balance between driving force and cofactor arrangement in achieving a hopping advantage with minimal losses in total free energy. Few biological ET reactions are driving force optimized ($-\Delta G° = \lambda$); therefore, they are sensitive to even small changes in cofactor arrangement and reduction potential. Finally, it is important to point out that those two-step systems with short hopping and overall distances have the fastest *absolute* rates; however, breaking a system with a long overall distance into two shorter steps produces a larger hopping *advantage*.

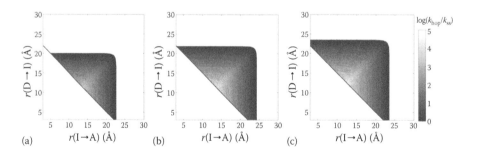

FIGURE 8.2 Hopping maps for a three-site ET system with $r_T = 25$ Å. The driving force for the first step is (a) −0.1 eV, (b) 0 eV, and (c) 0.1 eV. Each map has $-\Delta G°_{DA} = -0.2$ eV, $\lambda = 0.8$ eV, $\beta = 1.1$ Å^{-1}, and $H^{r0}_{AB} = 186$ cm^{-1}. The contours are plotted at 0.1 log unit intervals.

8.3   MULTISTEP REDOX CHAINS

Our first application of hopping maps was for $Re^I(CO)_3$(4,7-dimethylphenanthroline) (His124)–Trp122–Cu^I *Pseudomonas aeruginosa* azurin.[16] Oxidation of Cu^I by electronically excited Re^I is 100-fold faster than single-step ET, which is due to transient oxidation of Trp122. Dynamics and other interesting features of this system (and related Re^I–azurin systems) have been discussed extensively.[20–22] Importantly, we could resolve almost all of the rate constants for the hopping events and demonstrate that predictions from hopping maps were in accord with experiments.

We later analyzed hopping in ribonucleotide reductase, MauG, DNA photolyase,[17] and PRCs.[18] The key aspects of ET in those proteins are addressed in the following; however, these are only a few examples of proteins that rely on hopping for function. Cytochrome *c* oxidase, photosystems (PS) I and II,[23] hydrogenase,[24] nitrogenase,[25] cytochrome *c* peroxidase,[26] the cytochrome bc_1 complex,[27] and the cytochrome b_6f complex[28] are notable examples of proteins that rely on hopping. This handful of examples represents a diverse cross section of multistep ET reactions. These examples employ pure ET, such as ET along [FeS] clusters in hydrogenase, as well as ET that is coupled to chemical reactions, such as H^+ pumping in cytochrome *c* oxidase. Other examples require exogenous electron donors or acceptors for hopping (cytochrome *c* peroxidase). The complete catalytic cycles of PS I and PS II rely on all of the above scenarios (single ET reactions, coupled redox, and bimolecular ET) for function. In the following, we analyze a few protein systems that illustrate the power of hopping analysis; then, we return to the challenges associated with understanding function in multicomponent systems.

8.3.1   RIBONUCLEOTIDE REDUCTASE (RNR)

Ribonucleotides are converted to the corresponding deoxyribonucleotides required for DNA synthesis by RNRs.[29] The mechanism of RNR from *Escherichia coli* is probably the best understood,[30] although some features of the catalytic cycle translate between enzymes from different organisms.[29,31] *E. coli* RNR is a heterodimeric protein consisting of α2 and β2 subunits. Important here is that catalysis in RNRs is triggered 35 Å away from the ribonucleotide binding site. X-ray structures of the α2 and β2 subunits are known, and a binding model from the active complex leads to the proposed ET pathway shown in Figure 8.3.[32] Tyr356 is in a disordered part of the protein and has not been crystallographically characterized; however, other studies have shown that it is necessary for function.[33] The role of Trp48 in the hole transfer reaction is still unclear.[30]

We constructed a hopping map for the putative Tyr122–Trp48–Tyr356 redox chain in β2 of *E. coli* RNR. First, we must evaluate the distances between cofactors, which is challenging because Tyr356 has not been structurally resolved. A conservative (although speculative) approach places Tyr 356 about halfway between the α and β interface. We estimate that this makes the edge-to-edge Trp48–Tyr356 distance about 12 Å, and the total distance (Tyr122 to Tyr 356) is about 23 Å. The edge-to-edge Tyr122–Trp48 distance is 8 Å from the X-ray structure. The absolute reduction potentials for each hopping cofactor are not known in RNR; however, recent EPR

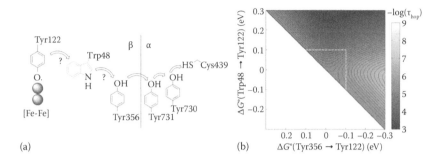

FIGURE 8.3 Proposed hole transfer pathway in *E. coli* RNR (a) and the hopping map for hole hopping in the β subunit (b). The map was calculated with $r_1 = 8$ Å, $r_2 = 12$ Å, $r_{total} = 23$ Å, $\lambda = 0.8$ eV, $\beta = 1.1$ Å^{-1}, and $H_{AB}^{r0} = 186$ cm^{-1}. The dashed white lines are drawn at $\Delta G°(\text{Trp48} \rightarrow \text{Tyr122}) = 0.1$ eV and $\Delta G°(\text{Tyr356} \rightarrow \text{Tyr122}) = -0.1$ eV.

measurements in modified RNRs illustrated that all of the driving forces are most likely in the ±100 meV range.

The calculated hopping rate constant from the map is between 10^5 and 10^6 s^{-1}. The calculated single-step rate constant at $\Delta G° = -0.1$ eV is 7 s^{-1}, which we estimate is near an upper limit based on our assumptions. RNR turns over with $k = 2$–10 s^{-1}, and this rate is controlled by conformational gating, not ET.[30,34] Therefore, analysis using hopping maps suggests that a redox intermediate (potentially Trp48) situated between Tyr122 and Tyr356 facilitates charge transfer in RNR such that redox events are not rate limiting.

An important limitation of our analysis using hopping maps is that experiments are not yet available that report on the exact position and environment of the redox cofactors at the conformationally gated transition state. Subtle changes in nuclear reorganization or electronic coupling could facilitate single-step ET directly from Tyr356 to Tyr122. Furthermore, redox reactions of tyrosine are inherently proton-coupled processes, so the application from semiclassical ET theory is not rigorous. These challenges are discussed in Section 8.4.

8.3.2 MauG

Methylamine dehydrogenase catalyzes the conversion of alkylamines to ammonia and aldehydes using a unique tryptophan–tryptophanyl–quinone (TTQ) cofactor.[35] This cofactor is synthesized using a hydroxylated Trp residue (Trp57) and Trp108 from the inactive "preMADH." The oxidizing equivalents for TTQ formation come from the di-heme protein MauG. This is an especially interesting system for analysis using hopping maps because the arrangement of cofactors is very well defined, and extensive data are available concerning the mechanism and ET pathway.[36–38]

The arrangement of cofactors in the MauG:preMADH complex is shown in Figure 8.4. During catalysis, H_2O_2 reacts with the five-coordinate, low-spin heme, which rapidly (>300 s^{-1}) generates a di-Fe^{IV} state, where both hemes are oxidized by one electron.[39] PreMADH is oxidized by two electrons, and in two additional cycles, is ultimately converted to the catalytically competent TTQ cofactor. The role of Trp93, situated between the two hemes, is not fully understood; however,

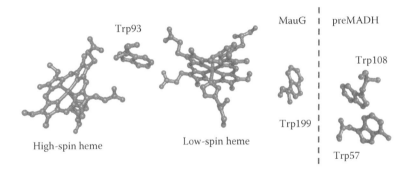

FIGURE 8.4 Redox cofactors in the MauG:preMADH complex. Image generated from PDB ID 3L4M.

it could serve as a hopping waystation that facilitates charge delocalization in the $Fe^{IV}Fe^{IV}$ state of the enzyme, as implicated in a structurally related cytochrome c peroxidase.[40] Single-step ET [$\Delta G° = 0$ eV and $r_T = 16$ Å (edge-to-edge)] is sufficient to account for the rapid formation of the $Fe^{IV}Fe^{IV}$ state [k_{calc}(single step) $= 2.5 \times 10^3$ s^{-1}]; however, hopping through Trp93 is a factor of 1000 times faster. Additional work is needed to understand this aspect of the MauG mechanism.

The case for hopping via Trp199 is more compelling. Importantly, site-directed mutagenesis of Trp199 to Phe or Lys generates MauG proteins that support redox chemistry and preMADH binding, but not catalysis.[35] The distance from the Fe^{IV} in MauG to the edge of Trp108 is 20 Å, and the edge of Trp57 is 23 Å from Fe^{IV}. Single-step ET with $\Delta G° = 0$ eV has $k_{calc} = 30$ and 1 s^{-1}, respectively. These rates are barely rapid enough to support catalysis ($k \sim 0.8$ s^{-1}), in qualitative agreement with the lack of reactivity in Trp199Phe and Trp199Lys mutants.

Hopping maps for Trp57 → Trp199 → Fe^{IV} and Trp108 → Trp199 → Fe^{IV} are shown in Figure 8.5. Again, we use $\lambda = 0.8$ eV and $\beta = 1.1$ Å^{-1} in these maps as a

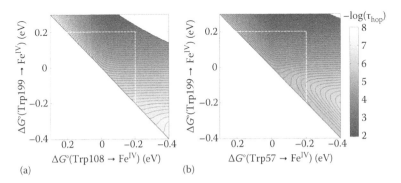

FIGURE 8.5 Hopping maps (a, b) for ET across the protein interface in the MauG:preMADH complex. The maps were constructed with $\lambda = 0.8$ eV, $\beta = 1.1$ Å^{-1}, and $H_{AB}^{r0} = 186$ cm^{-1}. The white dashed lines at $\Delta G°$(Trp199 → Fe^{IV}) $= 0.2$ eV and $\Delta G°$(Trp108/Trp57 → Fe^{IV}) $= -0.2$ eV are approximations of relevant driving force regimes for oxidation of preMADH by MauG. The contours are plotted at 0.1 log unit intervals.

starting point for rationalizing ET in MauG. Hopping is calculated to be at least 1000 times faster than single-step ET at low (biologically relevant) driving forces, as indicated by the white dashed lines. There is no clear evidence for which Trp (108 or 57) is oxidized first. The same hopping advantage is predicted for both cases, although the shape of the maps is slightly different at driving forces >0.3 eV. The overall redox process in MauG could involve movement of heavy atoms (bond breaking/forming); thus, analysis using hopping maps is not necessarily quantitative. In any case, addition of a redox waystation in the redox chain can substantially accelerate ET such that it is not rate limiting. This is important for MauG where the $Fe^{IV}Fe^{IV}$ intermediate can be generated, and is stable outside of the MauG:preMADH complex. Fast ET allows for reactivity only when the correctly oriented transient precursor complexes are formed.

8.3.3 DNA Photolyase

DNA photolyases use light to repair cyclobutane pyrimidine dimers (e.g., thymine) in bacterial DNA.[41] Understanding hopping in systems that use electronically excited states is of broad interest, especially in the context of building artificial light-harvesting systems. Furthermore, investigating mechanisms in proteins like photolyase and the photosynthetic reactions centers (Section 8.3.4) can provide natural inspiration for artificial systems. In the case of photolyase, there is no net redox change in a complete catalytic cycle. FADH$^-$ absorbs a photon of the appropriate energy, and an electron is transferred to the thymine dimer. The resulting cyclobutane pyrimidine dimer radical fragments, and the electron is ultimately transferred back to FADH$^•$. We constructed hopping maps for DNA photolyase to rationalize ET in this closed system.[17]

A notable feature of photolyases is the bent conformation of the FADH$^-$ cofactor, where the adenine ring is oriented almost perpendicular to the flavin ring (Figure 8.6).[42] The orientation of these cofactors has led to much speculation about the mechanism of ET between FADH and T–T dimers.[43,44] Very detailed investigations

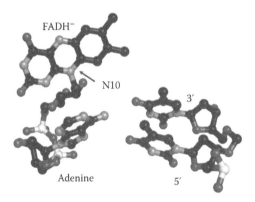

FIGURE 8.6 Arrangement of FADH$^-$ (flavin + adenine rings) and a bound T–T containing DNA photolyase. Image generated from PDB ID 1TEZ. Color codes: black, carbon; dark gray, oxygen; light gray, nitrogen; and white, phosphorus.

of the kinetics for the entire catalytic cycle were recently described.[45,46] The authors unambiguously demonstrate that the electron is transferred from *FADH⁻ to the 5′-T, then the T–T dimer fragments and the electron is returned to FADH⁺ via the 3′-T.[42] Hopping is not directly observed but adenine is required for function. An ET superexchange mechanism is suggested with an enhanced electronic coupling (β = 0.8 Å⁻¹) and increased reorganization energy (λ = 1.2 eV) relative to our empirical values (1.1 Å⁻¹ and 0.8 eV, respectively).[43] We note that this analysis relies on several assumptions, and while it is reasonable, our analysis using hopping maps offers an alternate picture.

One argument against hopping is that ET from *FADH⁻ to adenine is estimated to be ~0.1 eV uphill. However, if ET is isoergic, the tunneling time calculated from semiclassical ET theory (using our empirical λ and β) is in accord with the observed luminescence time constant for *FADH⁻ → FADH⁻ in substrate-free DNA photolyase (~1 ns). ET analysis in many proteins, including photolyase, relies on estimation of driving forces from solution reduction potentials, complicating quantitative analysis. The driving force for the first step is probably 0 ± 0.1 eV, which we calculated from the fluorescence spectrum of substrate-loaded photolyase,[45] the $E^{\circ\prime}$ value[47] for FADH⁺/⁻, and an estimate of the adenine reduction potential.[48] Subtle hydrogen bonding or electrostatic interactions can dramatically affect redox properties (including E°, reorganization energy, and electronic coupling) of embedded cofactors. Indeed, substrate-induced shifts in photolyase flavin redox chemistry are known.[47] Thus, we think that it is premature to preclude a hopping mechanism in DNA photolyase.

Hopping maps for DNA photolyase are shown in Figure 8.7. The map in Figure 8.7a is calculated using the parameters derived for a superexchange pathway and that in Figure 8.7b is calculated using β = 1.1 Å⁻¹ and λ = 0.5 eV, which we think is appropriate for ET involving organic cofactors.[49] Again, subtle changes to λ and β can easily offset each other, and it is typically very difficult to deconvolute these contributions. We used edge-to-edge distances of 5 Å (N10-FADH⁻ to adenine), 5 Å (adenine to 5′-T), and 9 Å (N10-FADH⁻ to 5′-T).

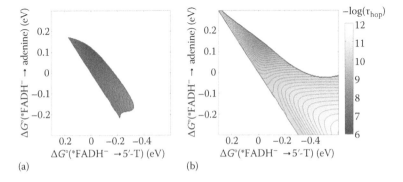

FIGURE 8.7 Hopping maps for DNA photolyase. The maps were constructed with (a) λ = 1.2 eV, β = 0.8 Å⁻¹ and H_{AB}^{r0} = 186 cm⁻¹, and (b) λ = 1.2 eV, β = 0.8 Å⁻¹ and H_{AB}^{r0} = 186 cm⁻¹. The contours are plotted at 0.1 log unit intervals.

The driving force for the first step is around 0 eV and that for the overall reaction is about −0.4 eV. The map made using the superexchange parameters has rate constants that are too low to account for the observed ET times (250 ps, $k_{obs} = 4 \times 10^9$ s^{-1}). The map made using our parameters has $k_{calc} \sim 10^{10}$ s^{-1} ($\tau \sim 100$ ps), in accord with experiments; however, the driving forces place the calculated point very near the hopping/single-step border. This analysis supports our suggestion that it is difficult to definitively conclude a mechanism for ET in DNA photolyases. Very small changes in any of the *estimated* ET parameters could alter the conclusion. Clearly defined reaction driving forces, reorganization energies, and electronic couplings would provide the parameters necessary for a firm conclusion.

8.3.4 Photosynthetic Reaction Centers

Purple sulfur bacteria use PRCs to generate the proton gradients required for ATP synthesis. The light-harvesting architecture used in these bacteria is structurally and functionally similar to that found in PS I and PS II; however, the proteins are far simpler.[50] As such, these were the first membrane proteins structurally characterized.[51] The light-harvesting events, and associated ET steps, are very well understood.[52–54] ET in PRCs has several unique features that we explored by constructing and analyzing hopping maps.[18]

The cofactors of a prototypical PRC (from *Rhodobacter sphaeroides*) is shown in Figure 8.8. A photon is absorbed by the special pair of chlorophylls (P), generating an oxidizing electronic excited state (*P), which transfers an electron to B_L. B_L^- reduces H_L, which ultimately reduces Q_A. The role of B_L in PRC ET was not clear from some of the earliest time-resolved studies of these proteins. It was apparent that H_L was reduced within 3 ps of P excitation, and a superexchange mechanism

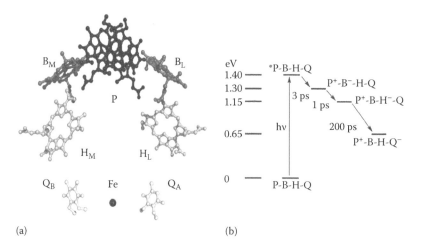

FIGURE 8.8 Arrangement of cofactors in *R. sphaeroides* PRCs (a) and energetics of ET through the L-branch (b). Image generated from PDB ID 3I4D.

was favored.[55] As time resolution and instrument sensitivity developed, additional investigation revealed that B_L is transiently oxidized in the overall reaction.[56] One recent report, employing single-crystal polarized spectroscopy at 100 K, suggests that *P reduction of B_L is the fast step (~1 ps) and subsequent reduction of H_L by $B_L^{\cdot -}$ has a time constant of 2 ps.[57] Note that ET occurs exclusively through the L-branch in wild-type PRCs.[50] This behavior is thought to arise from asymmetry of the amino acids surrounding the L- and M-branch cofactors.[58] This is a remarkable example of how subtle changes in the medium surrounding ET cofactors can lead to substantial changes in ET reactivity.

We restrict ourselves to analysis for net ET from *P to H_L; however, numerous ET events have been observed and investigated in varying levels of detail for PRCs.[12,18] The absolute reduction potentials are not known, but the driving forces for ET between *P and B_L,[59,60] and between B_L and H_L,[61,62] have been benchmarked with some certainty by different researchers (Figure 8.8). The energy gaps must be approximately equal to the reorganization energies (λ), since rates of ET in PRCs are essentially temperature independent.[63] We constructed hopping maps for PRCs (Figure 8.9) using these experimentally validated parameters.

The calculated hopping times constant for net *P $\rightarrow$ H_L ET in PRCs is about 1 ps, in good agreement with experiments. It is notable that these maps, derived solely from the semiclassical ET theory, faithfully reproduce the experimentally measured ET rates. As the temperature is lowered, the range of calculated ET rates increases, evidenced by the greater number of contour lines for the map at 100 K. A wider range of ET rates is calculated as the driving forces diverge from optimized ($-\Delta G^\circ = \lambda$) values. This analysis underscores the very careful balance of reorganization energy, driving force, and electronic coupling in PRCs. We note that this is not the only interpretation of ET function in PRCs; advanced simulations suggest that nuclear tunneling and protein dynamics could also play important roles.[64-66] The deep understanding of functional ET in PRCs highlights the complementarity of experiment and theory.

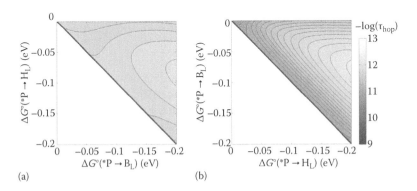

FIGURE 8.9 Hopping maps for PRCs at 298 K (a) and 100 K (b). The maps were made using $r_1 = 5$, $r_2 = 5$, and $r_{total} = 10$ Å; $\lambda_1 = 0.1$, $\lambda_2 = 0.15$, and $\lambda_{total} = 0.15$ eV; $H_{AB}^{r0} = 186$ cm^{-1}; and $\beta = 1.1$ Å. The contour lines in both maps are spaced at 0.1 log unit intervals.

8.4 CHALLENGES: STRUCTURE, DYNAMICS, AND MECHANISM

Analysis using the semiclassical ET theory and hopping maps are useful for ratio-nalizing functional electron/hole flow in biological and artificial systems. However, the examples that we analyzed here are "best-case" scenarios, where data gleaned from decades of research can be used to estimate the parameters required for appli-cation of hopping maps. There is much work still to be done to fully understand functional electron/hole flow in biological systems. This section highlights some of the biggest challenges to experimentalists and theoreticians.

8.4.1 STRUCTURE

Our analysis of hole transfer in RNR highlights an essential challenge in every ET analysis: location of the redox cofactors. As of early 2013, there were nearly 100,000 entries in the Protein Data Bank (PDB), and submissions were almost 10,000 in 2012 alone.[67] Remarkable advances in X-ray crystallography, including easier access to synchrotron radiation, assure that those numbers will continue to grow. However, even when high-quality structures are available, portions can be disordered, as in RNR. Analyses, including the use of hopping maps, can be suggestive but are not quantitative. Improved characterization of ET pathways will be possible as structural details of macromolecules continue to emerge.

One notable advance in crystallographic characterization is the discovery that some processes can be monitored *in crystallo*.[33,68] This type of structural characteriza-tion also aids in addressing the mechanistic challenges discussed below. Importantly, such *in crystallo* monitoring of redox transformations validates models of functional protein complexes. Recent structural work on PS II has highlighted the potential for X-ray damage to protein-embedded redox cofactors.[69,70] The emergence of X-ray absorption and X-ray emission as tools for metalloprotein characterization is advanc-ing our knowledge of redox cofactor structure.[68] Advances in the use of such comple-mentary techniques will provide both structural data and cofactor characterization that lie at the heart of ET and mechanistic analyses.

8.4.2 DYNAMICS

The cellular environment is a dynamic one, where proteins constantly sample dif-ferent interfaces and conformational space.[71] Interfacial electron/hole transfer is necessary for function in many central biological processes, such as respiration and photosynthesis. X-ray or nuclear magnetic resonance structures for a few redox protein complexes are known, including cytochrome c peroxidase:cytochrome c,[72] cytochrome c_{552}:Cu_A (Cu_A is the soluble portion of cytochrome ba_3 oxidase),[73] and MauG:preMADH.[36] These models greatly inform ET analysis, but they are a static picture of a dynamic process. Furthermore, in some proteins (e.g., RNR), electron/hole transfer can be triggered by substrate-induced conformational changes, so crys-tal structures does not necessarily represent ET transition states.[30] Such dynamics also could play a role in modulation of λ and/or $\Delta G°$, as in DNA photolyases where substrate binding perturbs $E°(\text{FADH}^{\bullet/-})$.

Intraprotein dynamics are also of central importance to ET in proteins. A great deal of computational work points to the importance of considering electron/hole transfer pathways (through bonds, hydrogen bonds, or space) and protein dynamics.[74,75] Likewise, dynamics in ET systems have been probed experimentally (e.g., PRCs,[76] DNA,[77] and flavin reductase[78]). However, both approaches have limitations. Treating large, dynamic proteins is still computationally intensive. Experimental approaches are limited by time resolution, sensitivity, and protein modification techniques, and protein modification could alter some dynamic aspects of ET reactions. This challenge highlights the limits of experimental and theoretical approaches, underscoring the value of collaborative efforts and continued advances in both fields.

8.4.3 MECHANISM

It is increasingly appreciated that a great many redox reactions are coupled to movement of heavy atoms. The most common reaction of this type probably is proton-coupled electron transfer (PCET), in which a change in cofactor redox state is coupled to proton transfer (PT). In the biological context, canonical examples include tyrosine redox cycling in RNR[30] or PS II[79] and C–H hydroxylation by cytochrome P450 compound I.[80] In these systems, H^+ and e^- transfer are tightly coupled, often being referred to as hydrogen atom transfer. PCET in other systems is much less coupled, and stepwise ET–PT is possible (e.g., in some of the proton-pumping steps in cytochrome c oxidase).[81]

Application of semiclassical ET theory to reactions where heavy atoms move on the same timescale as electrons is not rigorous, which has led to the development of new theories.[82–84] These theories have greatly enhanced our understanding of the factors important to PCET, but they still suffer from computational limitations (e.g., expense for large systems) and additional benchmarks against experiment are necessary. A key conclusion from both experiment[85] and theory[80] is that PCET, much like ET, depends strongly on reaction-driving forces and effective reorganization energies. Nature likely relies on both pure ET and coupled ET reactions in many different functions. For example, ET through [FeS] clusters in hydrogenase likely is uncoupled to H^+ movement, but at the catalytic site $2H^+$ and $2e^-$ are converted to H_2,[24] so ET and PT are inherently coupled. Experimental and theoretical approaches that reconcile the two distinct classes of reactions are needed if we are to advance our perception of entire biological redox cascades.

8.5 CONCLUDING REMARKS

The semiclassical ET theory provides an excellent starting point for understanding redox flow in biological systems. Predictions from the ET theory can reproduce experimental ET rates, as for simple reactions where reliable values for ET (λ, β, $-\Delta G°$) and structural (r) parameters are available. These are the parameters where nature exhibits amazing and subtle control. Assumptions for any of those parameters can compromise the predictive reliability of semiclassical treatments, including hopping maps. Advances in both experimental methods and theoretical approaches will continue to address structural, dynamic, and mechanistic challenges, ultimately improving our quantitative understanding of natural redox cascades.

ACKNOWLEDGMENTS

Our work was supported by the Arnold and Mabel Beckman Foundation and by the National Institutes of Health (DK019038 to HBG and GM095037 to JJW).

REFERENCES

1. Marcus, R. A., and Sutin, N. "Electron transfers in chemistry and biology." *Biochim. Biophys. Acta* 811 (1985): 265–322.
2. Mauk, A. G. "Biological electron-transfer reactions." *Essays Biochem.* 34 (1999): 101–24.
3. DeVault, D., and Chance, B. "Studies of photosynthesis using a pulsed laser: I. Temperature dependence of cytochrome oxidation rate in chromatium. Evidence for tunneling." *Biophys. J.* 6 (1966): 825–47.
4. DeVault, D., Parkes, J. H., and Chance, B. "Electron tunnelling in cytochromes." *Nature* 215 (1967): 642–4.
5. Meyer, T. J., and Henry, T. "Electron transfer reactions." In *Comprehensive Coordination Chemistry*, edited by G. Wilkinson, 331–84. New York: Pergamon, 1987.
6. Holwerda, R. A., Wherland, S., and Gray, H. B. "Electron transfer reactions of copper proteins." *Annu. Rev. Biophys. Bioeng.* 5 (1976): 363–96.
7. Kendrew, J. C., Bodo, G., Dintzis, H. M., Parrish, R. G., Wyckoff, H., and Phillips, D. C. "A three-dimensional model of the myoglobin molecule obtained by X-ray analysis." *Nature* 181 (1958): 662–6.
8. Perutz, M. F., Muirhead, H., Cox, J. M., and Goaman, L. C. G. "Three-dimensional Fourier synthesis of horse oxyhaemoglobin at 2.8 Å resolution: The atomic model." *Nature* 219 (1968): 131–9.
9. Swanson, R., Trus, B. L., Mandel, N., Mandel, G., Kallai, O. B., and Dickerson, R. E. "Tuna cytochrome *c* at 2.0 Å resolution. I. Ferricytochrome structure analysis." *J. Biol. Chem.* 252 (1977): 759–75.
10. Colman, P. M., Freeman, H. C., Guss, J. M., Murata, M., Norris, V. A., Ramshaw, J. A. M., and Venkatappa, M. P. "X-ray crystal structure analysis of plastocyanin at 2.7 Å resolution." *Nature* 272 (1978): 319–24.
11. Winkler, J. R., Nocera, D. G., Yocom, K. M., Bordignon, E., and Gray, H. B. "Electron-transfer kinetics of pentaammineruthenium(III)(histidine-33)–ferricytochrome *c*. Measurement of the rate of intramolecular electron transfer between redox centers separated by 15 Å in a protein." *J. Am. Chem. Soc.* 104 (1982): 5798–800.
12. Gray, H. B., and Winkler, J. R. "Electron tunneling through proteins." *Q. Rev. Biophys.* 36 (2003): 341–72.
13. Gray, H. B., and Winkler, J. R. "Long-range electron transfer." *Proc. Natl. Acad. Sci. U.S.A.* 102(10) (2005): 3534–9.
14. Lieb, E. H., and Wu, F. Y. "The one-dimensional Hubbard model: A reminiscence." *Physica A* 321 (2003): 1–27.
15. Eley, D. D., and Spivey, D. I. "Semiconductivity of organic substances. Part 9—Nucleic acid in the dry state." *Trans. Faraday Soc.* 58 (1962): 411–5.
16. Shih, C., Museth, A. K., Abrahamsson, M., Blanco-Rodríguez, A.-M., Di Bilio, A. J., Sudhamsu, J., Crane, B. R., Ronayne, K. L., Towrie, M., Vlček, Jr., A., Richards, J. H., Winkler, J. R., and Gray, H. B. "Tryptophan-accelerated electron flow through proteins." *Science* 320 (2008): 1760–2.
17. Warren, J. J., Ener, M. E., Vlček, Jr., A., Winkler, J. R., and Gray, H. B. "Electron hopping through proteins." *Coord. Chem. Rev.* 256 (2012): 2478–87.
18. Warren, J. J., Winkler, J. R., and Gray, H. B. "Hopping maps for photosynthetic reaction centers." *Coord. Chem. Rev.* 257 (2013): 165–70.

19. Moser, C. C., Chobot, S. E., Page, C. C., and Dutton, P. L. "Distance metrics for heme protein electron tunneling." *Biochim. Biophys. Acta* 1777 (2008): 1032–7.

20. Blanco-Rodríguez, A.-M., Busby, M., Grâdinaru, C., Crane, B. R., Di Bilio, A. J., Matousek, P., Towrie, M., Leigh, B. S., Richards, J. H., Vlček, Jr., A., and Gray, H. B. "Excited-state dynamics of structurally characterized $[Re^I(CO)_3(phen)(HisX)]^+$ (X = 83, 109) *Pseudomonas aeruginosa* azurins in aqueous solution." *J. Am. Chem. Soc.* 128 (2006): 4365–70.

21. Blanco-Rodríguez, A.-M., Busby, M., Ronayne, K., Towrie, M., Grâdinaru, C., Sudhamsu, J., Sykora, J., Hof, M., Záliš, S., Di Bilio, A. J., Crane, B. R., Gray, H. B., and Vlček, A. "Relaxation dynamics of *Pseudomonas aeruginosa* $Re^I(CO)_3(diimine)(HisX)^+$ (X = 83, 107, 109, 124, 126)Cu^{II} azurins." *J. Am. Chem. Soc.* 131 (2009): 11788–800.

22. Blanco-Rodríguez, A.-M., Di Bilio, A. J., Shih, C., Museth, A. K., Clark, I. P., Towrie, M., Cannizzo, A., Sudhamsu, J., Crane, B. R., Sykora, J., Winkler, J. R., Gray, H. B., Záliš, S., and Vlček, A. "Phototriggering electron flow through Re^I-modified *Pseudomonas aeruginosa* azurins." *Chem. Eur. J.* 17(19) (2011): 5350–61.

23. Nelson, N., and Yocum, C. F. "Structure and function of photosystems I and II." *Annu. Rev. Plant Biol.* 57 (2006): 521–65.

24. Mulder, D. W., Shepard, E. M., Meuser, J. E., Joshi, N., King, P. W., Posewitz, M. C., Broderick, J. B., and Peters, J. W. "Insights into [FeFe]-hydrogenase structure, mechanism, and maturation." *Structure* 19 (2011): 1038–52.

25. Seefeldt, L. C., Hoffman, B. M., and Dean, D. R. "Electron transfer in nitrogenase catalysis." *Curr. Opin. Chem. Biol.* 16 (2012): 19–25.

26. Poulos, T. L. "Thirty years of heme peroxidase structural biology." *Arch. Biochem. Biophys.* 500 (2010): 3–12.

27. Crofts, A. R. "The cytochrome *bc*1 complex: Function in the context of structure." *Annu. Rev. Physiol.* 66 (2004): 689–733.

28. Baniulis, D., Yamashita, E., Zhang, H., Hasan, S. S., and Cramer, W. A. "Structure–function of the cytochrome b_6f complex." *Photochem. Photobiol.* 84 (2008): 1349–58.

29. Nordlund, P., and Reichard, P. "Ribonucleotide reductases." *Annu. Rev. Biochem.* 75 (2006): 681–706.

30. Stubbe, J., Nocera, D. G., Yee, C. S., and Chang, M. C. Y. "Radical initiation in the class I ribonucleotide reductase: Long-range proton-coupled electron transfer?" *Chem. Rev.* 103 (2003): 2167–201.

31. Jiang, W., Yun, D., Saleh, L., Bollinger, J. M., and Krebs, C. "Formation and function of the manganese(IV)/iron(III) cofactor in *Chlamydia trachomatis* ribonucleotide reductase." *Biochemistry* 47 (2008): 13736–44.

32. Uhlin, U., and Eklund, H. "Structure of ribonucleotide reductase protein R1." *Nature* 370 (1994): 533–9.

33. Yee, C. S., Seyedsayamdost, M. R., Chang, M. C. Y., Nocera, D. G., and Stubbe, J. "Generation of the R2 subunit of ribonucleotide reductase by intein chemistry: Insertion of 3-nitrotyrosine at residue 356 as a probe of the radical initiation process." *Biochemistry* 42 (2003): 14541–52.

34. Yokoyama, K., Uhlin, U., and Stubbe, J. "A hot oxidant, 3-NO_2Y_{122} radical, unmasks conformational gating in ribonucleotide reductase." *J. Am. Chem. Soc.* 132 (2010): 15368–79.

35. Wilmot, C. M., and Yukl, E. T. "MauG: A di-heme enzyme required for methylamine dehydrogenase maturation." *Dalton Trans.* 42 (2013): 3127–35.

36. Jensen, L. M. R., Sanishvili, R., Davidson, V. L., and Wilmot, C. M. "*In crystallo* post-translational modification within a MauG/pre-methylamine dehydrogenase complex." *Science* 327 (2010): 1392–4.

37. Tarboush, N. A., Jensen, L. M. R., Feng, M., Tachikawa, H., Wilmot, C. M., and Davidson, V. L. "Functional importance of tyrosine 294 and the catalytic selectivity for the bis-Fe(IV) state of MauG revealed by replacement of this axial heme ligand with histidine." *Biochemistry* 49 (2010): 9783–91.

38. Tarboush, N. A., Jensen, L. M. R., Yukl, E. T., Geng, J., Liu, A., Wilmot, C. M., and Davidson, V. L. "Mutagenesis of tryptophan199 suggests that hopping is required for MauG-dependent tryptophan tryptophylquinone biosynthesis." *Proc. Natl. Acad. Sci. U.S.A.* 108 (2011): 16956–61.

39. Lee, S., Shin, S., Li, X., and Davidson, V. L. "Kinetic mechanism for the initial steps in MauG-dependent tryptophan tryptophylquinone biosynthesis." *Biochemistry* 48 (2009): 2442–7.

40. De Smet, L., Savvides, S. N., Van Horen, E., Pettigrew, G., and Van Beeumen, J. J. "Structural and mutagenesis studies on the cytochrome *c* peroxidase from *Rhodobacter capsulatus* provide new insights into structure–function relationships of bacterial di-heme peroxidases." *J. Biol. Chem.* 281 (2006): 4371–9.

41. Sancar, A. "Structure and function of DNA photolyase and cryptochrome blue-light photoreceptors." *Chem. Rev.* 103 (2003): 2203–38.

42. Mees, A., Klar, T., Gnau, P., Hennecke, U., Eker, A. P. M., Carell, T., and Essen, L.-O. "Crystal structure of a photolyase bound to a CPD-like DNA lesion after *in situ* repair." *Science* 306 (2004): 1789–93.

43. Prytkova, T. R., Beratan, D. N., and Skourtis, S. S. "Photoselected electron transfer pathways in DNA photolyase." *Proc. Natl. Acad. Sci. U.S.A.* 104 (2007): 802–7.

44. Acocella, A., Jones, G. A., and Zerbetto, F. "What is adenine doing in photolyase?" *J. Phys. Chem. B* 114 (2010): 4101–6.

45. Liu, Z., Tan, C., Guo, X., Kao, Y.-T., Li, J., Wang, L., Sancar, A., and Zhong, D. "Dynamics and mechanism of cyclobutane pyrimidine dimer repair by DNA photolyase." *Proc. Natl. Acad. Sci. U.S.A.* 108 (2011): 14831–6.

46. Liu, Z., Guo, X., Tan, C., Li, J., Kao, Y.-T., Wang, L., Sancar, A., and Zhong, D. "Electron tunneling pathways and role of adenine in repair of cyclobutane pyrimidine dimer by DNA photolyase." *J. Am. Chem. Soc.* 134 (2012): 8104–14.

47. Gindt, Y. M., Schelvis, J. P. M., Thoren, K. L., and Huang, T. H. "Substrate binding modulates the reduction potential of DNA photolyase." *J. Am. Chem. Soc.* 127 (2005): 10472–3.

48. Seidel, C. A. M., Schulz, A., and Sauer, M. H. M. "Nucleobase-specific quenching of fluorescent dyes. 1. Nucleobase one-electron redox potentials and their correlation with static and dynamic quenching efficiencies." *J. Phys. Chem.* 100 (1996): 5541–53.

49. Eberson, L. "Electron transfer reactions in organic chemistry." In *Advances in Physical Organic Chemistry*, edited by V. Gold, and D. Bethell, 79–185. London: Academic Press, 1982.

50. Hohmann-Marriott, M. F., and Blankenship, R. E. "Evolution of photosynthesis." *Annu. Rev. Plant Biol.* 62 (2011): 515–48.

51. Deisenhofer, J., Epp, O., Miki, K., Huber, R., and Michel, H. "Structure of the protein subunits in the photosynthetic reaction centre of *Rhodopseudomonas viridis* at 3 Å resolution." *Nature* 318 (1985): 618–24.

52. Boxer, S. G. "Mechanisms of long-distance electron transfer in proteins: Lessons from photosynthetic reaction centers." *Annu. Rev. Biophys. Biophys. Chem.* 19 (1990): 267–99.

53. LeBard, D. N., and Matyushov, D. V. "Energetics of bacterial photosynthesis." *J. Phys. Chem. B* 113 (2009): 12424–37.

54. Blankenship, R. E., Tiede, D. M., Barber, J., Brudvig, G. W., Fleming, G., Ghirardi, M., Gunner, M. R., Junge, W., Kramer, D. M., Melis, A., Moore, T. A., Moser, C. C., Nocera, D. G., Nozik, A. J., Ort, D. R., Parson, W. W., Prince, R. C., and Sayre, R. T. "Comparing photosynthetic and photovoltaic efficiencies and recognizing the potential for improvement." *Science* 332 (2011): 805–9.

55. Martin, J.-L., Breton, J., Hoff, A. J., Migus, A., and Antonetti, A. "Femtosecond spectroscopy of electron transfer in the reaction center of the photosynthetic bacterium *Rhodopseudomonas sphaeroides* R-26: Direct electron transfer from the dimeric bacteriochlorophyll primary donor to the bacteriopheophytin acceptor with a time constant of 2.8 ± 0.2 psec." *Proc. Natl. Acad. Sci. U.S.A.* 83 (1986): 957–61.

56. Kirmaier, C., and Holten, D. "Evidence that a distribution of bacterial reaction centers underlies the temperature and detection-wavelength dependence of the rates of the primary electron-transfer reactions." *Proc. Natl. Acad. Sci. U.S.A.* 87 (1990): 3552–6.

57. Huang, L., Ponomarenko, N., Wiederrecht, G. P., and Tiede, D. M. "Cofactor-specific photochemical function resolved by ultrafast spectroscopy in photosynthetic reaction center crystals." *Proc. Natl. Acad. Sci. U.S.A.* 109 (2012): 4851–6.

58. Marchanka, A., Savitsky, A., Lubitz, W., Möbius, K., and van Gastel, M. "B-branch electron transfer in the photosynthetic reaction center of a *Rhodobacter sphaeroides* quadruple mutant. Q- and W-band electron paramagnetic resonance studies of triplet and radical-pair cofactor states." *J. Phys. Chem. B* 114 (2010): 14364–72.

59. Bixon, M., Jortner, J., and Michel-Beyerle, M. E. "A kinetic analysis of the primary charge separation in bacterial photosynthesis. Energy gaps and static heterogeneity." *Chem. Phys.* 197 (1995): 389–404.

60. Katilius, E., Babendure, J., Lin, S., and Woodbury, N. "Electron transfer dynamics in *Rhodobacter sphaeroides*: Reaction center mutants with a modified ligand for the monomer bacteriochlorophyll on the active side." *Photosynth. Res.* 81 (2004): 165–80.

61. Goldstein, R. A., and Boxer, S. G. "The effect of very high magnetic fields on the reaction dynamics in bacterial reaction centers: Implications for the reaction mechanism." *Biochim. Biophys. Acta* 977 (1989): 78–86.

62. Hartwich, G., Lossau, H., Michel-Beyerle, M. E., and Ogrodnik, A. "Nonexponential fluorescence decay in reaction centers of *Rhodobacter sphaeroides* reflecting dispersive charge separation up to 1 ns." *J. Phys. Chem. B* 102 (1998): 3815–20.

63. Kirmaier, C., Holten, D., and Parson, W. W. "Temperature and detection-wavelength dependence of the picosecond electron-transfer kinetics measured in *Rhodopseudomonas sphaeroides* reaction centers. Resolution of new spectral and kinetic components in the primary charge-separation process." *Biochim. Biophys. Acta* 810 (1985): 33–48.

64. Balabin, I. A., and Onuchic, J. N. "Dynamically controlled protein tunneling paths in photosynthetic reaction centers." *Science* 290 (2000): 114–7.

65. Warshel, A., Chu, Z. T., and Parson, W. W. "Dispersed polaron simulations of electron transfer in photosynthetic reaction centers." *Science* 246 (1989): 112–6.

66. Wang, H., Lin, S., Allen, J. P., Williams, J. C., Blankert, S., Laser, C., and Woodbury, N. W. "Protein dynamics control the kinetics of initial electron transfer in photosynthesis." *Science* 316 (2007): 747–50.

67. RCSB Protein Data Bank. Available at http://www.rcsb.org/pdb/static.do?p=general _information/pdb_statistics/index.html (accessed May 15, 2013).

68. Kovaleva, E. G., Neibergall, M. B., Chakrabarty, S., and Lipscomb, J. D. "Finding intermediates in the O_2 activation pathways of non-heme iron oxygenases." *Acc. Chem. Res.* 40 (2007): 475–83.

69. Umena, Y., Kawakami, K., Shen, J.-R., and Kamiya, N. "Crystal structure of oxygen-evolving photosystem II at a resolution of 1.9 Å." *Nature* 473 (2011): 55–60.

70. Kern, J., Alonso-Mori, R., Tran, R., Hattne, J., Gildea, R. J., Echols, N., Glöckner, C., Hellmich, J., Laksmono, H., Sierra, R. G., Lassalle-Kaiser, B., Koroidov, S., Lampe, A., Han, G., Gul, S., DiFiore, D., Milathianaki, D., Fry, A. R., Miahnahri, A., Schafer, D. W., Messerschmidt, M., Seibert, M. M., Koglin, J. E., Sokaras, D., Weng, T.-C., Sellberg, J., Latimer, M. J., Grosse-Kunstleve, R. W., Zwart, P. H., White, W. E., Glatzel, P., Adams, P. D., Bogan, M. J., Williams, G. J., Boutet, S., Messinger, J., Zouni, A., Sauter, N. K., Yachandra, V. K., Bergmann, U., and Yano, J. "Simultaneous femtosecond X-ray spectroscopy and diffraction of photosystem II at room temperature." *Science* 340 (2013): 491–5.

71. Orrit, M. "The motions of an enzyme soloist." *Science* 302 (2003): 239–40.

72. Pelletier, H., and Kraut, J. "Crystal structure of a complex between electron transfer partners, cytochrome *c* peroxidase and cytochrome *c*." *Science* 258 (1992): 1748–55.

73. Muresanu, L., Pristovšek, P., Löhr, F., Maneg, O., Mukrasch, M. D., Rüterjans, H., Ludwig, B., and Lücke, C. "The electron transfer complex between cytochrome c_{552} and the Cu_A domain of the *Thermus thermophilus* ba_3 oxidase." *J. Biol. Chem.* 281 (2006): 14503–13.

74. Skourtis, S. S., Waldeck, D. H., and Beratan, D. N. "Fluctuations in biological and bio-inspired electron-transfer reactions." *Annu. Rev. Phys. Chem.* 61 (2010): 461–85.

75. Beratan, D. N., Skourtis, S. S., Balabin, I. A., Balaeff, A., Keinan, S., Venkatramani, R., and Xiao, D. "Steering electrons on moving pathways." *Acc. Chem. Res.* 42 (2009): 1669–78.

76. Wang, H., Hao, Y., Jiang, Y., Lin, S., and Woodbury, N. W. "Role of protein dynamics in guiding electron-transfer pathways in reaction centers from *Rhodobacter sphaeroides*." *J. Phys. Chem. B* 116 (2011): 711–7.

77. Schuster, G. B. "Long-range charge transfer in DNA: Transient structural distortions control the distance dependence." *Acc. Chem. Res.* 33 (2000): 253–60.

78. Yang, H., Luo, G., Karnchanaphanurach, P., Louie, T.-M., Rech, I., Cova, S., Xun, L., and Xie, X. S. "Protein conformational dynamics probed by single-molecule electron transfer." *Science* 302 (2003): 262–6.

79. McEvoy, J. P., and Brudvig, G. W. "Water-splitting chemistry of photosystem II." *Chem. Rev.* 106 (2006): 4455–83.

80. Green, M. T. "CH bond activation in heme proteins: The role of thiolate ligation in cytochrome P450." *Curr. Opin. Chem. Biol.* 13 (2009): 84–8.

81. Kaila, V. R. I., Verkhovsky, M. I., and Wikström, M. "Proton-coupled electron transfer in cytochrome oxidase." *Chem. Rev.* 110 (2010): 7062–81.

82. Hammes-Schiffer, S., and Stuchebrukhov, A. A. "Theory of coupled electron and proton transfer reactions." *Chem. Rev.* 110 (2010): 6939–60.

83. Kretchmer, J. S., and Miller, III, T. F. "Direct simulation of proton-coupled electron transfer across multiple regimes." *J. Chem. Phys.* 138 (2013): 134109–20.

84. Cukier, R. I., and Nocera, D. G. "Proton-coupled electron transfer." *Annu. Rev. Phys. Chem.* 49 (1998): 337–69.

85. Mayer, J. M. "Understanding hydrogen atom transfer: From bond strengths to Marcus theory." *Acc. Chem. Res.* 44 (2010): 36–46.

9 Modeling of Ligand Binding to Metalloproteins

Art E. Cho

CONTENTS

9.1 BACKGROUND AND INTRODUCTION

To model binding of ligands to proteins, one has to go beyond simple molecular mechanics. This is because of the relative orientations of ligands and the target proteins, which must be sampled with special care. For this purpose, protein-docking methods have been developed. During the last few decades, protein docking has become one of the most important molecular modeling methods used in the pharmaceutical industry.[1–5] A number of academic and commercial programs have been developed, and their performance has been improved steadily. The state-of-the-art docking programs can reproduce a set of experimentally determined structures within 2 Å of root mean square deviation (RMSD) close to 80% of the time. To achieve such accuracy, docking programs exploit various empirical scoring schemes, as well as energy-based ones. For the electrostatic energy portion of scoring, most of the current docking methods use force field-based fixed electric charges for both protein and ligand atoms.[2,6–9] It has been shown that a variable charge model using the combined quantum mechanics/molecular mechanics (QM/MM) methods for docking can be effective. The correction of the atomic charges on ligand atoms according to the polarization of the binding site environment improves prediction of the correct binding modes of ligand–protein complexes.[10] With a more extensive test on a larger set of samples, however, it was found that there still remain some cases in which even the new method, which we will call QM/MM docking henceforth, fails to improve the results of conventional docking practice. An analysis to understand the failure of QM/MM docking revealed that there are certain classes of proteins for which it failed more often than others in redocking experiments. One of them was the metalloprotein group, which contain metal ions in the binding sites. Careful analysis of these specific cases led us to conclude that charges on the metal ions determined by the force field parameter are not appropriate for docking.[11]

Although metalloproteins are an important target protein group playing major roles in physiological processes, receptor binding of potential drugs, and drug metabolism,[12] docking studies involving metalloproteins pose a serious challenge since the ligand interactions with transition metals can be treated appropriately only at the quantum mechanical level.[13–15] Attempts have been made to improve docking accuracy on metalloproteins, especially zinc complexes, by deviating from force field-based atomic charges.[16,17] The authors of these works emphasize on the practicality of the methods and therefore attempt to improve the docking accuracy and binding affinity prediction by reparametrization of the metal ion force field. Khandelwal et al., in particular, devised a method for predicting the binding affinity of metalloproteins by combining a series of QM/MM and force field-based molecular dynamics (MD) calculations. In another work, Strynadka et al. used a fluctuating atomic charge model of force field, which is parameterized by a semiempirical quantum chemical method, to study zinc complexes.[18] It should also be mentioned that recent efforts on empirical parameter fitting of scoring functions have brought significant improvements in docking of metalloproteins.

In an attempt to further improve docking of metalloproteins and understand the mechanism behind it particularly, we have extended the idea of QM/MM docking that was used for the general docking problem. The extension was aimed at including protein atoms surrounding the binding sites along with metal ions located nearby in addition to ligand atoms as a quantum region for QM/MM calculations. Except for the issue of quantum region assignment, we followed all the basic procedures of the original QM/MM docking protocol. Although employment of quantum mechanical calculations for an extended number of atoms in protein–ligand complexes may not be suitable for fast docking applications because of the demand on computational power, it would be useful to check how it performs in the prediction of the binding modes of metalloprotein complexes. In this regard, our research was done to see whether one can incorporate quantum mechanical calculations successfully into docking that can accommodate proteins with metal ions in the binding sites. Although such implementation can be utilized as a procedure in lead optimization, the purpose of this work is mainly to elucidate the true physical/chemical workings behind docking involving interactions of ligands with metal ions, which a conventional approach with empirical parameters would probably overlook. Within the confinement of native docking of metalloproteins, we also tried to answer the question if the extension of earlier QM/MM docking could be achieved so as to come up with an industry-applicable algorithm.

This chapter is organized as follows. We start by describing the detailed method of the extended QM/MM docking along with other computational details. As we use Schrödinger's QSite, which is an implementation of the frozen orbital method, for QM/MM calculations, there is an associated technical barrier that has to be overcome. We describe how to solve that problem here. With this modified version of QM/MM calculation in place, we then proceed to test the idea of extended QM/MM docking by performing a proof-of-concept experiment: to run QM/MM calculations at native poses to generate new sets of charges, which, in turn, are utilized in the subsequent docking. It is again recognized that this procedure is not for realistic industrial settings, in which one does not have the knowledge of native poses of ligands. Therefore, we test a modified version of the "survival of the fittest (SOF)" protocol, devised as a QM/MM docking method, in which one assumes no knowledge of native poses. Keeping the basic structure of the protocol, we make suggestions for improvement of the application of the method in general. All of the proposed methods are tested with 10 metalloprotein complexes for which Glide 4.0 failed 90% of the time, and previous QM/MM docking algorithm failed 50% of the time. Our results show that for this class of proteins, the extension of QM/MM docking enables better description of the binding sites by including metal ions and the surrounding atoms in quantum calculations. It should also be noted that our method differs from previous works on metalloprotein binding studies employing QM/MM calculations, such as the one discussed earlier in which we focused on the rescaling of atomic charges coming from QM calculations rather than QM/MM optimization. Finally, we discuss the future directions in the conclusion.

9.2   METHODS

9.2.1   DOCKING METHOD

We employed the Glide 4.0 program[19] throughout our study. In particular, we used the SP (standard precision) mode of Glide. The docking algorithm in Glide utilizes a hierarchical search protocol. Selection of the final ligand pose for the regular Glide is done with Glidescore, which is an extension of an empirically based Chem-Score function of Eldridge et al.[20] For our QM/MM docking protocol, however, final ligand selection is primarily determined by the total Coulomb–van der Waals energy (Ecvdw), with the Coulomb energy screened by a distance-dependent dielectric constant, since we have, in effect, emphasized the Coulombic energy portion of the scoring function by recalculating the atomic charges. The effectiveness of this choice of scoring function was demonstrated in the previous work.[10] The ligand structures and the corresponding receptor protein structures were prepared manually using utilities provided in the Schrödinger Glide suite. We started all the test cases from the raw Protein Data Bank (PDB) files. During this preparation, protein–ligand complexes were minimized by up to 0.3 Å RMSD in coordinates to the crystal structures. We calculated the RMSD in coordinates of the resulting ligand configurations to these structures to determine the accuracy of binding mode prediction. After we generated ligand poses, we discarded those that were within 0.6 Å of RMSD values to any previously accepted poses, a process we call "clustering." For regular Glide, we kept

5000 poses for each ligand docking from initial generation for refinement. After the refinement, we kept 400 poses for minimization using grids, during which a maximum of 100 steps were imposed. Finally, we scored 10 poses and ranked them after minimizations. These settings along with clustering parameters were modified for the initial stage of the SOF algorithm, as will be explained later.

9.2.2   QM/MM DOCKING

In the earlier work of QM/MM docking, a docking protocol coined as "survival of the fittest[10]" was implemented. In this work, an initial docking with regular force field was performed to produce a set of poses that were fed into QM/MM one-point energy calculations regarding only ligands as the quantum mechanical region, which, in turn, were fitted to generate a new set of atomic charges based on density functional theory (DFT)[21] quantum mechanical calculations. Using this new set of charges, a new generation of docking runs was performed and, in the end, the best scoring pose was selected. The key idea in this protocol was that at a pose that is close to the native structure, quantum mechanical calculation will produce atomic charges that give rise to lower Coulombic energy values at the native pose. For a validation study, one simply performs QM/MM calculation at the native pose of each complex to adjust the atomic charges on ligand atoms. This calculation would generate a set of charges in a given binding site environment that is theoretically the best possible one for docking. For QM/MM calculations, we used QSite version 4 of Schrödinger's 2006 suite, which combines IMPACT 4.0 and Jaguar 6.5. All of our calculations for the QM part were done with DFT. The 6-31G* basis set of Pople and coworkers,[22–24] and hybrid DFT functional B3LYP,[25–28] which has been shown to yield excellent results for atomization energies and transition states in a wide range of chemical systems, were used.

9.2.3   CHARGE FITTING

To calculate atomic charges after single-point energy QM/MM calculations, we used electrostatic potential (ESP) fitting, which is done by first calculating the molecular electrostatic potential (MEP) and fitting atomic charges to it. For the frozen orbital method[29] as implemented in the QSite program, ESP atomic charge fitting becomes problematic for atoms near the QM/MM boundaries because of the extra electron planted in the frozen orbital. This issue was addressed by the following remedy. To calculate charges on QM/MM systems containing frozen orbitals, we first optimized the geometry (or generate binding mode prediction by docking) and then performed a single-point QM/MM energy calculation with QSite. This generated a file with a basis functions list. We subsequently modified this file (a Jaguar restart file) by replacing the MM atoms having basis functions in the QM/MM calculations with dummy atoms having the same basis functions. The total charge and multiplicity of the system were also adjusted at this time such that each frozen orbital was occupied by only one electron. We also ensured that the frozen orbitals were taken to be the highest occupied orbitals in generation of the guess wavefunction. This guess wavefunction was subsequently provided to Jaguar and, without performing any self-consistent field (SCF) iteration to keep the same wavefunction as in the QM/MM calculation, we generated a grid of electrostatic potential values. This grid was then used to generate the QM/MM ESP charges. To generate the atomic charges, we used the RESP program. In the RESP grid file, we first duplicated the position of the atoms directly involved in a frozen orbital. We constrained the charges on these duplicated atoms to between −0.25e and +0.25e to correct for the dipolar effects on side-chain cuts, and −0.3e +0.3e or +0.1e −0.1e for backbone cuts. All charges on dummy atoms in the MM region were constrained to be 0e. Finally, the charges obtained using these constraints were used as the QM/MM ESP charges. It should also be noted that since our method is based on the frozen orbital method rather than the link atom method, hydrogen atoms near the boundary are treated just like any other atoms.

9.2.4 CHOICE OF LIGAND–PROTEIN COMPLEXES

For illustration of the method, we select 10 metalloproteins, which contain Zn, Fe, Mg, and Mn. For this set of complex structures, the Glide 4.0 SP mode failed to predict the binding mode within 2.0 Å RMSD of the native pose as the top-scoring pose in eight of them. All of the examples were downloaded from PDB and prepared according to the procedure described earlier. The ligands for these complex structures are depicted in Figure 9.1. For a quantitative estimate of how the extension of the QM/MM docking performs, we selected eight MMP complexes. These

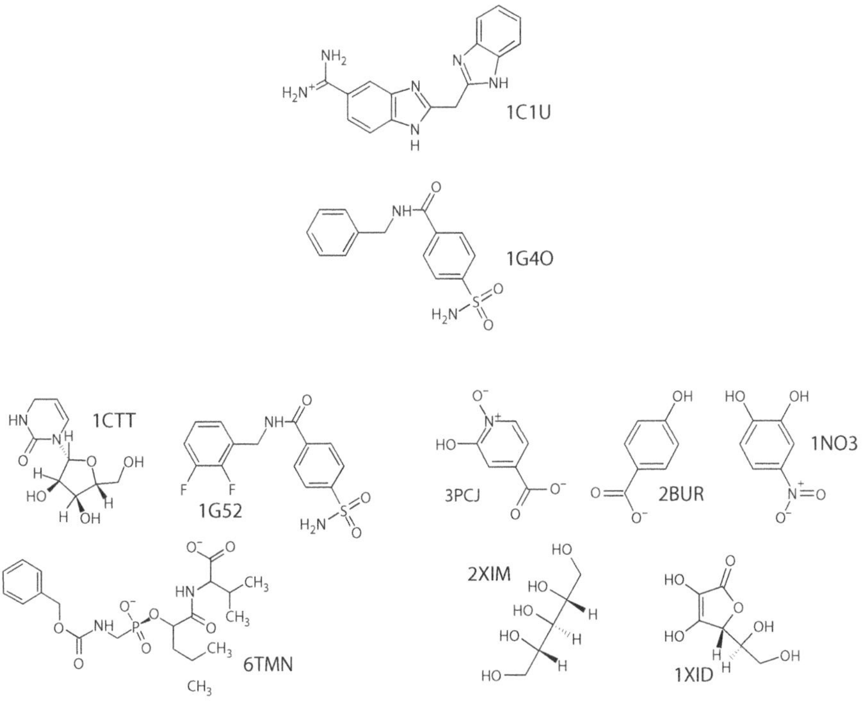

FIGURE 9.1 Structures of ligands used for the redocking experiment.

complexes are high-quality crystal structures with resolution better than or equal to 2.0 Å, and diverse in terms of ligands and receptors.

9.3 RESULTS AND DISCUSSION

9.3.1 INCLUDING ONLY METAL IONS IN THE QM REGION

In the case of carbonic anhydrase II with PDB ID 1G52, the Glide 4.0 SP mode gave the top-scoring pose to be 4.54 Å RMSD from the native structure. We ran an earlier version of the QM/MM docking protocol with the QM/MM charges computed at the native pose for the same complex. Only the ligand was considered in the QM region for this computation. This setting should give the best possible answer for the given protocol. It turned out, however, that the QM/MM docking protocol also gave a wrong binding mode prediction, yielding 3.79 Å RMSD for the top-scoring pose. It is well known that force field charges for metal ions are not quite as optimized as real partial charges. Usually, the formal charge values are used for partial charges without modification. The charge assigned by force field on Zn ion is thus +2e. As can be seen in Figure 9.2a, this strong positive charge attracts negative charges on fluorine atoms in the upper phenyl ring, drawing the whole ring toward the ion center. This explains the high RMSD value for the top-scoring SP pose. Calculating QM/MM charges on ligand atoms with the same Zn atomic charge changed the atomic charge on N atom significantly but did little to the negative charge on the fluorine atoms. Performing docking with this setting, the predicted pose was found to be in the folded ligand structure form similar to the one from the SP run (Figure 9.2b). The natural extension of QM/MM docking at this point would be to include metal ions in the QM region.[30] After fitting the new charges based on QM calculation on the Zn atom along with the ligand, we found that the charge on the Zn atom was reduced to +1.90e. This loss of charge is small and did very little to alter the charge configuration

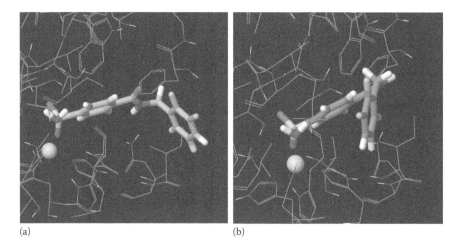

(a)   (b)

FIGURE 9.2 Docked poses for 1G52. (a) Glide SP docking result and (b) QM/MM docking result. Both poses have upper ring with fluorine atoms bent toward the zinc atom.

on ligand atoms. Again with docking with this new set of charges on ligand atoms and Zn atom, we still found the top-scoring pose to be in the folded configuration.

9.3.2 Extended QM/MM Docking Protocol

We thus further extended the QM region in the calculations to include protein atoms surrounding the binding site in an attempt to simulate a more realistic picture of the binding site. We included in the QM region side chains of those residues that are within 4 Å of the ligand with a restriction that only the residues for which QSite has parameters can be at the boundary of the QM and MM regions. If there exists no parameter for a selected residue side chain, then we have to include the whole residue and make a cut in the adjacent residues. Figure 9.3 shows the proximity of the ligand in the binding site of 1G52, which were treated as the quantum region. As was noted in Section 9.2, locating QM/MM boundaries between atoms of the protein complicates the ESP fitting. We applied the procedure described earlier to correct the artifacts created by the frozen orbitals on the fitted atomic charges. In the extended QM/MM docking, the charge on the Zn atom now reads +1.16e, which is a significant reduction. Using this new charge configuration, we performed docking and found the top-scoring pose to be 0.91 Å RMSD from the crystal structure (Figure 9.4). Reducing the charge on Zn atom prevented the upper phenyl ring to bend toward the ion center. We also checked in this example how the selection of residues in the QM region affects charge calculation on the Zn atom. The immediately neighboring residues of Zn atom are three histidine residues (HIS). These residues are obviously included in the QM region. However, addition of other residues in the QM region

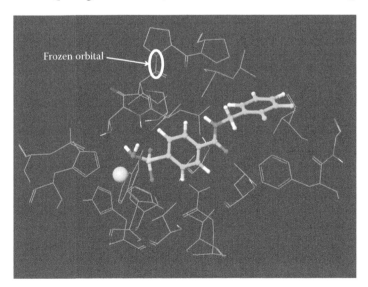

FIGURE 9.3 Proximity of the ligand in the binding site of 1G52. Residues that are within 5 Å of the ligand are shown. Residues in pink are in the quantum region. Frozen orbitals are located at the boundary between the QM and MM regions. An example of the frozen orbital is marked in the figure.

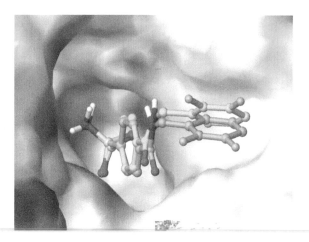

FIGURE 9.4 Superposition of the ligand docked by extended QM/MM docking and native conformation. Extended QM/MM docking result is the top one and native ligand is the bottom one.

beyond these three changed the charge value on Zn atoms by <0.05. Inspection of the binding site geometry led us to conclude that residues other than three HISs are too far from the metal ion to have any influence. Our method still followed the rule of selection "residues that are within 4 Å of the ligand" since residues other than HISs can affect the polarization of ligand atoms.

For another example, protocatechuate 3,4-dioxyenase 1 (PDB ID: 2BUR), a similar argument applies. The ligand of this complex structure has a carboxylate group on one end and a hydroxyl group on the other. Even if the hydroxyl group becomes deprotonated, because of the larger overall negative charge on the carboxylate group, the ligand tends to be oriented in such a way that the carboxylate group points to the ion center. This is because of the strong positive charge on the Fe atom, which is assigned as +3e by the force field. Including only the Fe atom along with the ligand did not solve this problem; however, including surrounding protein atoms as well reduced the positive charge on the Fe atom significantly and therefore helped in finding the right binding mode by docking.

The results of the 10 complex structure redocking experiment with extended QM/MM docking protocol are listed in Table 9.1. The new docking protocol succeeded in finding the right binding mode, defined by a pose with <2 Å RMSD from the crystal structure, in all but one of them. It found poses with 1 Å RMSD or lower in seven of 10 cases. Of these 10 cases, Glide 4.0 SP succeeded only in two of them, while the original QM/MM docking protocol succeeded in five of them. For the case of 2XIM, examination of the binding site tells us that there are a few water molecules that mediate interactions between the ligand and the receptor. The presence of crystal waters that were completely eliminated for our study probably has an effect on the final outcome of the binding mode prediction. We believe that to treat this problem correctly, one needs to carry out more extensive quantum mechanical calculations, which should also address the issue of water molecule orientation.

TABLE 9.1

RMSD (in Å) of Top-Scoring Poses from Native Poses for Glide SP 4.0, QM/MM Docking, and Extended QM/MM Docking

PDB	Glide SP 4.0		QM Dock		Ex-QM Dock		Charge on Metal Ion[a]
	Ecvdw	RMSD	Ecvdw	RMSD	Ecvdw	RMSD	
				Zn			
1C1U	−45.5	5.11	−74.3	0.62	−55.8	0.41	1.59
1G4O	−42.6	4.35	−46.9	0.33	−32.7	1.41	1.23
1CTT	−33.5	4.85	−50.9	0.59	−53.7	0.74	0.84
6TMN	−48.9	8.17	−89.7	0.35	−61.2	1.37	1.68
1G52	−44.8	4.54	−48.9	3.79	−41.2	0.91	1.16
				Fe			
3PCJ	−48.6	0.33	−58.6	0.35	−30.0	0.07	1.49
2BUR	−32.2	4.14	−31.4	4.05	−22.3	0.39	1.69
1NO3	−23.3	3.89	−26.6	3.54	−27.1	0.27	1.78
				Mg			
2XIM	−34.9	4.25	−36.5	4.39	−29.8	4.01	1.35:1.30
				Mn			
1XID	−26.0	4.07	−32.7	4.48	−30.3	0.45	1.12:1.32

[a] Charge values on metal atoms calculated with the Ex-QM Dock method. The force field-based charges on these atoms were +2e for Zn, Mg, and Mn, and +3e for Fe. There are two Mg atoms in the 2XIM protein binding site and two Mn atoms in the 1XID protein binding site.

9.3.3 TEST OF "PRACTICAL" PROTOCOL

As pointed out earlier, the above studies are only for validation of the idea, since the method cannot be used in an industrial setting, when one does not have a solved complex structure. In the earlier work of QM/MM docking, we set forth of the method coined as the "survival of the fittest" algorithm to address this issue. The method relied on the fact that in an initial round of regular force field-based docking, within the top five to 10 poses obtained, one can usually find a pose that is close enough (<1 Å RMSD) to the native configuration, which will, in turn, give rise to the "good" charge configuration. This becomes even more critical for our extended QM/MM docking protocol. We tested a few possibilities for improvement, in which the final prediction of poses is as diverse as possible with at least one of them being close to the native structure. In the settings for Glide, this can be achieved by controlling parameters in two directions. In the "Advanced Settings" window on the "Settings" tab, there are three parameters for the selection of initial poses section. The default parameters are 5000 for "how many poses per ligand to keep for the initial phase," 100.0 for "scoring window for keeping initial poses," and 400 for "how many poses per ligand to keep for energy minimization." We raised each of these parameters up

TABLE 9.2

**Binding Mode Prediction of Extended
"Survival of the Fittest" (SOF) Algorithm**

	Initial Docking		Ex-SOF
PDB	Min. RMSD	Rank	Final RMSD
		Zn	
1G4O	0.93	4	1.67
		Fe	
3PCJ	0.24	7	0.30
1NO3	0.78	5	1.02
		Mn	
1XID	0.57	3	0.98

to 20,000, 200.0, and 800, respectively. The intention is not to throw away any of the potentially "good" poses in the initial filtering process. For the second direction, we focused on the clustering of output poses. In the "Advanced Settings" window of the "Output" tab, the default value for clustering of the output pose is set to 0.5 Å. We have argued in an earlier work that within minimization, poses with 1.0 Å RMSD from each other can be regarded as identical.[7] What we are trying to accomplish by controlling this parameter is to ensure the presence of a "good" pose within five to 10 ranked structures without listing hundreds of them, for which one has to perform expensive QM/MM calculations. For our purpose, we found, after examination of a few examples, that 1.5 Å would be a reasonably good number that forces the final docking poses to be diverse enough so that low RMSD poses are included in the list of the top 10 ranked poses. In the same window, we have also a final filtering condition based on Coulomb–van der Waals energy. We increased this number from 0.0 to 20.0 kcal/mol not to err in filtering out "good" poses. Table 9.2 is the result of our test of SOF for the extended QM/MM docking method on four examples. It shows that with our settings of Glide parameters, we found a pose that is <1.0 Å RMSD from the native pose in the initial docking within the top 10 scoring poses for all of them. These low RMSD poses eventually led to correct binding mode predictions.

9.3.4 TEST OF DOCKING ACCURACY AND SCORING RELIABILITY ON MATRIX METALLOPROTEINASE

To further estimate quantitatively how the extension of QM/MM docking performs, we tested our method on a set of eight matrix metalloproteinases (MMPs). The ligands of these metalloprotein complexes are depicted in Figure 9.5. Table 9.3 shows the result of our test. In this table, we see that except for the 1HFS case, the RMSD values are <2.0 Å. In fact, in six of eight cases, we found the final prediction of docking poses to be <1.0 Å RMSD. Examination of incorrect docking of 1HFS indicates a potential reason for the failure to find the correct binding mode. Although

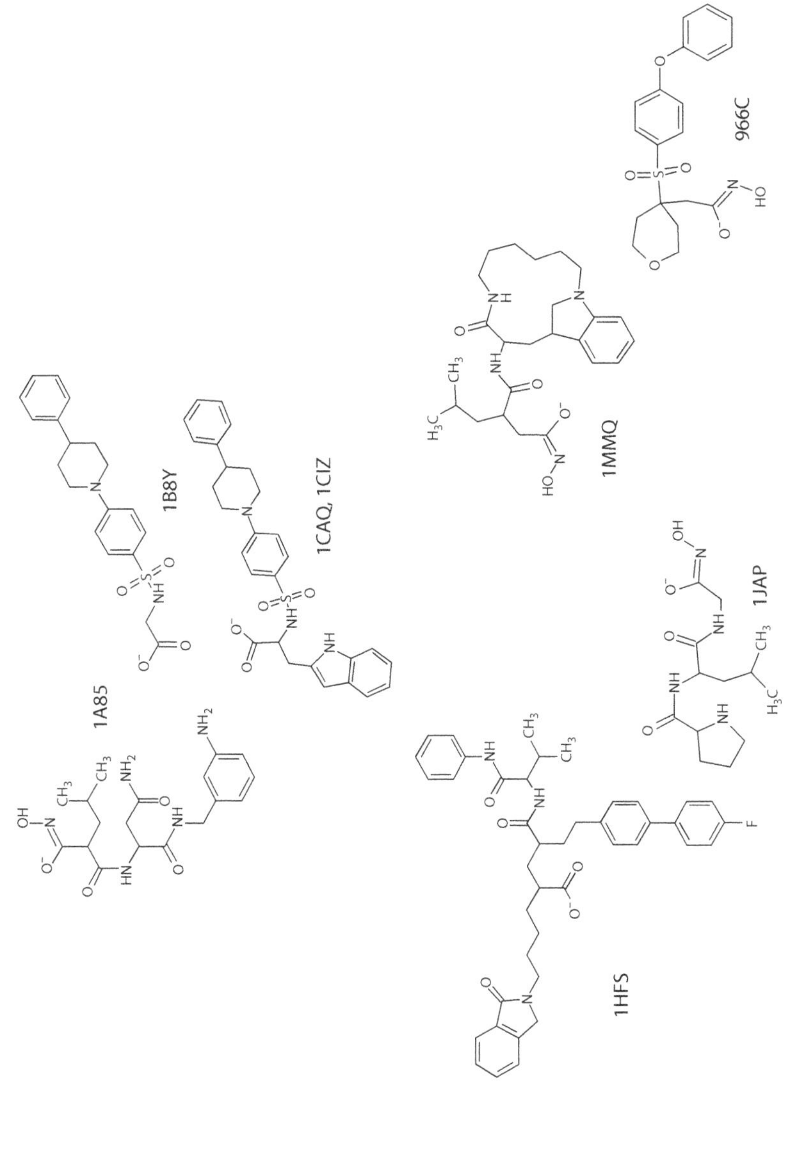

FIGURE 9.5 Structures of the eight MMP ligands.

TABLE 9.3
Binding Free Energy Calculations of Eight MMP
Inhibitors Using Extended QM/MM Docking

PDB	$\Delta G_{experimental}$ (kcal/mol)	RMSD (Å)	$\Delta G_{predicted}$ (kcal/mol)	Glidescore
1A85	−6.26	1.37	−7.32	−7.79
1B8Y	−10.71	0.74	−10.04	−11.05
1CAQ	−10.15	0.82	−9.96	−10.94
1CIZ	−10.53	0.66	−9.46	−10.92
1HFS	−11.87	3.24	−11.75	−13.16
1JAP	−6.44	0.50	−6.34	−7.30
1MMQ	−10.26	0.40	−9.53	−8.65
996C	−10.42	0.31	−9.05	−9.08

the carboxylate group correctly binds to the zinc atom in the binding site of 1HFS, one of the three arms with phenyl ring points to the wrong direction. This part of the ligand is solvent exposed and surrounded by hydrophobic residues. We feel that to accurately describe such a motif, one needs a different approach. A possible solution to this problem is being pursued in our laboratory at the moment. In the table, we also list the modified Glidescore as our prediction of binding free energy along with experimentally observed free energy. This energy value is different from the usual Glidescore given by the Glide program in that it was calculated with the rescaled charge values from our new method. The usual Glidscores were calculated by in-place scoring with the same pose predicted by our method. The coefficient of determination (R^2: square of the correlation coefficient) from regression analysis between the predicted free energies and experimental free energies for seven MMP complexes, which excludes one case with a wrong binding mode, is 0.89 with a standard error of 0.51 kcal/mol. This result is significantly better than the usual Glidescore result, which gave 0.65 for R^2 and 1.02 kcal/mol for the standard error. It seems apparent that except for 1MMQ and 966C, there is a tendency that the usual Glidescore overestimates the binding free energies whereas the modified Glidescore underestimates them. This is understandable since the modified Glidescore is evaluated with reduced zinc atom charges. Overall, we still feel the need for further development in the calculation of binding free energies especially because the Glidescore given here, albeit with improved charge values, does not account for solvent effects properly.

9.4   CONCLUSION AND FUTURE DIRECTIONS

The results presented here certainly demonstrate the importance of including protein atoms in the QM region for QM/MM docking on metalloproteins to succeed. Whether this new method can be used as a practical docking solution remains a question, however. Including more atoms in the QM region simply means more computational time. Compared with the earlier QM/MM docking protocol, the extended

QM/MM docking can take >20 times in computational time for the same system. This reflects the fact that >100 protein atoms are added to the QM region. In SOF implementation of the extended QM/MM docking, this lengthy QM/MM calculation has to be repeated 5–10 times. Although the process is "embarrassingly" parallelizable, without parallel computing, the whole calculation can be deemed prohibitively expensive. However, since the transfer of charge from the metal ions is predominantly to the surrounding protein atoms, it is conceivable to devise a protocol in which such calculation is performed as a preliminary procedure, akin to grid generation of current docking algorithms. This procedure will undoubtedly shorten the computational time and can even be applied to a fast docking (virtual ligand screening) environment. It is more sensible than universally rescaling the charge on metal ions. Another source of complication for the extended QM/MM docking is the use of grids that is prevalent in current docking programs. Pregenerated grids are used in docking to speed up the ligand screening under the assumption that the receptor is rigid. In our case of extended QM/MM docking, after each QM/MM charge calculation on protein atoms, grids must be regenerated. Grid generation is also ill suited for docking with flexible proteins. It is our contention that for future development of docking methods, this practice of using grids for scoring should be abandoned. Instead, on-the-fly scoring without grids should be implemented as suggested in our past work.[7] It is our belief that the computational power of today can handle such implementation.

REFERENCES

1. Halperin, I.; Ma, B. Y.; Wolfson, H.; Nussinov, R. *Proteins* 2002, 47(4), 409–443.
2. Gschwend, D. A.; Good, A. C.; Kuntz, I. D. *J Mol Recognit* 1996, 9(2), 175–186.
3. Abagyan, R.; Totrov, M. *Curr Opin Chem Biol* 2001, 5(4), 375–382.
4. Kuntz, I. D. *Science* 1992, 257(5073), 1078–1082.
5. Drews, J. *Science* 2000, 287(5460), 1960–1964.
6. Brooijmans, N.; Kuntz, I. D. *Annu Rev Biophys Biomol Struct* 2003, 32, 335–373.
7. Cho, A. E.; Wendel, J. A.; Vaidehi, N.; Kekenes-Huskey, P. M.; Floriano, W. B.; Maiti, P. K.; Goddard, W. A., 3rd. *J Comput Chem* 2005, 26(1), 48–71.
8. Kramer, B.; Metz, G.; Rarey, M.; Lengauer, T. *Med Chem Res* 1999, 9(7–8), 463–478.
9. Verdonk, M. L.; Cole, J. C.; Hartshorn, M. J.; Murray, C. W.; Taylor, R. D. *Proteins* 2003, 52(4), 609–623.
10. Cho, A. E.; Guallar, V.; Berne, B. J.; Friesner, R. *J Comput Chem* 2005, 26(9), 915–931.
11. Cho, A. E. *BioChip J* 2007, 1(1), 70–75.
12. Lipscomb, W. N.; Strater, N. *Chem Rev* 1996, 96(7), 2375–2433.
13. Deerfield, D. W.; Carter, C. W.; Pedersen, L. G. *Int J Quant Chem* 2001, 83(3–4), 150–165.
14. Dudev, T.; Lim, C. *J Phys Chem B* 2000, 104(15), 3692–3694.
15. Rulisek, L.; Havlas, Z. *J Am Chem Soc* 2000, 122(42), 10428–10439.
16. Shurki, A.; Warshel, A. *Adv Protein Chem* 2003, 66, 249–313.
17. Khandelwal, A.; Lukacova, V.; Comez, D.; Kroll, D. M.; Raha, S.; Balaz, S. *J Med Chem* 2005, 48(17), 5437–5447.
18. Strynadka, N. C. J.; Eisenstein, M.; Katchalski-Katzir, E.; Shoichet, B. K.; Kuntz, I. D.; Abagyan, R.; Totrov, M.; Janin, J.; Cherfils, J.; Zimmerman, F.; Olson, A.; Duncan, B.; Rao, M.; Jackson, R.; Sternberg, M.; James, M. N. G. *Nat Struct Biol* 1996, 3(3), 233–239.

19. Friesner, R. A.; Banks, J. L.; Murphy, R. B.; Halgren, T. A.; Klicic, J. J.; Mainz, D. T.; Repasky, M. P.; Knoll, E. H.; Shelley, M.; Perry, J. K.; Shaw, D. E.; Francis, P.; Shenkin, P. S. *J Med Chem* 2004, 47(7), 1739–1749.
20. Eldridge, M. D.; Murray, C. W.; Auton, T. R.; Paolini, G. V.; Mee, R. P. *J Comput Aid Mol Des* 1997, 11(5), 425–445.
21. Kohn, W. *J Comput Chem* 1999, 20(1), 1.
22. Rassolov, V. A.; Pople, J. A.; Ratner, M. A.; Windus, T. L. *J Chem Phys* 1998, 109(4), 1223–1229.
23. Rassolov, V. A.; Ratner, M. A.; Pople, J. A.; Redfern, P. C.; Curtiss, L. A. *J Comput Chem* 2001, 22(9), 976–984.
24. Tran, J. M.; Rassolov, V. A.; Ratner, M. A.; Pople, J. A. *Abstr Pap Am Chem S* 1999, 217, U443.
25. Becke, A. D. *J Chem Phys* 1993, 98(2), 1372–1377.
26. Kim, K.; Jordan, K. D. *J Phys Chem* 1994, 98(40), 10089–10094.
27. Perdew, J. P.; Emzerhof, M.; Burke, K. *J Chem Phys* 1996, 105(22), 9982–9985.
28. Stephens, P. J.; Devlin, F. J.; Chabalowski, C. F.; Frisch, M. J. *J Phys Chem* 1994, 98(45), 11623–11627.
29. Philipp, D. M.; Friesner, R. A. *J Comput Chem* 1999, 20(14), 1468–1494.
30. Cho, A. E. *BioChip J* 2008, 2(2), 61–66.

Index

Page numbers followed by f, t and n indicate figures, tables and notes, respectively.